Beaches and Dunes of Developed Coasts

This volume discusses the role of humans in transforming the coastal landscape. The book details the many ways beaches and dunes are eliminated, altered and replaced and the differences between natural landforms and the human artifacts that replace them. A distinguishing feature of the book is the emphasis on the importance of retaining naturally functioning beaches and dunes in ways that achieve natural values while accommodating development and use.

The issues dealt with in this monograph are important in coastal management. It will therefore be of interest to practicing coastal engineers and scientists working on applied research topics, as well as to planners and managers of coastal resources at all levels of government. The book will be of particular value to investigators planning for the future of coastal development under accelerated sea level rise. It will also be useful as a supplementary reference text for graduate and advanced undergraduate courses in geography, geology, ecology and other disciplines dealing with issues in environmental management, environmental ethics and the interaction between science, technology and society.

KARL F. NORDSTROM is a professor of marine and coastal sciences at Rutgers University. He is a member of the editorial board of the Journal of Coastal Research. His previous books include *Living with the New Jersey Shore* (Duke University Press), *Coastal Dunes: Processes and Morphology* (John Wiley), *Estuarine Beaches* (Elsevier Applied Science), *Estuarine Shores: Evolution, Environments and Human Alterations* (John Wiley).

Beaches and Dunes of Developed Coasts

KARL F. NORDSTROM
Institute of Marine and Coastal Sciences
Rutgers University
New Brunswick, USA

CAMBRIDGE
UNIVERSITY PRESS

PUBLISHED BY THE PRESS SYNDICATE OF THE UNIVERSITY OF CAMBRIDGE
The Pitt Building, Trumpington Street, Cambridge, United Kingdom

CAMBRIDGE UNIVERSITY PRESS
The Edinburgh Building, Cambridge CB2 2RU, UK
40 West 20th Street, New York NY 10011–4211, USA
477 Williamstown Road, Port Melbourne, VIC 3207, Australia
Ruiz de Alarcón 13, 28014 Madrid, Spain
Dock House, The Waterfront, Cape Town 8001, South Africa

http://www.cambridge.org

First published 2000
First paperback edition 2003

Typeface Lexicon (*The Enschedé Font Foundry*) 10/14 pt *System* QuarkXPress® [SE]

A catalogue record for this book is available from the British Library

Library of Congress Cataloguing in Publication data
Nordstrom, Karl F.
  Beaches and dunes of developed coasts / Karl F. Nordstrom.
    p. cm.
  Includes bibliographical references and index.
  ISBN 0 52147013 7 (hardback)
  1. Coastal zone management. 2. Coastal ecology. 3. Coasts.
I. Title.
HT391.N58 2000
333.91'7–dc21 99–13766 CIP

ISBN 0 521 47013 7 hardback
ISBN 0 521 54576 5 paperback

Contents

Preface

A considerable proportion of the coastline of many countries is now developed with buildings and transportation routes or protected against flooding and erosion, and many shorefront communities that are only partially developed or stabilized are well on the way to total transformation. Some coastal areas have been designed and built as human artifacts and bear little resemblance to the coast that formerly existed. Strip development dominates the land conversion process in many coastal communities, placing the location of much of the development where it is readily affected by wave and wind processes. Shore protection structures emplaced to protect these developments are often placed where they have the greatest impact on shoreline processes and the most dynamic landforms. Shorefront property owners alter the coastal landforms to suit their needs, and beach users alter the vegetation and landforms without even knowing it.

There is no indication that the trend toward increasing development will be reversed. Economic evaluations indicate that many locations could justify even greater expenditures for construction of new buildings and protection projects, and these expenditures will be forthcoming despite an increase in the rate of sea level rise and potential for storm damage. Major storms cause considerable property damage but little lasting effect on the landscape compared to human activities. Communities that have been severely damaged by coastal storms are often rebuilt to larger proportions. Increasing population pressure, combined with the value of shorelines for human use, makes the occupation of the coast widespread and inevitable under present management practice.

Our understanding of the relationship between human activities and coastal evolution is inadequate, despite the growing body of literature on human-induced changes in coastal environments. Geomorphologists, who are specialists in the study of landforms, have been reluctant to study developed coastal systems. Those who study developed systems often examine them in the context of what is happening in the present without considering how the developed landform arrived at the present stage of evolution. Scientists may examine landform evolution through time by showing how erosion rates or vegetation patterns change but do not explain this change in the the context of social processes. Some investigators are deterred by the magnitude of the problem of isolating cause and effect in complex developed systems, or they are trained to look at landscape evolution at vastly greater time scales than occur under developed conditions. The net result of the lack of focus on developed coastal systems is that human alterations most often are viewed as an aberration, rather than an integral component of landscape evolution. Use of the terms interference and intervention (rather than alteration) to describe human action underscores the lack of appreciation of the role of humans as landscape agents or a lack of objectivity in assessing their effects.

This book identifies the way coastal landforms are transformed by human action and the reasons why this transformation occurs. The context is geomorphology, and ecological and human factors (such as economic or social constraints) are presented in terms of how they affect landform size, shape, location and mobility. The focus is on beaches and dunes because human alteration of these features and the natural processes that shape them is so prevalent in the coastal zone. The book is not an examination of issues related to fields outside geomorphology, but it will have value to practitioners in those fields as a reference document. The conflict between natural and human functions and values and static and dynamic approaches to management are principal threads throughout the book, as is the issue of whether humans should be considered intrinsic or extrinsic agents of landscape evolution. My intention is to minimize discussion of purely natural processes, except to illustrate some of the basic differences in the way natural and human-altered landforms are created and evolve. Basic information on natural processes is available in several excellent texts (e.g., Carter 1988; Komar 1998), and there is no reason to repeat that information here.

This book is an attempt to provide an objective, scholarly scientific treatise of human-induced landscape conversion that can serve as a basis for environmental debate, a starting-point for an approach to restoration of endangered living and non-living resources, and a reference volume for those wishing to conduct more extensive research on these topics. The discussion is intended to be of interest to coastal engineers and scientists as well as to planners and managers of coastal resources at all levels of government, students, shorefront managers and users of the beaches and dunes of the coast. The book is intended to help provide the basis for a management program for beaches and dunes, but it is not intended to discuss basic principles of management and policy. Guidelines for scientists, outlines for future studies and research methods and opportunities are presented, but the book is not designed as a handbook.

The proportion of publications devoted explicitly to human-altered coastal landforms is relatively small compared to so-called natural landforms, considering the scale of human impact. Articles on the geomorphology or sedimentology of developed coasts are rare in basic research journals, with some important exceptions, notably the *Journal of Coastal Research*. Articles in environmentally oriented journals and publications with emphasis on planning, management or societal aspects of coastal change are plentiful and contain valuable insight to the human processes operative in developed coastal systems and the way shorelines have been modified from natural conditions. The literature in engineering forums on the relationship between structures, coastal processes and beach change is truly vast. The proceedings of conferences sponsored by the American Society of Civil Engineers are particularly valuable, although much of the research is concerned with design of structures rather than the effects of structures after they are emplaced. Books oriented toward planning, management or societal aspects of coastal change often do not provide detailed assessments of mechanisms of landform evolution in developed physical systems. Proceedings of multidisciplinary conferences dealing with management issues contain valuable case studies of developed systems, but synthesis of these studies is not possible in those forums. There are numerous books and reports with a regional orientation that discuss developed portions of coastlines and collections of case studies that include developed coasts, but these volumes do not provide a synthesis of data on landforms or vegetation of developed systems or models of shoreline change specifically for developed coasts. Books that have addressed the nature of human alterations to beaches and dunes in some detail include Carter (1988), Viles and Spencer (1995) and Bird (1996). I have attempted to make this book complementary to Carter (1988) by focussing almost exclusively on the vast literature that has been produced since his publication and complementary to Bird (1996) by presenting human alterations in a basic research context with implications for modeling landform evolution rather than managing landforms. The book is complementary to Viles and Spencer (1995) in its concentration on human activities in beaches and dunes.

The focus of this book is on landforms of exposed ocean coasts. Space constraints preclude inclusion of the vast literature on estuarine environments, although human alterations of the coasts of estuaries have been profound.

Modifications in streams and estuaries and human actions outside the portions of the coastal zone occupied by beaches and dunes, such as ocean dumping, pollution of air, water and sediments and alterations to flora and fauna, are not addressed unless the changes directly affect sediment budgets and landforms on open-coasts. The emphasis is on results of field investigations and observations rather than physical (scaled down) models or theoretical/mathematical models, although results of those types of studies are used where they provide the only meaningful assessment of human alterations of the coastal landscape.

There is no tradition of research on the geomorphology of developed coasts, and the structure and emphasis of this book could be handled in a number of ways. I have elected to lead off with a historical perspective of the economic and social forces leading to coastal development. The ways that humans alter coastal landforms to achieve specific needs other than shore protection are presented in chapter 2. Alterations resulting from projects designed to replenish sediments in beaches and dunes are presented in chapter 3. Chapter 4 focuses on interaction of waves and winds with coastal structures that have been designed for a variety of purposes, including shore protection, habitation, recreation and transportation. The ways that human alterations are compatible or incompatible with natural processes and the ways their physical characteristics and temporal scales of evolution differ are then discussed in chapters 5 and 6. The reasons why natural values do not fare well when pitted against traditional human values are identified in chapter 7, where management programs are examined in the specific context of beaches and dunes. Restoration and nature development is discussed principally in chapter 8, along with guidelines for restoring or maintaining natural landform characteristics in developed areas. The book concludes with a discussion of future research programs that will provide a basis for understanding the role of beaches and dunes in developed landscapes and serve as the basis for programs for their planning and management.

Acknowledgements

Financial assistance and administrative support for the numerous projects that led to results published in this book were provided by: the NOAA Office of Sea Grant and Extramural Programs, US Department of Commerce, under Grant Nos. R/S-95002 and R/D-9702; the German Academic Exchange Service (DAAD), the Department of Physical Geography and Soil Science of the University of Amsterdam and the Institute of Marine and Coastal Sciences, Rutgers University.

I am grateful to the following people who responded to my request for information on developed coasts, directed me to sources of information, or guided me to interesting locations in the field: J.R. Allen, E. Anthony, S.M. Arens, D. Assendorp, D. Bennett, C. Braun, M.S. Bruno, R.W.G. Carter, D.M. Chapman, R.G.D. Davidson-Arnott, I. Eliot, P. Fabbri, J. Garofalo, J. Gebert, S. Hotta, D. Jenkins, G. Keller, T. Keon, N.C. Kraus, R. Lampe, R. Levy, M.N. Mauriello, A. Miossec,, R. Paskoff, O.H. Pilkey, E. Pranzini, N.P. Psuty, E. Sanjaume, D.J. Sherman, H.M. Sterr, T. Sykes, T.A. Terich, J. van Boxel, F. van der Meulen, D. van der Wal, H.J. Walker, A.T. Williams, K. Wright.

Special thanks for the support provided by J.F. Grassle and J.M. Verstraten. Thanks also to N.L. Jackson and L.M. Vandemark, who provided comments on a draft of the book. This is Contribution Number 98–15 of the Institute of Marine and Coastal Sciences, Rutgers University and Sea Grant Publication NJSG-98–388. The US Government is authorized to produce and distribute reprints for government purposes notwithstanding any copyright notation that may appear hereon.

1

The developed coastal landscape: temporal and spatial characteristics

Introduction

Human-altered coasts vary greatly in appearance, from landscapes where human impact is significant but barely perceptible (Figure 1.1) to landscapes where cultural features visually dominate the landscape (Figure 1.2). No one would deny the prominent role that humans play in altering the coastal landscape (Walker 1984). The more difficult tasks involve identifying: (1) how human-altered landforms may be defined; (2) whether humans are or should be the dominant agent in landscape evolution; and (3) whether human needs or actions should determine the characteristics and values of the resulting landforms. These broad issues can be separated into several areas of investigation (Table 1.1) that are examined in detail. It is assumed here that the landforms of interest (beaches and dunes) and the habitats within them are desirable for both their natural and human values, and that it is better to have these landforms than not to have them. Human actions are then evaluated in terms of loss, gain or conversion of these landforms.

This chapter addresses the first two areas of investigation identified in Table 1.1 by providing a historical perspective on the human forces that drive coastal development and the processes and stages of alteration from natural shorelines to artifacts in attempts to maximize human values. The focus is on long-term and large-scale transformations of landscapes within which individual landforms are altered. The most obtrusive human modifications are highlighted, along with some of the economic and social reasons for the conversion. Evolution at the scale of individual landforms and at shorter temporal scales are evaluated in greater detail in subsequent chapters.

The historical perspective is based largely on activities in western Europe and the USA, because of the availability of information for those locations. Recent economic and social forces are evaluated to show how improvements in communication and transportation, increases in expendable income and

Figure 1.1 Manzanita Oregon, USA. The natural-appearing dune in the foreground has been transformed from its natural appearance as a result of the introduction of European beach grass (*Ammophila arenaria*). This vegetation was first planted in the nineteenth century. It spread rapidly along the Pacific coast and provided a more complete trap to blown sand than the native vegetation and created a higher, more linear, and better vegetated foredune than existed previously.

Figure 1.2 Arma di Taggia, Italy, showing onshore and offshore structures and beach grading operations that have changed the appearance and function of the coast by altering the energy of the waves, the topography of the beach and the natural vegetation and habitat.

Table 1.1 *Major areas of investigation addressed in this book*

How do developed coasts evolve?
How do landforms and landscapes change due to changes in social and economic processes?
How are landforms altered to achieve specific human needs?
How are landforms enhanced to retain their utility?
What are the physical characteristics of the resulting landforms?
What are the temporal scales of evolution of coastal landforms?
Under what conditions are humans intrinsic or extrinsic agents of landscape evolution?
How effective are regulations affecting landforms?
What are the viable alternative approaches to restoring landforms?
How can human alterations be made compatible with natural processes?
What are the ways that natural values can be maintained while accommodating human use?
What is the significance of using static or dynamic approaches to management?
What are the research requirements for beaches and dunes on developed coasts?

creation of global economies have made human alterations an international phenomenon, contributing to an exponential increase in the pace and scale of development. Much emphasis is placed on the evolution of tourism, because it has been a driving force in altering beaches and dunes, and on the implementation of shore protection projects that have extended the impact of humans beyond the initial modifications to the landscape that were designed to accommodate tourism activities.

Perspective on some of these trends is provided in an assessment of a case study of the shoreline of New Jersey, USA, a location that is examined in some detail throughout the rest of the book. The New Jersey coast has the longest history of stabilized barrier island shoreline in North America; it has the most developed coastal barriers and the highest degree of stabilization in the USA; and it has been identified as a template by which developing barriers can be evaluated to show the incompatibility of shorefront development (Pilkey 1981; Nordstrom 1987a; Mitchell 1987; Pilkey and Wright 1988; Hall and Pilkey 1991).

The impact of humans through time

Finding detailed information on the impacts of humans on beaches and dunes through time is a difficult task. There is less information on these landforms than on many other aspects of the environment because these features were of little value in traditional economies. There is little doubt that some human impact on coasts dates back tens of thousands or even hundreds of thousands of years in some areas. Human presence along the coast of Italy, for example, is described as occasional, beginning about 300 000 years BP, and the first settlement is documented about 40 000 years BP (Torresani 1989).

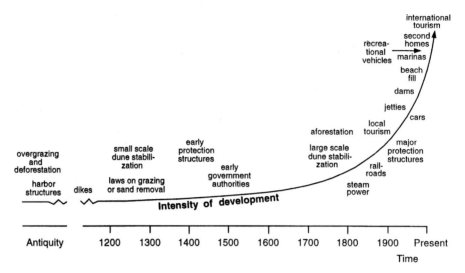

Figure 1.3 Idealized representation of the intensity of human development through time with contributing human actions or features.

Evidence of human influence in the past is obscured due to sea level rise. Most coastal and nearshore archaeological sites are less than 6000 years old because of the rapid rise in sea level prior to that time (Walker 1981a; Bird and Fabbri 1987). Real influence on the coastal landscape in Italy apparently started with the Etrurians from the ninth century BC (Torresani 1989), and there is a rich assemblage of archaeological and geomorphological evidence in Italy and other parts of the Mediterranean in the past 2000 years (Bird and Fabbri 1987).

Figure 1.3 presents an idealized representation of the intensity of development in historical times (revealed in the trend line) along with the principal human actions or features for which there is adequate documentation. The figure is based largely on reports of activities in western Europe and North America. The time of initiation of human activities is generalized because data prior to about 1800 are spotty and site-specific, but the figure does reveal the general change from incidental or accidental actions to direct modification in response to changes in population pressures, perception of resources, income, leisure time and technological advances. Other locations may have gone through similar phases of landscape conversion, but the phases may have been at different dates or have had different durations.

The trend line in Figure 1.3 is portrayed as smooth because it represents a global curve. A site-specific curve would show considerable short-term fluctuations. The impacts of humans on the coast have always undergone periods of greater and lesser impact. In the distant past, periods of declining human use have been related to piracy, war and disease (Torresani 1989). Pronounced cycles of shoreline advance and retreat are associated with changes in social and demographic characteristics on the Mediterranean coast (Innocente and Pranzini 1993), and at least one and perhaps several phases of active deposition of sediment and advance of the shoreline occurred on the Mediterranean coast in classical antiquity as a result of human activity (Paskoff 1987).

There is a pronounced change in the slope of the curve beginning about two centuries ago corresponding to the availability of steam power that enabled large modifications to the landscape (Marsh 1885; De Moor and Bloome 1988; Terwindt *et al.* 1988; Meyer-Arendt 1992). The slope of the curve has increased with increases in the size and availability of machinery, with development of the internal combustion engine, and with the growth of tourism as a major industry.

Overgrazing and deforestation

Overgrazing and deforestation of drainage basins leads to increased quantities of sediment delivered to the coast. These human actions were likely the earliest causes of major changes to coasts (Walker 1985), and they have undoubtedly had great impact in the Mediterranean, where there has been a long history of human settlement. Deforestation in Italy occurred at a modest scale during the Roman period (Fabbri 1985a; Postma 1989). Deforestation there accelerated in the Middle Ages when people migrated from the coast to higher and drier regions and needed new land for agriculture and settlement; and large areas were deforested by the middle of the nineteenth century when farming reached its maximum extent (Postma 1989). Deforestation associated with settlement in the USA led to delivery of considerable volumes of sediment in the nineteenth and twentieth centuries, leading to locally high rates of accretion on the Pacific coast where rivers flowed directly to the ocean.

Overgrazing and deforestation had a more direct effect on coastal landforms when these activities were practiced in the dunes themselves. Historical perspective on problems of dune destabilization are presented in Sherman and Nordstrom (1994). Human activity in dune fields in Cornwall, England exists from the Neolithic that occurred from about 6500 to 4500 BP (Lewis 1992). There is evidence of Bronze Age occupation of dune fields in Europe (Higgins 1933). Actual reports of wind blown sand and sand drift date back to the tenth century in continental Europe (Klijn 1990) and prior to 1066 in Wales (Higgins 1933). Dune mobility has been increased by human activities associated with grazing, such as burning plant species to produce more desirable vegetation cover and cutting wood in trees to supply shelter or fuel for shepherds (Corona *et al.* 1988). Problems of dune destabilization appeared to reach their greatest extent in Europe in the eighteenth and nineteenth centuries, when actions at the national scale were taken to stabilize the dunes.

Dune stabilization

Stabilization of drift sands with vegetation plantings was practiced in Great Britain after the stormy periods of the fourteenth and fifteenth centuries (Ranwell 1972). Aforestation occurred as early as the early part of the sixteenth century in Japan (Hotta *et al.* 1991). Aforestation took place in the Doñana Dunes in Spain in 1737 using *Pinus pinea*, but, since then, many other species were used all over the Mediterranean and Portugal using mainly non-indigenous species (van der Meulen and Salman 1996). Aforestation led to commercial activities to make use of the new resources, including lumbering and charcoal-making (Corona *et al.* 1988). The relative success of dune aforestations led to the belief that dunes had to be stabilized with trees whether they were mobile or not (van der Meulen and Salman 1996). Large-scale measures were taken to stabilize drift sand in the eighteenth century in Prussia, Denmark and France (Marsh 1885) and in the nineteenth century in Poland (Piotrowska 1989) and The Netherlands (Klijn 1990). Foredunes for controlling wind blown sand were widely built in European countries in the eighteenth and nineteenth centuries in Europe and in the 1920s in Japan (Hotta *et al.* 1991). Active measures to stabilize drift sand in the USA occurred at Cape Cod, USA before 1775 (Marsh 1885). Stabilization of present-day Golden Gate Park in San Francisco began in 1869 Lamb (1898). Large-scale stabilization projects began on the Oregon coast of the USA in the early twentieth century.

Water regulation activities

Water regulation activities affect coastal sediment budgets, leading to erosion and accretion at the shoreline. These activities may include reclamation, stream channel diversion and stream damming. It is likely that attempts to reclaim land and control flooding have occurred for millennia in population centers near ports. Fabbri (1985a) notes that patterns of canals of Roman Age (2200 to 2000 BP) still exist on the landscape in the plain of the Po River, and he speculates that artificial levees built by the Romans along channels of the Po River affected sedimentation rates on adjacent beaches. A dramatic increase in human influence on European coasts occurred on the Dutch coast between 1100 and 1300 (Berendsen and Zagwijn 1984). This is about the time when dikes are identified on maps in France (Lahousse *et al.* 1993) and when dikes became common in Britain and Germany (Doody 1996; Garniel and Mierwald 1996).

Large-scale stream diversion occurred near the end of the sixteenth century in northeastern Italy when the Venetians diverted several rivers near their city, including the Po and the Piave (Zunica 1990; Bondesan *et al.* 1995). Diversion of the Po was extensive, involving relocation of a portion of the channel more than 5 km long (Bondesan *et al.* 1995).

Dramatic change in coastal sediment budgets occurred in the twentieth century owing to reduction in river sediment supply because of upstream dams and mining of sediment from river beds (Ferrante *et al.* 1992; Marabani and Veggiani 1993). Major periods of dam building affecting coastal sediment budgets occurred prior to World War II in many industrial countries and after World War II in many others (Paskoff 1992). The recent increases in sedimentation rates at dams have reduced the amount of sediment delivered to the coast and reversed the accretional trends that formerly occurred in many areas due to overgrazing and deforestation.

Navigation improvements

Navigation improvement structures are among the earliest human features used to directly control coastal processes, and they are among the earliest structures identified in inventories of human alterations (Leidersdorf *et al.* 1994). Harbor works that are still functioning may be traced back to antiquity (Inman 1974). These structures have great longevity because of their scale (that results in a high degree of survivability) or because of their importance (that justifies rebuilding if they are damaged).

The most profound changes to coasts due to navigation improvements have occurred since the mid nineteenth century, when advances in power machinery facilitated opening and closing of inlets and enabled construction of massive jetties and channel dredging projects to stabilize them. Many jetties were constructed at inlets on the coast of the USA from the late nineteenth century to present. Nearly every harbor in southern California is artificial, being either dredged in low sandy areas, followed by jetty construction, or created by building breakwaters in the nearshore (Wiegel 1994). The coast of New South Wales has over 30 jettied river entrances (Druery and Nielsen 1980). The effects of these human actions extend far beyond the limits of the navigation channels themselves because the alterations affect the sediment exchanges and shoreline fluctuations at a larger spatial scale than the site of the navigation improvement.

Early regulations

Early laws to preserve littoral defences by prohibiting cattle on dikes or removing sand and vegetation date from the thirteenth century in Italy (Franco and Tomasicchio 1992). Laws to control migrating sand date from the thirteenth century in The Netherlands (van der Meulen and van der Maarel 1989) and the sixteenth century in Denmark and Great Britain (Marsh 1885; Gray 1909; Ranwell 1959; Jensen 1994). Government authorities designed to deal with coastal hazards and erosion problems date back to the fifteenth century with formation of the Dutch water boards and the Venetian Water Committee (Franco and Tomasicchio 1992). Laws regulating activities on dikes and in dunes appear to have been the principal actions taken to control activities in coastal landforms prior to the twentieth century and were driven by the need to control specific types of coastal hazard, rather than to protect natural components from irreversible losses. The late twentieth century saw passage of many laws that attempt to reduce losses from coastal hazards and protect the natural environment in the face of dramatic increases in the pace of coastal development.

The growth of tourism

There was little or no interest in direct use of the exposed part of the coastal zone in many countries up to the mid nineteenth century due to the difficulty of traversing lagoons and marshes and the occurrence of malaria (Cencini and Varani 1989). The second half of the nineteenth century saw the beginning of relatively large-scale coastal tourism and development of seaside resorts in many locations (Meyer-Arendt 1990; Ehlers and Kunz 1993; Grechischev *et al.* 1993; Kelletat 1993; Nordstrom 1994a; Fabbri 1996). Contributors to increased use of the coast and change in the character of resorts during that time period were expansion of steamship service and railroad systems, changes in the organization of time between working and non-working hours, reduction in the number of hours in the work week, reduction of time devoted to religious practices on Sundays, and formation of an urban middle class with money and mobility (Fabbri 1990).

Mass tourism occurred after World War II, due to a general increase in national incomes and its distribution to different social levels, combined with increased free time (Cencini and Varani 1989; Ridolfi 1989). The diffusion of tourism has turned many ports and fishing villages into resorts (Fabbri 1989; Meyer-Arendt 1991; Anthony 1997). Automobile access has been the primary stimulus for development in many areas, extending the zone of development beyond centers of mass transit. For example, construction of the Trans-peninsular Highway in Baja California, Mexico in 1973 increased tourist arrivals 500 percent (Fermán-Almada *et al.* 1993).

Second homes have become more popular in recent decades, resulting in a greater amount of environmental degradation and exposure to coastal hazards than occurred in formerly clustered hotel-dependent activities

(Ridolfi 1989; Good 1994). Spaniards vacationing in Cantabria, for example, prefer to buy or rent flats in new high-rise hotels rather than stay in hotels or camping areas, contributing to the extension of urbanization into areas formerly occupied by farms (Fischer *et al.* 1995). Many new homes are detached single-family or duplex structures that use considerably greater space than condominiums. Many homes often are used only for a few weeks a year, raising questions as to whether their cost is worth their use (Fabbri 1989). In other cases, older weekend cottages are torn down and replaced by larger homes (Griggs *et al.* 1991b).

A large number of marinas have been built in recent years, particularly on the Atlantic coast of France and in the Mediterranean, where they are considered a means of drawing income to a municipality and contributing to its prestige (Miossec 1988; McDowell *et al.* 1993; Anthony 1994, 1997). Pleasure boating in Italy was limited and elitist up to the end of the 1960s but has expanded greatly since then as a result of increased leisure time, leveling of standards of living, widespread increase in income and perception of yachting as a less exclusive pastime (Rizzo 1989). The average distance between harbors on the northern Adriatic coast is now 6 to 7 km (Rizzo 1989), yet demand often exceeds mooring capacity (Ridolfi 1989). There are up to 50 marinas on the French coast, where they are considered a threat to the natural environment (Miossec 1993). There are over 4000 harbors in Japan, or about one for every 8 km of shoreline; many of them are built out from shore into the open sea (Walker 1985).

Mobile homes (or caravans) have increased in numbers over the past several decades. These units may be static and rented on site, towed or self-propelled. The direct impact of vehicles varies according to whether they are confined to regulated camping sites or allowed to drive to undeveloped sites where their subsequent use is uncontrolled. Their indirect impact is related to: (1) the increased access of a larger number of beach and dune users to relatively undeveloped areas; (2) increases in the number of pedestrian trips across dunes as users make return trips to their caravans; and (3) damage to surface cover and substrate as users create burial pits for disposal of refuse (Mather and Ritchie 1977).

Off-road vehicle use has increased as well. Small off-road vehicles, "dune buggies," were a novelty at Sand Lake, Oregon, USA until the 1960s, but they numbered in the thousands on weekends in 1979 (Wiedemann 1990). Off-road vehicle registrations at Cape Cod National Seashore USA grew from 966 in 1964 to 5843 in 1978, and 33 378 vehicle passes were made through access points in this seashore between June and September 1976 (Godfrey and Godfrey 1981). The Aberffraw sand dune system in Anglesey, Wales had 3.2

km of vehicle tracks and 2.2 km of footpaths in 1960 and 11.7 km of vehicle tracks and 16.5 km of footpaths in 1970 (Liddle and Grieg-Smith 1975a).

Stages of landscape conversion through tourism

The relationship between natural coasts and human-modified coasts may be presented in terms of generic scenarios, representing the results of applying different human values and levels of investment. Most coastal communities undergo incremental development that progresses through stages. The stages in many resorts in Thailand, for example, include: (1) construction of simple low-budget visitor dwellings; (2) upgrading of these structures as visitor numbers increase; (3) selling of land to developers; (4) construction of hotels to meet increasing demand; and (5) expanding buildings and infrastructure while ignoring legislation on zoning and land use (Chou and Sudara 1991). Models for growth of European and North American resorts present growth in terms of: (1) an exploration stage; (2) a period of commercial involvement and infrastructure development; (3) a settlement-expansion stage; and (4) an increase in intensification of sites already developed (Butler 1980; Meyer-Arendt 1990, 1993a). The end of the settlement stages is a maturation stage when all potentially developable land has been developed as either low or high density, and levels of tourist visits have stabilized (Meyer-Arendt 1993a). By this stage, human-induced environmental degradation is often recognized and translated into government controls (Meyer-Arendt 1993a). Levels of maturation vary from site to site, reflecting a combination of physical and cultural attributes, land use regulations and market demands; locations that developed prior to restrictive legislation can mature within the stage of land use intensification, whereas locations that developed more slowly may have been halted in earlier stages (Meyer-Arendt 1993a).

Meyer-Arendt (1985, 1990, 1991, 1992, 1993a) reviews the characteristics of seaside resorts and how their form has changed since the nineteenth century. Resorts in the nineteenth century reflected an urban spatial structure attributed largely to the mode of access (steamship or railroad), resulting in a concentration of tourists at nodes. Concentric zones of human activity and infrastructure emanated from locations where transportation routes ended, either near the shorefront or at docks or railroad stations somewhat inland from the beach. These zones (Figure 1.4) have been subject to redefinition through time, but they include: (1) the core central business district (CBD), often of compact shape; (2) the recreational business district (RBD), usually of linear shape, corresponding to the orientation of the beach and swash/surf zone that is the principal attraction of coastal resorts; (3) an accommodation

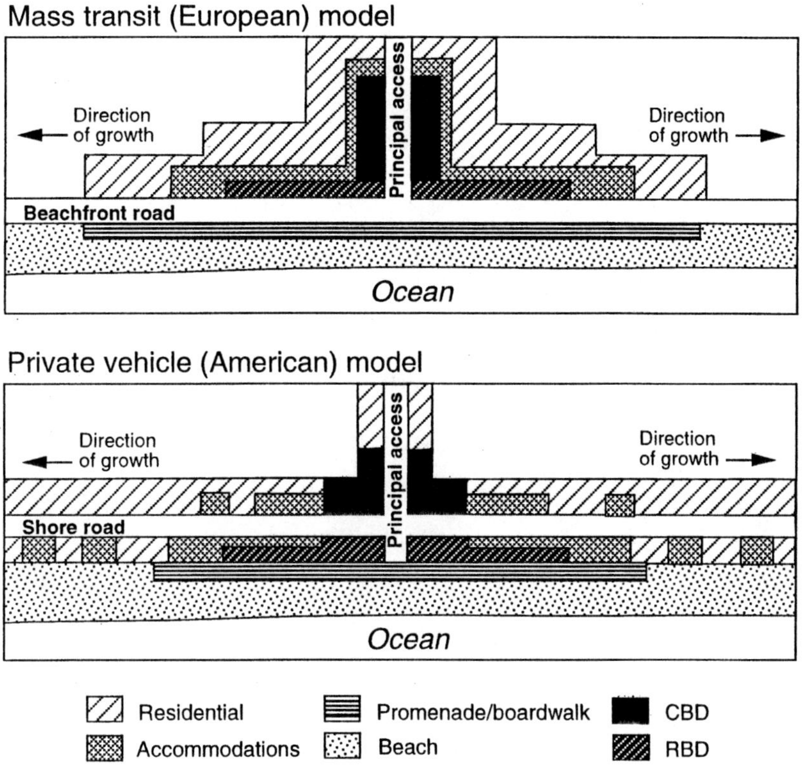

Mass transit (European) model

Private vehicle (American) model

∅ Residential	≡ Promenade/boardwalk	■ CBD
▨ Accommodations	⠿ Beach	▧ RBD

Figure 1.4 Relationship of recreational business district and central business district to beach and transportation routes. Synthesized from models presented in Meyer-Arendt (1990) and visual observations of seaside resorts.

area (hotels, boarding houses) around both the central business district and recreational business district; and (4) a residential zone for non-tourists or (more recently) owners of second homes. The recreational business district may be distinguished from the central business district in both form and function. The central business district may offer a full range of retail goods and be centrally located with respect to permanent residents; the recreational business district is located close to the beach with maximum accessibility to lodging facilities and travel routes (Stansfield and Rickert 1970; Meyer-Arendt 1990).

The tourist structures in the recreational business district were often (as now) separated from the beach itself by a pedestrian promenade, a highway, or both (Figure 1.5). Many resorts on the northeast coast of the USA have an elevated boardwalk instead of a ground-level promenade. Promenades (and boardwalks) may be built out onto the landward portion of the beach (Fernandez-Rañada 1989), and they usually end at the longshore termini of

Figure 1.5 Nice, France showing linear recreational business district, beachfront promenade and thoroughfare.

the recreational business district and do not extend along the shoreline fronting isolated residential structures.

The nodes of developments were T-shaped where reliance on mass transit produced a limited number of routes or disembarcation points (Figure 1.4). Initially, automobile access created a T-pattern of development, with the focus being one or two central beach hotels with amusement facilities. The recreational business districts expanded by increasing concentration of businesses in these core areas and lateral elongation along the shoreline as a recreational frontier (Meyer-Arendt 1990, 1991). Expanded use of private vehicles resulted in elimination of railroads as a principal mode of access, and many rail lines to seaside resorts have closed (Gale *et al.* 1995). The use of air travel by international tourists has created a large-scale pattern where high-value, low-density leading-edge development in the form of villas are farthest removed from the airport, with low-value, high-density properties in high-rise apartments and condominiums closer to it (McDowell *et al.* 1993).

Lateral elongation can occur as contiguous development or leap to new nodes that represent *loci* of speculative development or conversion of pre-existing coastal communities that add tourism to traditional economies. Development and expansion of resort communities is strongly tied to developments in transportation. The new transportation links can be built under speculation, specifically for tourism, such as at Atlantic City, New Jersey (Nordstrom 1994b) or they can be built for economic values that precede demand for tourism, such as at Progresso, Mexico (Meyer-Arendt 1991).

Seaside resorts have changed considerably when compared with their nineteenth-century counterparts in that they have diversified to incorporate retail residential and industrial functions akin to inland towns and cities. Increased mobility allowed by private vehicles and proliferation of shopping chains into resorts have blurred the sense of place and quality of resorts as an exclusively coastal experience (Gale *et al.* 1995).

The user population can come from local non-coastal urban areas or can be international, and much of the character of coastal resorts reflects the mix of user populations. Former characteristics associated with folk culture and specific characteristics of the natural environment are altered by the imprint of popular culture, creating a new landscape layer (Meyer-Arendt 1990). For example, ports that were part of the tourist folklore and considered part of the landscape have been replaced by massive developments driven by tourism and fueled by competition among seaside resort communities trying to capture readily available private investment (Anthony 1997).

Stages of shore protection

Towns threatened by erosion in the distant past were allowed to succumb when their time came, or they essentially migrated inland with the new coastline (Paskoff 1987). Written reports of shore protection works date back to AD 537 in Italy (Franco and Tomasicchio 1992). Seawalls built of timber piles and stones were common on the coastal barriers near Venice in the seventeenth century and were built to large scale in the mid eighteenth century (Franco and Tomasicchio 1992; Marchi 1992). Protective works on the Nile Delta coast began as early as 1780 (Fanos *et al.* 1995). Groins were built more than 200 years ago on the Dutch coast (Pluijm *et al.* 1994). Groin building activities began at Atlantic City, New Jersey between 1857 and 1876 (Weggel and Sorensen 1991), at Norderney, Germany beginning in 1857 (Kunz 1990), the west coast of Denmark in 1875 (Laustrup 1993), Hawaii in 1900 (Fletcher *et al.* 1997).

Large-scale shore protection projects became common in the past 100 years as availability of heavy machinery and the benefit–cost ratio of protecting infrastructure of ever-increasing scale became more favorable. The first large-scale shore protection project in the USA occurred early in the twentieth century with construction of the Galveston seawall and raising of the height of the island near the beach (Wiegel 1991).

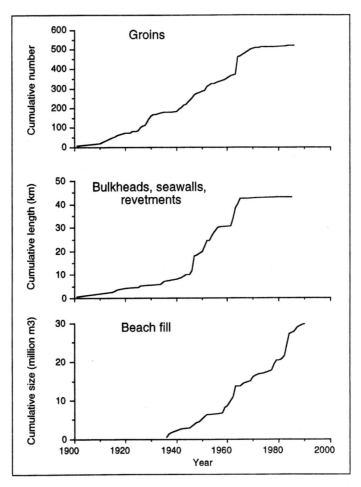

Figure 1.6 Trends in the implementation of shore protection projects in New Jersey since 1900. Modified from Nordstrom (1994a).

Shore protection methods change in frequency of implementation, depending on changing preferences of residents and attitudes of planners and engineers. Examination of these changes (Figure 1.6) often reveals an early preference for groins, followed by a period of construction of shore-parallel structures, to a period of beach nourishment that is currently favored (Caputo *et al.* 1991; Kana 1991; Paskoff and Kelletat 1991; Nordstrom 1994a). In some locations, such as Italy and Japan, offshore breakwaters have been in vogue, with submerged breakwaters becoming popular relative to traditional emergent structures (Lamberti and Mancinelli 1996). The number of detached breakwaters in Japan increased from 2305 in the early 1980s (Toyoshima 1982) to about 4800 by 1989 (Silvester and Hsu 1993). Sawaragi (1988) reports a growth rate of offshore detached breakwaters at about 10 times the growth rate for construction of dikes and seawalls from 1970 to 1985.

In some locations, different types of structures are added to the protective infrastructure through time as erosion proceeds. The shoreline of Kaike, Japan has undergone the following stages: (1) a series of groins built from 1947 to 1955; (2) a seawall built from 1955 to 1961; (3) emplacement of concrete blocks in front of the seawall in the 1960s; and (4) a detached breakwater system built seaward of all of these structures from 1971 to 1981 (Toyoshima 1982). In other cases, a similar type of structure is used, but its dimensions and cost increase through time as each structure fails (Pilkey 1981; Kana 1983; Reynolds 1987). Structures built initially as backup protection are replaced by structures designed to provide the principal defense, as has occurred on the east coast of the USA, where bulkheads have been replaced by seawalls.

Relatively large-scale beach nourishment projects were used to improve urban resorts in the USA as early as 1922 (Dornhelm 1995). Nourishment became popular in the USA in the 1950s (Domurat 1987). The first large-scale beach nourishment project in Europe was on Norderney, Germany in 1951–52 (Kunz 1993a). The first planned nourishment in France was in Cannes in 1960–61 (Anthony and Cohen 1995). Nourishment became popular at the end of the 1970s in Poland (Zawadzka-Kahlau 1995) and in the 1980s in Georgia (Zenkovich and Schwartz 1987). Since the 1970s, 90 percent of the US federal appropriation for shore protection has been for beach nourishment (Sudar *et al.* 1995). Over 60 percent of all sand emplaced on the Gulf Coast of the USA was emplaced during the past 16 years, with approximately 25 percent of all sand emplaced within the past 6 years (Trembanis and Pilkey 1998). Beach nourishment is increasingly favored because of the decreasing real cost of sand (Laustrup 1993).

Use of shore protection structures has increased along with the increase in construction of buildings and support infrastructure. A total of 80.9 km of seawalls, 11 km of breakwaters and 914 groins over a length of 54.8 km have been constructed on the Russian Black Sea coast between Tuapse and Psou since the 1950s (Grechischev, *et al.* 1993). 178 groins were built over an 8.8 km reach in Yucatan in Mexico (Meyer-Arendt 1991). 1300 groins extend over a distance of 125 km in France (Monadier *et al.* 1992). Over 130 groins were built on the 4.4 km long Tybee Island, Georgia since the late 1800s (Olsen 1996). The percentage of the California, USA coastline protected by structures increased from 42.9 km in 1971 to 210.6 km in 1989 (Griggs 1995).

The proportion of coastline that is protected against flooding and erosion or that is significantly affected by human development varies from high estimates of about 100 percent for Belgium (De Moor and Bloome 1988; De Wolf *et al.* 1993) to low estimates of about 12.9 percent for Italy (Cencini and Varani 1988). The national averages may obscure the intensively developed and protected shorelines where low coastal formations and desirable beaches have been foci for resort development. Coastal defense structures have been built along 60 percent of the coast of northeastern Italy from Grado to Rimini, including 56 km of groins and breakwaters and 55 km of seawalls (Bondesan *et al.* 1995).

The coastline of Japan has been drastically altered by massive engineering projects (Nagao 1991; Watanabe and Horikawa 1983). As many as 10 043 groins were reported in Japan by the early 1980s, along with 5579 km of seawalls and 2838 km of dikes (Toyoshima 1982). The total length of protection structures along the coast of Japan is over 10 000 km (Nagao 1991). Many coastal areas in Japan have been built as human artifacts and bear little resemblance to the coast that formerly existed (Kawaguchi *et al.* 1991; Nagao and Fujii 1991).

Case study: development of an intensively developed coast

The preceding sections provide a brief survey of only a few of the large-scale landscape conversions that have occurred over millennia in a variety of different locations characterized by different user populations and different economic and social constraints. This section identifies trends in one segment of coast, the State of New Jersey, USA, focussing on activities that have occurred since human alterations became dominant (corresponding to the steep rise in the curve on Figure 1.3).

Figures 1.7 and 1.8 synthesize changes that occurred to the barrier islands of New Jersey since early development initiatives in the mid nineteenth century. Beaches at the New Jersey shoreline had some limited use as a recreational resource as early as 1790 (Domurat 1987), but only a few hunting cabins, homesteads and boarding-houses existed on the islands prior to the mid and late nineteenth century (Sea Isle City 1982; Koedel 1983). This period corresponds to the exploration stage, when access is initially difficult, and human impact on the physical environment is limited to access paths to the beach (Meyer-Arendt 1993a). The barriers were low and narrow prior to intensive human modification, and they were backed by salt marsh and fronted by foredunes that formed broad-based ridges in relatively stable portions of the barriers (Transect A, Figures 1.7 and 1.8 at Phase 1) and hummocks in more dynamic areas, such as near inlets (Nordstrom 1994a). The upland portions of the barriers had lush growth of cedar, holly and other trees and a variety of grasses (Sea Isle City 1982). Several New Jersey barriers appeared to be highly mobile prior to human development; one of these barriers migrated across its

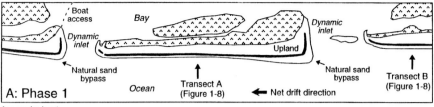

A: Phase 1

Access by boat
Scattered buildings
Paths to water

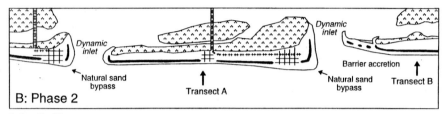

B: Phase 2

Railroad building
initial development at access nodes
Local elimination of dunes

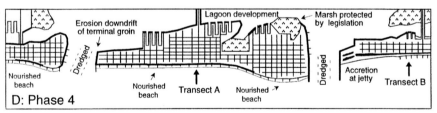

C: Phase 3

Construction of road networks and new settlement nodes
Widespread elimination of dunes
Beginning of marsh filling
Dredging of inlets
Emplacement of groins and bulkheads at eroding settlements

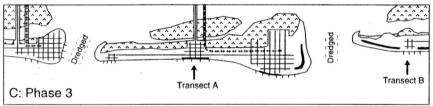

D: Phase 4

Construction of lagoon residences
Widespread use of groins and bulkheads
Beach nourishment in intensively developed areas
Jetty construction at inlets
Restrictive environmental regulations

Marsh   Dunes   Railroad   Bulkhead
Upland   Roads   Groins   Jetty

Figure 1.7 Plan view of phases representing the development and protection of coastal barriers in New Jersey, USA through time. Transects A and B are presented in profile view in Figure 1.8 Modified from Nordstrom (1994a).

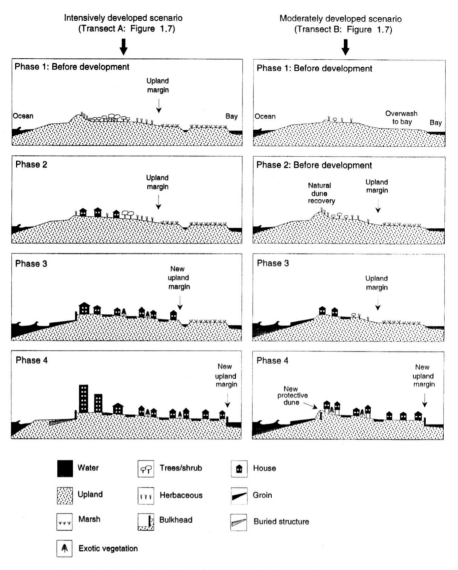

Figure 1.8 Profile view of phases representing the development and protection of coastal barriers in New Jersey, USA through time. Transects A and B are presented in plan view in Figure 1.7 Modified from Nordstrom (1994a).

width in no less than 200 years (USACOE 1957). Locations adjacent to inlets were especially mobile (Figure 1.7, Phases 1 and 2).

Railroad lines, built to support real estate speculation ventures, extended along 57.8 percent of the New Jersey shoreline by 1885–86 (Nordstrom 1994a). The resulting land use pattern reflected a clustering of recreational business districts near railroad stations on the upland portions of the barriers (Figure 1.7B). The infrastructure development stage is a time when landscaping often is accompanied by removal of geomorphic features and replacement of upland vegetation by introduced species (Meyer-Arendt 1993a). Modifications on the New Jersey barriers included grading dunes to a flatter form to facilitate construction of buildings and roads (Figures 1.7 and 1.8, Transect A in Phase 2 and Transect B in Phase 3) and destruction of the natural vegetation, that was often dense shrub.

The settlement expansion stage in growth of coastal resorts was rapid following construction of railroads and the first permanent business districts. The permanent population of Atlantic City was nearly 5500 by 1880 (Funnel 1975). Land values at Atlantic City grew from almost nothing in 1854 to $50 million in 1900 (Domurat 1987).

The barrier island resort industry in New Jersey developed rapidly in the early twentieth century, due primarily to increased use of private automobiles (Koedel 1983), accelerating the pace of the settlement expansion phase. New isolated communities appeared, in some cases linked by roads rather than railroads. Growth then extended outward from these locations, both alongshore on the upland portion of the barriers and bayward onto the marsh surface. Filling of the marsh behind the upland occurred on several barriers between 1886 and 1902 (Figure 1.7C), and the marsh at Ocean City was filled all the way to the edge of the backbay by 1907 (Nordstrom 1994a).

The settlement expansion stage involves the transformation of the physical environment into a cultural one (Meyer-Arendt 1993). Shore protection measures begin to be employed in the settlement expansion stage, although the intensity of development may be too limited early in this stage for public works to be cost-effective (Meyer-Arendt 1993a). Efforts were made to control erosion in New Jersey as early as 1847 at Cape May (USACOE 1957). The widespread use of protection structures came in the settlement expansion phase after 1900 (Figure 1.7C).

Dredging of channels into the backbarrier marsh to accommodate boats occurred in the larger settlements between 1905 and 1913. Large-scale conversion of the marsh to lagoon developments for private housing and their associated boat docks (Figure 1.7D) occurred on many of the barriers after World War II. Most of these projects placed materials dredged from the new waterways on the marsh surface to provide a substrate for houses and roads. Construction of lagoon housing is now severely restricted by regulations governing use of wetlands, but marshland is now found only in isolated enclaves on the back sides of the developed portions of barriers in New Jersey (Nordstrom 1994a).

Many inlets that existed in 1885–86 and all inlets that formed since that time were closed artificially by the US Army Corps of Engineers or kept from reopening after natural closure to eliminate undesirable shoreline fluctua-

Figure 1.9 Atlantic City, New Jersey, showing juxtaposition of high-rise buildings and beach. A geotextile tube that will form the core of an artificial dune is seen on the beach at lower right.

tions, facilitate land transportation or increase the hydraulic efficiency of nearby controlled inlets. Navigation improvement projects at inlets were undertaken to maintain a predictable navigation channel and a static shoreline. Five of the 12 New Jersey inlets that now exist are stabilized by jetties, constructed since 1911. Two of the unjettied inlets are maintained by dredging, changing the amount of sediment transferred across the inlets and the location of accretion and erosion on adjacent shorelines.

The stage of land use intensification (including construction of high-rise condominiums and hotels and replacement of pre-existing structures with structures of higher density) is not a requisite stage of resort evolution, but it is characteristic of resorts of high recreational demand (Meyer-Arendt 1993a). The proximity of the New Jersey shoreline to the large urban population centers of New York City and Philadelphia provided the stimulus for further intensification. Restrictions to the number of units per building in New Jersey have limited high-rise constructions to only a few locations, such as Atlantic City (Figure 1.9), where the buildings are so large and so close to the beach that they profoundly affect wind processes and aeolian transport. Other communities are characterized by a high density of multiple-unit low-rise structures. Most communities are moderately developed in detached houses that are fronted by a narrow dune (Figure 1.10), a bulkhead or both.

A level of development such as that portrayed along Transect A (Figure 1.8)

Figure 1.10 Long Beach Island, New Jersey, showing narrow beach and linear dune characteristic of shorefront developed in detached houses on an eroding barrier island.

has been sufficiently intensive to justify use of large-scale nourishment operations to protect buildings and infrastructure. Human actions along a coast with the level of development portrayed on Transect B indicate a reliance on less costly public expenditures and include dune building and flood proofing (accomplished by raising buildings) with a bulkhead providing back-up protection.

The greatest change in the dimensions of the New Jersey barriers due to development is the dramatic increase in the width of upland through filling of the marsh (Figures 1.7 and 1.8). The maximum elevations of isolated dunes in developed areas and in presently undeveloped areas are generally higher than they were in 1886 (the time of the earliest state topographic surveys), but they do not extend as far inland. Dunes do not exist in many areas that are now protected by bulkheads and seawalls. There are numerous locations where dunes exist but are no higher than 1.5 m above the elevation of the backbeach and no wider than 15 m (USACOE 1990). Despite this limitation to their human utility value as a form of flood protection, the dunes are a more continuous barrier than they were prior to development in 1886, and they are more likely to restrict barrier island mobility. Several communities in New Jersey are now protected by a combination of beach nourishment, groins, bulkheads and dunes. Many communities employ at least two of these methods of protection (Nordstrom 1994a). Over 43 km of the 205 km long

New Jersey shoreline is protected by shore-parallel structures, and there are over 300 groins (USACOE 1990).

Offshore depth contours up to 5.5 m below MLW have generally moved landward on the New Jersey coast since the beginning of development in the mid nineteenth century (USACOE 1957), but the upper foreshores have undergone less erosion since development, and several barriers have an accretional trend (Everts and Czerniak 1977; Dolan *et al.* 1978; Galvin 1983). As a result, there has been little onshore migration of the intertidal shoreline over the past century (Nordstrom 1994a), but the slope of the subtidal shoreline has probably steepened.

Major storms on the New Jersey coast have periodically eroded beaches, destroyed or damaged buildings, boardwalks, support infrastructure and protection structures, but post-storm reconstruction efforts have rebuilt these facilities, usually to larger proportions. In most cases, the seaward-most line of human structures is located no farther landward than where it was prior to the storm. The result of human activity on the New Jersey shore has been an increase in the number of buildings subject to storm hazards, accompanied by a decrease in the role of natural processes in altering the extent of natural landforms and their mobility or the configuration of the islands or the likelihood that they will migrate (Nordstrom 1994a).

Recent growth trends

Economic and social forces

There is no reason to expect that the pace and scale of coastal development will decrease on a global basis. The primary determinant of development in beachfront communities is growth demand, based on rising income and employment in inland areas (Hillyer *et al.* 1997). Coastal resources and coastal tourism are foundation blocks supporting economic and social development programs (Sorensen and Brandani 1987; Huber and Meganck 1990). International tourism is now a rapid growth industry that plays a key role in developed countries (Houston 1996a) and is perceived as the most important economic sector to be exploited in many developing countries (Huber and Meganck 1990; Awosika and Ibe 1993; Pearsall 1993). This tourism development is often funded by foreign investors (Freestone 1991), increasing the economic base required to develop new coastal areas. National and state governments often become catalysts in resort developments by passing tourism-encouragement laws, altering economic policies (e.g., providing free trade status, income-tax relief, altering laws on foreign investments and land use), building or providing subsidies for access routes and infrastructure, creating development corporations or entering into partnerships with entre-preneurs (Atherley *et al.* 1991; Awosika *et al.* 1991; Chou and Sudara 1991; Awosika and Ibe 1993; Bringas Rábago 1993; Domroes 1993; McDowell *et al.* 1993; Schmahl and Conklin 1991; Wong 1993).

Developers construct buildings as close to the beach as possible, often in contrast to the locations preferred by former local residents (Kelletat 1993). In many locations, growth of the tourist industry has occurred with little direc-tion and control and lack of governmental guidance (Wong 1990; Fermán-Almada 1993). Construction is frequently incompatible with beach dynamics because selection of sites for resorts is based on non-geomorphological factors (Wong 1990). Environmental awareness in these locations may be lacking among tourists and entrepreneurs alike (Domroes 1993), and there may be little local expertise in the smaller countries to deal with environmental issues (Cambers 1993a).

Rapid development of tourism can result in many socio-economic disad-vantages, including crowding, pollution, lowered skill levels, diminished coastal access and alienation of the former resident populations (Atherley *et al.* 1991; Cervantes-Borja and Meza-Sanchez 1993; McDowell *et al.* 1993). Paradoxically, there are now low living standards in locations originally selected for quality tourism (Fabbri 1985b). Access to beaches may be restricted for local people, resulting in loss of traditional activities and func-tions (Knight *et al.* 1997).

Many countries are now facing major conflicts between economic recovery and sustainable development of coasts (Savov and Borissova 1993). Much of the natural beauty of the environment is subsequently destroyed through overuse and pollution (Bunpapong and Ausavajitanond 1991; Chou and Sudara 1991), and management plans that could prevent this destruction are hampered by limited funding levels and public and governmental under-standing and support (Premaratne 1991).

Development is often at the expense of physical features and ecological values that are not appreciated by developers who may favor activities that do not require a seashore setting (Demirayak 1995). One of the worst aspects of tourism development is the narrow conception of the value of beaches as rec-reation platforms rather than dynamic natural environments. Much of the scenic beauty of human-altered coastal environments is human-constructed (Chou and Sudara 1991), and the definition of beauty has been redefined by the nature of both the cultural facilities and the reshaped landforms.

The pace and scale of recent development

Construction of buildings and resort infrastructure resulting in coastal urbanization depends in large part on the levels of disposable income that are often tied to economic booms (Meyer-Arendt 1991). Major growth in Miami Beach USA occurred in the 1930s (Wiegel 1992b). Booms in the USA and Mexico occurred soon after World War II (Meyer-Arendt 1993a). Development of coastlines has accelerated in much of the world in the last few decades (Wong 1985, 1988; Koike 1988; Cencini and Varani 1988; Moutzouris and Maroukian 1988; Valentini and Rosman 1993). The Lido di Jesolo in Italy was almost completely uninhabited until the 1950s, but all the space was soon occupied, and the entire shoreline was protected by shore protection structures by 1978 (Zunica 1990). Marco Island, Florida was altered from a wilderness to a fully developed shoreline, including protection structures, in fewer than 20 years (Reynolds 1987). Coastal population densities in the USA went from 96 persons per square kilometer in 1960 to 132 persons per square kilometer by 1988, when density was four times the national average (Beatley *et al.* 1994). Approximately 80 percent of the population of California lives within 50 km of the coastline (Griggs 1994). The population of the Crimean Black Sea coast increased 25 percent from 1970 to 1981, and the simultaneous number of guests on the coast of the Black Sea is now more than 10 million persons (Aibulatov 1993).

The effort of Thailand to increase international tourism has produced spectacular economic results. Accommodations in Pattaya expanded from one hotel to over 25 000 tourist rooms from 1965 to 1990 (Branan *et al.* 1991). Ko Samui had an increase in visitor arrivals from 15 000 to 300 000 between 1980 and 1987 (Chou and Sudara 1991), while, in Patong, the number of guest rooms increased from 364 to 3067 between 1981 and 1989 (Bunpapong and Ausavajitanond 1991).

The coast of the Mediterranean has recently experienced accelerated construction to accommodate fast-growing tourist and residential populations, resulting in decreased space for both natural areas and recreation (Fabbri 1985b; Cencini *et al.* 1988; Haeseler 1989; Warne and Stanley 1993). Much of the alteration of the French Riviera shoreline occurred from 1965 to 1980 (Anthony 1997). About 3500 km or 50 percent of the Spanish coastline has been developed for tourism over the past 40 years (Figure 1.11), with the Costa del Sol experiencing at least a tenfold increase in visitors over the 30 years prior to 1990 (McDowell *et al.* 1993). The Mediterranean basin received more than 100 million international tourists during the peak season, and this number is expected to double by the year 2020 (Hawkins 1996). Almost 75 percent of the coastal dunes in the Mediterranean have been damaged or destroyed by tourism (van der Meulen and Salman 1996), and the western Mediterranean dune region has been described as "one of the most seriously endangered ecosystems of our time" (Haeseler 1989). Even in undeveloped

Figure 1.11 Benicasím, Spain, 1994, showing rapid transformation of coastal landscape through massive construction of dwelling units and promenade on landward side of the shoreline and artificial fill and groins on seaward side.

areas, the landscape has been profoundly altered by many different kinds of human activities in the past (Corona *et al.* 1988; van der Meulen and Salman 1996). There is an increasingly more uniform pattern of shore-based recreation in the Mediterranean (Anthony 1997), and the development process is obscuring the natural characteristics and the former cultural characteristics in areas now developed as resorts. Throughout the world, the old cultural landscapes and their relationship to natural features are being replaced by a new cultural landscape with a new emphasis on natural features as recreation platforms and high-density development as the best use of beach front land (Schmahl and Conklin 1991).

Prospect

The pace, types and locations of development are dictated by investment markets as well as by consumer demands. Booms in the late 1960s and mid 1980s in southern Spain were followed by downturns linked to the 1970s oil crises, fluctuating European currency regulations or exchange rates, bad publicity (health scares, crime) and changing fashions (McDowell *et al.* 1993). Competition from cheaper (in development and running costs) resorts have made economic problems more difficult on some well-established coasts (Anthony 1997). Despite these fluctuations, conversion of the shoreline is inexorable. Isolated shorefront communities may undergo periods of stagnation, disinvestment and economic decline due to obsolescence of facilities, diversion of investment interest elsewhere or loss of beach resources (Meyer-Arendt 1990; Stansfield 1990; Paternoster 1991; Hoffman 1992), but they often experience further growth when economic resources become available. Some shorelines that are difficult to alter, such as rockbound shores, may enjoy relative immunity for a time because of prohibitive costs involved in altering them, but, with increase in demand and the use of appropriate technology, these coasts eventually become altered (Anthony 1997).

The availability of technology and capital now enables protection of large portions of coastline that would not have been protected in the past. Efforts may now be devoted to protecting dynamic undeveloped shorelines. One example is the Dutch island of Rottumeroog, where spontaneous public discussion of the eroding island led to formation of a pressure group to raise money for continued maintenance to lengthen the life of the island, despite evidence that new islands develop in that location over the course of time (Koster and Hillen 1995). Efforts in The Netherlands to apply active human inputs to enhance nature conservation and to retain a dynamic system document the changing role of humans from consumers of environmental resources to regulators of environmental resources.

The ever-increasing pace and scale of human alterations due to technological advances, and the availability of funds to modify the landscape to achieve human goals, indicate that humans are or soon will be the dominant agent in landscape evolution in coastal locations that have human value. Even in communities that undergo economic decline, buildings and shore protection projects remain in place to affect coastal processes. These structures can affect processes and shoreline morphology millennia after they are constructed and have lost their human value. Paskoff and Oueslati (1991) note that 44 of 89 archaeological sites along the Gulf of Gabes in southern Tunisia are presently affected by marine erosion.

Human dominance does not necessarily have to be detrimental to beaches and dunes. Intensively managed coasts can evolve in dramatically different ways in response to differences in social processes and perceptions of human functions and values of coastal landforms. Environmentally friendly methods can be found to use and manage beaches and dunes. Subsequent chapters identify the significance of differences in management at the landform scale; the ways human actions affect the value of these landforms for natural and human use values; and the ways that human actions can be directed toward compatible uses.

2

Altering landforms to suit human needs

Introduction

Beaches and dunes are eliminated, altered through use, reshaped, remobilized, stabilized or created to suit human needs (Table 2.1). Many of these changes are designed for multiple purposes and may involve several different kinds of human processes occurring in several stages (Pye 1990). The alterations may be intentional (where the landform is considered less suitable than the alternative) or unintentional (where people change the sediment budget or vegetation characteristics that, in turn, change locations of deposition and erosion). Some uses may be considered consumptive, in that the landform is totally eliminated from a specific location (e.g., the dunes in the background of Figure 2.1). Other uses may reshape or even create landforms, but the form and genesis of the feature will be a function of human needs.

Many alterations are designed to create one kind of landform at the same time that they destroy a different kind. As a result, it is difficult to classify human actions in mutually exclusive categories. The major headings used in this chapter are based on degree of alteration, beginning with alterations that completely eliminate landforms or prevent them from responding to natural processes by converting their surfaces to cultural features. Subheadings are based on the rationale for the alteration (in terms of resource value) that is responsible for the landscape conversion. The alterations described in this chapter are of large scale and include changes in the form and surface characteristics. Creation of new landforms and increases in their dimensions through beach and dune nourishment are assessed in chapter 3. Changes due to the effects of hard structures (including sand fences) or fixed physical plants are discussed in chapter 4. The significance of departures of human-altered landforms from their natural equivalents in terms of physical characteristics and landform classification and evolution is discussed in chapters 5 and 6.

Table 2.1 *Ways that landforms are altered to suit human needs (modified from Nordstrom 1998)*

Eliminating for alternative uses
    Facilitating construction
        Transportation routes and terminals
        Parking areas
        Alternative recreation surfaces (e.g., golf courses)
        Buildings
    Non-coastal (agriculture, landfills)
    Mining
        Construction aggregate
        Minerals
        Liming material and substrate for crops
        For covering landfills

Altering through use
    Pedestrian trampling and vehicle use
        For access
        For direct recreation
    Harvesting
        Gathering wood, seaweed, flowers, fruits
        Removing vegetation for fuel, thatch
        Planting forests
        Grazing
    Waste disposal
        From day-use tourist activities
        Random disposal of cars, machinery
        From beach cleaning
        From commercial (mining) activities
    Extraction and recharge
        Drinking water
        Oil and gas
        Watering gardens and waste-water disposal
        Concentrating surface runoff
    Military
        Active uses (bombing, maneuvers)
        Fortresses and bunkers
        Harbor structures
    Cemeteries
    Inscribing graffiti and carving caves

Reshaping
    Increasing levels of protection
        Grading beaches
        Bulldozing dunes
        Breaching barriers to control flooding
        Dredging inlet channels to cause deposition
    Preventing or alleviating hazards
        Sand inundation
        Cliff failure

Table 2.1 (*cont.*)

Enhancing recreational use
    Widening beaches for recreation platforms
    Creating platforms for cabanas, pavilions
    Clearing the beach of litter
    Eliminating obstacles to access
    Providing or retaining views of the sea
    Maintaining navigation channels
Enhancing environmental values
    Creating more naturalistic landscapes
    Altering environments for wildlife

Altering landform mobility
    Introducing new sediments into beach, dune
    Closing, stabilizing or creating inlets
    Relocating channels or altering cycles
    Placing barriers to trap sand
    Stabilizing dunes
        Planting vegetation
        Armoring surfaces
    Remobilizing dunes
    Altering natural vegetation
        Controlling density (mowing, grazing, fires)
        Planting species to increase diversity
        Introducing exotics

Altering external conditions
    Damming streams
    Mining streams

Eliminating landforms

The emphasis of this book is on beaches and dunes that presently exist or that can be created. Accordingly, little emphasis is given to the processes that eliminate dunes in ways that prevent their re-formation at or near those sites. In many cases, the shape of the landform survives but its surface characteristics and natural and human functions may be altered (e.g., golf courses). In other cases, human actions totally remove all evidence of the original landform (bulldozing dune fields for airports). Often, the landform is truncated or re-created by artificial means at an adjacent location, such as when beaches are separated from dunes as a result of construction of a shore-front road or when dunes that are leveled for construction are rebuilt seaward of the new structures. The characteristics and significance of these new landforms is treated in subsequent chapters.

Figure 2.1 North of Pietrosanta, Italy, showing pronounced difference in characteristics of the shore under relatively natural conditions and altered to enhance recreation.

Facilitating construction

Dunes have been eliminated for construction of buildings and roads in nearly every country. Elimination of dunes close to the beach for construction of human facilities may now be prevented or highly regulated in many countries, but this elimination is still widespread in countries where environmental regulations are not as strict. Even in countries with strict controls, restrictions may not apply to inland portions of the dune where the threat of storm hazard or restrictions to coastal access are not perceived to be an issue.

Roads and railroads may be built on the beach or on dunes that are graded to facilitate construction. These features are common along coasts with mountainous interiors, such as along the Mediterranean (Cencini and Varani 1989; Peña *et al.* 1992). Parking lots have been built right on the beach, and they are a conspicuous feature in some regions, such as southern California. Parking lots eliminate only a portion of the beach environment, whereas transportation routes may create a continuous strip of development alongshore and truncate the beach, separating it from inland dunes.

Construction of golf courses occurs on the dunes (Mather and Ritchie 1977; Doody 1989), and thousands of hectares have been converted to this use along the Mediterranean coastline alone (van der Meulen and Salman 1996).

Creation of golf courses in dunes often involves destruction of pre-existing relief, soil type, vegetation and natural wildlife, accompanied by introduction of rich soil or compost, agricultural grass species, fertilizers, pesticides and water (van der Meulen and Salman 1996). Golf courses, like parking lots and landfills, eliminate only selective portions of the landforms.

Bluff tops may be graded for construction right up to the edge of the bluff, destroying surface vegetation cover, diverting drainage and loosening compacted surface materials (Kuhn and Shepard 1980). These alterations may initiate new slope failures (Kuhn and Shepard 1980), extending the impact to locations that are not altered directly. Alternatively, the slopes of these bluffs may be landscaped by property owners (Lee and Crampton 1980), changing both the initial appearance of the bluffs and their evolution.

Normally dunes are graded prior to development to facilitate construction. In most cases, the grading involves leveling the previous natural dune (and perhaps building a smaller dune seaward of the building). Dunes may be left relatively intact, by grading only those portions required to accommodate individual houses or parking areas. Occasionally, dunes are graded into creative forms, such as sequentially lower terraces to create a topography suitable for viewing the sea from inland buildings (Healy *et al.* 1990).

Grading of dunes to facilitate construction removes stabilizing vegetation and can initially increase sediment mobility and sand transport rates (Gusmão *et al.* 1993), but this mobility is short-lived. Subsequent replacement of the natural surface by a cultural one reduces the area of the original landforms, eliminates its value as habitat or source of sediment or seeds for adjacent natural environments, and truncates previous natural environments, reducing their size, mobility, and potential for growth or expansion. These secondary effects (discussed in chapters 5 and 6), as well as the effect of bringing the locus of human activity closer to natural environments, extend the zone of potential landform alteration beyond the locations where landforms are eliminated directly. Greater attention in this book is placed on these other impacts and the landforms generated by them than the landforms that have already been lost.

Agriculture

Agriculture and harvesting have been practiced in dunes (Mather and Ritchie 1977; Carter 1985; Tinley 1985), and there have been numerous attempts to cultivate dune fields all along the Mediterranean (van der Meulen and Salman 1996). References to use of dunes for crops rarely identify the amount of area converted or the extent that the topography and soil conditions are changed, but it is assumed that little of the form or function of dunes survives where crops are grown. Harvesting (discussed later) may occur with little modification to the landform. Crops grown in the machairs of Scotland include barley, potatoes and root crops (Mather and Ritchie 1977). Dunes have been used for paddy fields (Lin 1997), and tulips and asparagus are favored for growth in dune sands (Guilcher and Hallégouët 1991). In some cases, the areal extent of the landscape conversion can be great, such as in the I Macconi section of southern Sicily, where the entire dune system was destroyed to provide space for specialized farming (Cencini and Varani 1989).

Mining

Beaches, dunes, the shoreface and coral reefs have been mined for: (1) construction aggregate for ports, roads, railroads and houses; (2) minerals and liming material; (3) new substrate for crops; and (4) landfills. Mining is mentioned frequently in the literature (Mather and Ritchie 1977; Guilcher 1985; Tinley 1985; Cencini and Varani 1989; Chapman 1989; Haeseler 1989; Anctil and Ouellet 1990; Armah 1991; Postolache and Diaconeasa 1991; Paskoff 1992; Sanjaume and Pardo 1992; Aibulatov 1993; Cambers 1993a; Domroes 1993; Gusmão *et al.* 1993; McDowell *et al.* 1993; Anthony and Cohen 1995; Ciavola and Simeoni 1995; Knight *et al.* 1997). With notable exceptions (e.g., Mather and Ritchie 1977; Lewis 1992), the specific purposes, methods, and effects on landforms are rarely reported. In most cases, the sediment is used because it is cheaper and more convenient than the alternative. Mather and Ritchie (1977), for example, present a case where the cost of bagged liming material from an inland source is 6.5 times the cost of obtaining liming material from shell sand.

Scale of operations

Some sand has been extracted from almost every beach accessible by tractor in the beaches of the highlands and islands of Scotland, with significant amounts taken from about 1 beach in every 5 (Mather and Ritchie 1977). Guilcher and Hallégouët (1991) report mining of 80 000 m³ of sediment behind a single spit in Brittany during one year. Beach mining in Israel prior to 1964 removed sediment at an annual rate 10 to 20 times larger than the naturally occurring rate, causing an accelerated erosion of the beaches and nearby cliffs (Nir 1982). More than 5 million m³ of sand were mined from up to 40 km of the Ashdod coastline from 1949 to 1963 (Golick *et al.* 1996). Qu *et al.* (1993) project a loss of 20 000 tons km^{-1} of the entire shoreline of China in 10 years. An estimated 2 million m³ was removed from Varadero Beach, Cuba prior to 1978 (Martí *et al.* 1996). Mining along the Georgian Black Sea coast has resulted in a loss of 40 million m³ of sediment and has changed the sediment budget from positive to negative in many locations (Aibulatov 1993; Peshkov 1993). Aibulatov (1993) notes that the effect of material mined from one quarry affected surf zone morphodynamics and the ecosystem over a distance of 30 km. Removal of 10 million m³ from 4 to 5 m water depth for construction of the Ilyichevsk Port resulted in a loss of material into the quarry along 15 km of coast, reducing the width of the largest beaches 3 to 4 times, eliminating narrow beaches, tripling the occurrence of landslides, increasing erosion of the bottom 2 to 3 times out to a depth of 6 to 7 m, and increasing the rate of cliff retreat from 0.8 m to 3.0 m a^{-1} (Aibulatov 1993). The immediate geomorphic effect of one period of mining at the Pakiri-Mangawhi coast offshore site was a borrow pit about 2 m deep and tens of meters wide that was rapidly reworked over periods of hours to days; extractions at this site amounted to about 110000 m³ a^{-1} (Hesp and Hilton 1996).

Locations and types of mining

Changes in mining practice have occurred due to improvements in technology and changes in laws governing mining. Manual operations using small boats and carts have changed to mechanized operations using cranes and tractors (Guilcher 1985), although mining still may be done manually with shovels and baskets in places (Green and Cambers 1991). A switch to modern machinery for mining calcareous sand has favored erosion in the Bay of Goulven, France, at a location that was formerly accreting (Guilcher and Hallégouët 1991).

In some cases, the focus of activity has shifted from the beach to offshore. Mining along the Pakiri-Mangawhi coast of New Zealand, for example, progressed from use of a sailing scow beached on the foreshore at falling tide, to mechanical excavators mounted on barges working in the shallow nearshore, to the present operations using suction pumps mounted on barges working seaward of the offshore bar in depths of 3 to 8 m; the main reason that mining has not shifted farther offshore apparently has been the lax requirement of the regulations in New Zealand relative to other countries (Hesp and Hilton 1996).

Effects of mining

Hesp and Hilton (1996), citing others, report that mining of nearshore sediment may cause or exacerbate erosion of beaches by drawdown, interception of transported sediments, removal of sheltering bars and banks and alteration of wave refraction patterns. Destruction of reefs for coral mining has resulted in increased wave activity and erosion at the shoreline in

Bali (Knight *et al.* 1997). Removal of the boulder-armored surface of gravel barriers by mining can increase beach mobility by exposing a large volume of smaller clasts (Dickinson and Woolfe 1997).

Mather and Ritchie (1977) identify the effects of direct removal of sediment from coastal dunes in Scotland. Small diggings, at the scale of the individual farmer, initially exploit blowouts that are increased in size and depth until digging ceases due to difficulty of vehicular access or because the water table is reached. At that time, a new site is found nearby. The long-term effect of this process is creation of a surface with scattered pits at several different stages of development and restabilization. Abandoned sites may be used as rabbit-warrens or rubbish tips. At sites of larger commercial activity, the site is usually a single entity that proceeds along the most convenient dune ridge or sand hill, with the water table acting as the local base level of activity. Support equipment and infrastructure may include screening plants and conveyors and gravel-covered roads. Usually, little is done to remove the derelict structures or restore the landscape when the site is abandoned (Mather and Ritchie 1977). Revegetation programs have been successfully conducted in some areas, although the restored landscape may depart conspicuously from the pre-mined landscape because the sediment budget, size and location of the foredune will have changed (Chapman 1989).

Difficulties of preventing mining

There are now widely accepted criteria for conducting commercial dredging operations (Anctil and Ouellet 1990). Most international agencies now only approve sand mining seaward of the 18 to 25 m isobath (Nielsen *et al.* 1991, cited in Hesp and Hilton 1996). Mining is limited by law in some countries but effective control is often lacking (Postma 1989). Even where rates of removal of sediment from beaches has decreased, the removal from river mouths may have increased (Zenkovich and Aibulatov 1993). Deposits in dunes normally belong to the landowner, who may exploit them directly or lease them to a contractor (Mather and Ritchie 1977). Even local authorities may encourage mining by purchasing sediment from beaches for use in their roads (Mather and Ritchie 1977).

Altering landforms through use

Observations of alteration to soil characteristics and vegetation through pedestrian trampling or off-road vehicles are well represented in the literature (e.g., Liddle and Greig-Smith 1975a and b; Boorman and Fuller 1977; Eastwood and Carter 1981; Wiedemann 1984; Carter 1985; Anders and Leatherman 1987; Carlson and Godfrey 1989; Williams and Randerson 1989;

Guilcher and Hallégouët 1991; Sanjaume and Pardo 1992; Andersen 1995a; Apostolova *et al.* 1996). Most studies have been conducted on human activities in dunes, although trampling by animals in dunes occurs. Traffic and indiscriminate parking on tops of cliffs leads to destruction of cliff-top vegetation, and pedestrian access to beaches and surfing sites leads to incision of gullies on slopes of bluffs (Dias and Neal 1992; Bird 1993).

Previous studies of the effects of pedestrian trampling and vehicle use are summarized by Chapman (1989) and Andersen (1995a), who indicate that trampling can create paths; inhibit formation of incipient dunes; reduce the height of established dunes; erode the surface; compact the soil; change the bulk density, organic matter and moisture content of soil; reduce the cover and height of vegetation; decrease production of biomass; reduce the number of flowering species; cause disappearance of vulnerable species; introduce weeds and exotic plants; interfere in the natural succession; cause loss of biodiversity; and disrupt fauna. Disruption to fauna, although ecologically important (Watson *et al.* 1997), is not treated in detail here, but it can be an important rationale for controlling vehicle use.

The events leading to geomorphic or biologic change may be simple, such as the direct uprooting of vegetation, causing immediate formation of a bare patch, or they can be complex, involving an interrelated chain of events. For example, trampling may cause changes in plant cover, increases in bulk density of trampled sediments and associated reduction of pore space that can lead to decrease in the rate of gas exchange that affects microbial and decomposition processes and changes the hydrology of the sediments, affecting the composition of microarthropods (Koehler *et al.* 1996).

Impacts to beaches and dunes due to human uses are separated into pedestrian and vehicular types for convenience, although geomorphic effects may be caused by a combination of both processes, perhaps in association with trampling by animals. This section is confined to only a few examples of the most direct alterations to landforms.

Trampling by pedestrians

Trampling by virtually any faunal species of a size that can disturb individual grains of sand has a geomorphic effect, readily revealed by footprints in the sand. In most cases, the natural fauna that use the beaches and dunes are so small, and use is so infrequent, that the effects are eliminated by wave and wind transport. Humans can change the size and frequency of alteration to the point where this activity determines the evolution of the landform. Activities that have little immediate impact on large-scale geomorphic features can locally change vegetation and affect the likelihood of sediment entrainment or slope failure, leading to the categorical conclusion that the likelihood of occurrence of erosion increases in line with trampling impact (Mather and Ritchie 1977).

Thresholds and rates

Estimates of the occurrence of trampling damage on Scottish dunes made by Mather and Ritchie (1977) revealed that the onset of trampling damage occurs in the range of 26 to 100 visitors per day. They found that there could be trampling damage at some sites at lower intensities but it is invariably light. When more than about 60 users per day (2000 per season) occurs, there is almost certainly likely to be some trampling damage. Boorman and Fuller (1977) found that 2889 passages reduced the vegetation cover of dunes dominated by *Ammophila* to 50 percent. Rates such as these can be only rough estimates because it is difficult to separate the effects of pedestrians (both walking and running, with shoes and barefoot), car wheels, caravan wheels, sheep, cattle and dogs. Various thresholds are provided for trampling rates beyond which destruction of vegetation is critical, but many site-specific factors make a single threshold unrealistic (Mather and Ritchie 1977). Beach grass is especially sensitive to trampling, in part because it has no physical characteristics that deter pedestrians (Mather and Ritchie 1977). Accordingly, the threshold is lower where beach grass is present and is also lower on irregular ground than on level terrain. Trampling on sloping portions of the dune more readily exceeds the shear strength of the sand–vegetation–root complex (Mather and Ritchie 1977).

Susceptible locations

Path density and size decrease away from parking areas and access nodes (Eastwood and Carter 1981; Bonner 1988). Uncontrolled pedestrian access from parking areas is a particular problem (City of Stirling 1984), but, even where pathways are provided and maintained, damage to vegetation can occur if the pathways are not maintained at sufficient width to accommodate specific uses and equipment (Koltasz-Smith and Partners 1994). Mather and Ritchie (1977) note that caravan (camper) sites are more problematic than day-use sites from the standpoint of pedestrian trampling because residents in caravans are more likely to make several trips between their vehicles, toilets and the beach in the course of a day. They found that 92 percent of the beach complexes used for caravanning suffered some form of trampling damage, compared to only 32 percent of sites not used for that purpose. The existence of high trampling damage was found on 13 percent of the caravan sites compared to 1 percent on the other sites.

Effects on landforms and landscapes

Bonner's (1988) assessment of 40 walkovers at Fire Island National Seashore revealed a range of path sizes from almost imperceptible notches in the dune crest 0.1 m deep and 0.8 m wide to gaps 2.3 m deep and 22.9 m wide. She noted that path form changed as size increased, with top width increasing relative to bottom width. A landward displacement of form and mass occurs because the unvegetated paths act as conduits for sediment blown in from the beach. McCluskey *et al.* (1983) monitored a 3 m wide by 1.5 m deep notch in a dune crest 3.4 m above the back beach in the vicinity of Bonner's study and found that the rate of transport through the notch was 418 percent greater than across an equivalent length of dune crest 20 m away. These increases in bare ground and sediment transport are often viewed as negative by managers of dune reserves, but trampling can have positive impacts, in that it can keep vegetation communities in a dynamic stage and enhance growth of some vegetation types.

It is perhaps easier to identify the physical effects of trampling on dunes than it is to say categorically that trampling is adverse to dunes. The creation of bare sand surfaces and increases in mobility are not long-term threats to the natural value of dunes, but paths will retain their human origin in size, shape, location and function unless they develop into blowouts that eventually conform to the characteristics of dune-forming winds. The departure in appearance of a dune landscape with paths from a natural dune landscape is most noticeable from the air, and air photographs of dunes impacted by trampling (available to managers of dune reserves) reveal a most unattractive surface. Views from the ground (seen by users of dune reserves) may reveal a more aesthetically appealing landscape. Even where paths are linear, they may actually contribute to a more natural appearance (viewed from the ground) than an artificially revegetated dune would have in their absence. The result may be a dune of greater ecological value in that a light trampling impact of 5 to 10 visitors per day is a way to secure the continuation of an open vegetation for some areas (Andersen 1995a). The major difficulty of using trampling (or many other culturally induced impacts) to achieve desirable results in dunes is monitoring and controlling the activity to have the optimum effect.

Off-road vehicles

Off-road vehicles are used for fishing, touring, search and rescue, patrolling, utility maintenance and scientific study in addition to illegal "dune busting" activities (Godfrey and Godfrey 1981). Impacts of these vehicles on the mobile portion of the beach are difficult to assess because the habitat changes frequently under natural conditions, but these impacts are considered important because much of the life on the beach is concentrated in and around the organic drift lines (wrack lines) that are often in the way of vehicle use (Godfrey and Godfrey 1981). Results of unpublished studies summarized in Godfrey and Godfrey (1981) indicate that vehicles used on beaches can: (1) reduce populations of diatoms by 90 percent (although the next tidal cycle can restore populations); (2) reduce by one half the population of microbes under patches of organic detritus (presumably by desiccation when detritus is broken up); and (3) pulverize and disperse organic matter in drift lines, thereby destroying young dune vegetation and losing nutrients (Godfrey and Godfrey 1981).

Field experiments at Fire Island, USA (Anders and Leatherman 1987) indicate that even low-level vehicle impacts will severely damage dune vegetation. Vehicle impacts were conducted at the rate of one pass per week for 1 year followed by a 5.5 month recovery period and then at a rate of one-half the former level. Within one year, the dune vegetation in their impact zones was essentially eliminated, while the control areas increased in vegetation cover and the vegetation front advance seaward.

Anders and Leatherman (1987) also investigated conditions adjacent to a dune cut that was heavily used by vehicles entering and exiting from one direction and lightly used from the other direction. The vegetation front was closer to the dune crest and about five times less extensive on the heavily impacted site than on the adjacent lightly used site. The loss of vegetation resulted in limited sand deposition, and the dune landward of the impacted area accumulated sand at the expense of the area undergoing vehicle impact, resulting in a steepening of the foredune profile with greater potential for development of an erosional scarp (Anders and Leatherman 1987).

In a study of impacts at Cape Cod, USA, Godfrey and Godfrey (1981) found that 50 passes of a vehicle were sufficient to stop the seaward growth of the dune completely at Cape Cod, creating a scarped, rather than sloping dune front. They also reported that plant biomass in dunes was reduced to nearly zero after 70 to 175 vehicle passes. They conclude that no meaningful carrying capacity for vehicles can be assigned to dune vegetation because it makes little difference whether 100 or 10 000 vehicles use a path.

Not all damage by off road vehicles is due to wanton or excessive use. Anders and Leatherman (1987) note that, during winter when beaches are narrow, drivers do not hesitate in running over the dormant vegetation since it appears to be dead, resulting in unintentional impact on buds and emerging tillers. Drivers can think that they are not causing damage, even when they are, because they drive on underground rhizomes (Hoogeboom 1989).

Mather and Ritchie (1977) support the findings of Godfrey and Godfrey (1981) that vehicle damage on sloping ground is greater than on level ground. They also found that the effect of individual vehicles traversing vegetated surfaces can be considerable when they make turns. Car wheels tend to cut deeper grooves during turning, and these grooves develop directly into blowouts or can be exploited by animals, such as rabbits (Mather and Ritchie 1977).

Regrowth of vegetation in impacted sites indicates that rapid recovery is possible after termination of vehicle impacts (Godfrey and Godfrey 1981; Anders and Leatherman 1987; Judd *et al.* 1989). Godfrey and Godfrey (1981) concluded that it is better to have a few well-managed and well-patrolled dune tracks than to have many lightly used tracks. They report that dune biomass on the dune crest recovered after 2 years and the former impacts were virtually undetectable after 6 years, but that tracks through the older stabilized dunes to landward remained visible for at least 8 years of follow-up observations.

Vehicle impacts, like pedestrian impacts, decrease away from access points (Judd *et al.* 1989). Formal identification of edges of parking areas and well-signposted legal access points are effective means of controlling adverse effects of off-road vehicles (Western Australia Department of Planning and Urban Development 1993), but fences may be required in remote portions of beach and dune reserves where there is no conspicuous management presence. Visitor pressure no longer constitutes a serious threat to biodiversity or dune system stability in well-managed dune reserves (Mullarda *et al.* 1996), but even in some of these reserves there are still isolated locations where the landscape bears evidence of vehicle impact.

Harvesting

Gathering of wood on beaches may locally increase rates of aeolian transport and reduce the potential for incipient dune forms by eliminating small-scale obstacles. Removal of wood may also increase susceptibility of upland formations to attack by wave uprush, although evidence for this may be circumstantial (Komar and Shih 1991). Seaweed is harvested from the beach in some countries, contributing to increased incidences of trampling in addition to the effects of direct removal of sediment and biomass (Mather and Ritchie 1977; Bird 1996).

Picking or removing vegetation in dunes can be more directly destructive of landforms. Removal of cuttings can impoverish a site (Hewett 1985), as can removal of vegetation litter, even where the surface topography is not altered. Types of vegetation removed from dunes include decorative plants, such as sea-holly (*Eryngium maritimum*), soap grass (*Gypsophila paniculata*) and sea-grass (*Lathyrus japonicus*) (Olsauskas 1995); trees (Piotrowska 1989; Borowka 1990); and dune grass that is used for fuel and thatch (Randall 1983; Westhoff 1985; Skarregaard 1989).

Aforestation (planting trees) has been conducted to protect crops from sea winds (Cencini and Varani 1989) and to establish a forest industry and stabilize dunes (Blackstock 1985; Sturgess 1992). The practice was carried out on a relatively large scale in the past. For example, aforestation conducted in Italy between 1950 and 1970 increased the woodland area by 600 000 ha (Postma 1989). About 78 percent of the 101 000 ha of sand dune coast managed by the French National Forestry Service is aforested (Favennec 1996). Ecological problems associated with aforestation include loss of flora and fauna, changes to soil characteristics, lowering of the water table and seeding into adjacent unforested areas (Sturgess 1992; Janssen 1995). Aforestation, particularly with exotic species, is now considered less desirable than it was in the past, and attempts are being made to remove areas of woodland to restore dune habitat that has greater conservation value (Sturgess 1992). Despite these efforts in some locations, the practice is continuing on a large scale in other locations, such as Turkey (van der Meulen and Salman 1996).

Grazing

There is little useful information about the effect of grazing by indigenous species on dunes to provide a template against which human-induced species can be evaluated, but there is evidence of the effect of introduced species in many areas. For example, the introduction of hoofed fauna to the dunes in Australia, where they were not indigenous, had dramatic effect on remobilizing dunes, as did invasion by rabbits (Chapman 1989). Grazing has been reduced in many countries, but there is still serious overgrazing of dune systems in some countries (van der Meulen and Salman 1996).

Evaluations of grazing are presented in Garson (1985), Leach (1985), Oosterveld (1985), Corona *et al.* (1988), Doody (1989), Westhoff (1989), van Dijk (1992) and Kooijman and de Haan (1995). Grazing can result in increased availability of light at the ground and decreased stock of organic matter and nutrients due to a decreased input of litter and accelerated rates of decomposition (Kooijman and de Haan 1995). Grazing can affect aeolian transport directly by reducing the average length of the vegetation, with a resulting decrease in its effect on diminishing wind speed near the surface.

The indirect effects of trampling and rubbing may be more critical at some sites than the direct effects of grazing (Mather and Ritchie 1977). Sheep, for example, rub against and under the edges of small blowouts to seek shelter from wind in winter or shade in summer, leading to direct landform change (Mather and Ritchie 1977; Angus and Elliott 1992).

Rabbit grazing in infested areas may change the vegetation characteristics, creating a flora that is less rich and less attractive than locations that are less severely grazed; their trampling pressures may be low, but they produce numerous small bare-sand faces that can be exploited by wind erosion and grazing by rabbits and domestic stock (Mather and Ritchie 1977). Rabbit grazing has resulted in greater sand drift, disappearance of a number of species and dominance of species avoided by rabbits, although a moderate rabbit stock stimulates the pattern of diversity of the vegetation cover (Westhoff 1985). Boorman (1989) studied the grazing status of 777 quadrats in 48 dune sites in Britain and found that over 70 percent of the quadrats were directly grazed by rabbits and over 98 percent of the sites appeared affected by rabbit grazing to some extent, leading to the conclusion that the dominant grazing animal in British sand dunes is the rabbit.

Domestic animals are easier to control and can be used to maintain specific dune conditions. From the human perspective, grazing can be considered detrimental or beneficial depending on type and intensity of use that vary in their effect depending on where they occur in the dune environment. Cows, horses and sheep have different ways of grazing and result in different vegetation assemblages (Hewett 1985). Boorman (1989) found that dune slacks appear to be best grazed by cattle, and grazing intensities applied to slacks can be greater than those applied to dune grassland, although excessive grazing is likely to cause serious damage, especially in the wetter areas; the plant community in the seaward-most mobile dunes is open with more bare ground for colonization and plant competition is more intense, diminishing the need for control of plant growth through grazing (Boorman 1989).

Temporal effects of grazing are important in that the present state of vegetation may reflect the management of the past rather than current practice, and a change in grazing management often takes many years to effect a permanent change in the vegetation (Boorman 1989). The grazing intensities of the British dunes are much less than in the past, and there has been a wholesale development of coarse vegetation and the invasion of scrub, with the loss of species-rich short-turf communities (Boorman 1989). Light grazing can be used to remove prolific growth from the most vigorous plants to reduce competition and increase the amenity and conservation values in dunes by increasing the number of species, although the control of grazing animals may require careful investigation to determine adequate stock density, followed by investment of considerable human resources for management and capital for fencing (Hewett 1985; Westhoff 1985; Boorman 1989).

Waste disposal

Beaches and dunes have been used as convenient places to dispose of isolated bits of refuse, including products associated with day-use tourist activities and larger items unassociated with beach use, such as old cars or farm equipment. The hummocky characteristics of dunes make them especially appealing sites for disposal of undesirable materials (and for use as toilets) because the disposal activity is inconspicuous from other users of the coast. Refuse disposal introduces a new type of beach material into the coastal landscape; it reduces aesthetic appeal; it introduces dangers for people and grazing animals; and it results in ecologically undesirable plant species (Mather and Ritchie 1977). Disposal of refuse is common at caravan sites, but it can also be common in residential areas where residents see no great aesthetic or recreational uses for the beach (Mather and Ritchie 1977).

The disposal of clean fill may be viewed as non-consumptive, because the added materials help reduce erosion rates and because the waves may rework the dumped material to a size compatible with a beach location, but these materials can change the composition of beach materials, change the size of the landforms or create new landforms. As much as 1.8 million tons of mine waste were dumped at Rapid Bay, South Australia between 1942 and 1982, creating a gravel beach (Bourman 1990). About 100 million m³ of mine waste were dumped on the shore of Chañal Bay, Chile from 1938 to 1962, followed by a dumping of 35000 metric tons daily at a new location north of there (Paskoff and Petiot 1990). Progradation in Chañal Bay between 1938 and 1975 was 20 to 25 m a^{-1}, resulting in a shoreline advance up to 900 m over this period (Paskoff and Petiot 1990). An estimated 360000 tonnes a^{-1} of colliery spoil dumped on the northeast coast of England accumulate on the local beaches (Humphries and Scott 1991). Dumping at one site has pushed the mean high-water line about 200 m seaward, leaving a cone of spoil extending from beach level to the cliff top and a relatively cohesive raised beach forming a step in the beach profile (Humphries and Scott 1991). Over the past 35 years, more than 80 million m³ of sand waste from land-based amber mines has been deposited on the shore near Kaliningrad, resulting in formation of beaches 50 to 100 m wide and creation of foredunes at distances up to 50 km away from the dumping ground; the resulting coastal morphology contrasts markedly with the previous sediment-starved coast that had been eroding at rates of 0.5 to 0.7 m a^{-1} (Boldyrev et al. 1996). Discharge of spoil from a diamond processing plant onto a local beach in Namibia has resulted in 300 m of local beach accretion, and a proposed mining operation

Figure 2.2 Disposal ridge occupying the location of a foredune at Brigantine, New Jersey, showing litter and coarse-grained materials that reveal its non-aeolian origin.

will discharge approximately 26 million m³ onto the neighboring beach (Smith *et al.* 1994).

Disposal operations that make use of at least some sediment similar to that found on local beaches or dunes may be perceived as multipurpose win–win operations. One example is using sediment dredged from lagoons in southern California (to restore tidal volume) to rebuild the dunes (to provide protection and restore the historic landscape and natural values) (Tippets and Jorgensen 1991). Other disposal operations may simply recycle beach materials as a by-product of the waste material. Sediment and litter removed from the surface of the beach in cleaning operations, for example, is placed in the foredune in some communities. The resulting deposits may be in isolated hummocks or in ridges that mimic the location and form of dunes (Figure 2.2) (Nordstrom and Arens 1998). Kelp berms (Hotten 1988) are another of these "disposal landforms."

Extraction and recharge

Subsidence

Withdrawal of oil and gas or groundwater (for drinking, agricultural or industrial use) are the most common forms of extraction leading to subsidence, local increases in sea level rise and increased rates of beach loss (Inman *et al.* 1991; Awisoka and Ibe 1993; Flick 1993; Wiegel 1994; Bondesan *et al.* 1995; Nicholls and Leatherman 1996). Lowering of the beach profile by subsidence reduces wave energy damping and results in correspondingly greater wave energy at the shoreline (Dean 1997). It also brings portions of the coast under the influence of storm waves more frequently than otherwise would have occurred. The impact of subsidence usually is highly regional in nature (Postma 1989), but it can greatly exceed local rates of sea level rise. Notable locations of human-induced subsidence include the coast of southern California and the northeastern coast of Italy. Rates of induced subsidence in southern California range from 0.03 to 0.20 m a^{-1} or 30 to 200 times the rate of present sea level rise (Inman *et al.* 1991). Subsidence rates of up to 0.06 m a^{-1} occurred near Codigoro, Italy between the 1940s and 1966 due to extraction of methane-bearing water (Bondesan *et al.* 1995).

Changing water levels in dunes

Groundwater extraction, artificial recharge and artificial drainage have all had undesirable effects on dune systems (Westhoff 1985; van der Meulen and Jungerius 1989a; van Dijk 1989; Llamas 1990; Angus and Elliott 1992; van Beckhoven 1992; Geelen *et al.* 1995). The western part of The Netherlands has relied on the dunes as a source of drinking water for the past 100 years because of the polluted nature of surface water (Louisse and van der Meulen 1991). Of special significance is the threat to wet dune slacks that are among the most seriously threatened natural habitats in The Netherlands (Geelen *et al.* 1995).

Artificial recharge of dunes has been applied in The Netherlands since the early 1950s (Louisse and van der Meulen 1991). Westhoff (1985) reports infiltration rates of recharged water of about 150 million m^3 a^{-1}. Pre-treated eutrophic river water is infiltrated into the dunes in open reservoirs and allowed to percolate through the dune sand and later recovered by means of shallow wells, altering the groundwater regime and producing a water quality that differs from the original oligotrophic dune water (Louisse and van der Meulen 1991). Recharge with purified water from the Rhine river has resulted in an unnatural water quality (richer in nutrients), disturbance in the direction and volume of groundwater flow and unnatural water table fluctuation (Geelen *et al.* 1995).

Effects of watering surface vegetation

Maintenance of exotic vegetation on lawns, gardens and golf courses in arid and semi-arid environments uses greater volumes of water than would otherwise occur. The quantities used may be considerable. Modern 18-hole golf courses in a Mediterranean sand dune may require up to 500 000 million m³ of fresh water per year (van der Meulen and Salman 1996). Kuhn and Shepard (1980) report that landscape irrigation alone adds the equivalent of 1.25 m to 1.52 m of additional rainfall each year to lawns and gardens on coastal bluffs in southern California. These locations might otherwise receive only about 0.25 to 0.40 mm of rainfall (Griggs 1994). The rise in water table associated with watering can occur even during dry periods (Kuhn and Shepard 1980).

Watering of lawns and gardens (and waste-water disposal) above coastal bluffs contributes to a slow steady rise in the water table that adds weight to the cliff materials, weakens cliff materials (by solution in some cases) and lubricates surfaces along which slides develop (Kuhn and Shepard 1980; Griggs *et al.* 1991b; Dias and Neal 1992; Ledesma Vasquez and Huerta Santana 1993). Irrigation of land southwest of Odessa since 1983 has resulted in a landslide frequency 3 times the previous rate and an increase in shore erosion at 2.5 times the previous rate (Shuisky and Schwartz 1988).

Concentration of surface runoff at locations where it is discharged into existing gullies from culverts at the top or face of the bluffs leads to acceleration of erosion rates in gullies (City of Stirling 1984; Griggs *et al.* 1991a; Griggs 1994). Watering can make gullies active in the dry season in locations where natural precipitation is minimal (Dias and Neal 1992). Flow can be increased at locations where shore-normal channels carry water from shore-parallel flows. Kuhn and Shepard (1980) report cutting of a new canyon by this means at a location where no canyon previously existed; initial growth was slow, but during a subsequent one-day event the gully eroded landward a distance of 72 m, delivering about 38 000 m³ of sediment to the beach. Flow may also be concentrated where fences, stairways and drain pipes are located along former drainage avenues, causing erosion of the bluff face and collapse of stairs and storm drains during periods of heavy rains (Kuhn and Shepard 1980; Lee 1980).

Military use

Military use of beaches and dunes and coastal cliffs includes bombing, shelling from naval vessels, amphibious landings and construction of fortresses and bunkers in addition to building support infrastructure required to house and move personnel (Mather and Ritchie 1977; Doody 1989; Demos 1991; Guilcher and Hallégouët 1991; Jensen 1995; Binderup 1997). Military activities may be consumptive at specific sites in the short term, but they may provide the opportunity for long-term evolution by natural processes in other areas. The direct impact on the vegetation in bombing ranges, for example, may be limited to the immediate vicinity of targets that may be widely separated, allowing intervening areas to develop naturally (Doody 1989). Military use of dunes for maneuver or target areas can prevent urbanization or agricultural uses in them while they are in use and allow for natural development of dunes after this use has ceased (Doody 1989; Demos 1991; Guilcher and Hallégouët 1991). A large part of the 7000-year-old dune system at Morrich More, UK (representing one of the the most complete sequences of sand dune vegetation in Britain) remains intact because of bombing activities (Doody 1989).

There has been little reason or opportunity to assess specific effects of military activity on beach and dune change, but there has been considerable study of the impact of permanent fixtures associated with military use, such as harbor structures or bunkers (chapter 4). It appears that greater long-term beach change and greater long-term degradation of the dune occur where static structures (including buildings and roads) are in place (Doody 1989).

Other uses and activities

There are many other local or small-scale uses of beaches and dunes that alter the characteristics of these landforms (e.g., creating "sand art," throwing stones on gravel beaches), but there are no assessments of impacts of these uses because the results appear inconsequential to landform development. There are reports of dunes being used as cemeteries (Western Australia Department of Planning and Urban Development 1994a; Lin 1997). This use may lead to preservation of some of the character of the landform, but it also may lead to the use of static protection structures to prevent exhumation (Mather and Ritchie 1977). Inscribing graffiti and carving caves into sea cliffs is another activity that can have important local impact, especially where natural sea cliff recession rates amount to a few centimeters a year (Komar 1979; Lee 1980; Lee and Crampton 1980). Dias and Neal (1992) report on excavation of caves 1 to 2 m deep, representing an amount of change that would take many decades to achieve by natural processes on some cliffed coasts.

Reshaping landforms

Reshaping landforms by grading (bulldozing, scraping) of beaches and dunes by earth-moving equipment (Figure 2.3) is one of the most ubiquitous alterations, but it is also one of the least well studied, and the practice is still controversial in terms of its benefits or adverse impacts to geomorphic systems. Beaches and dunes are reshaped primarily to: (1) provide shore protection; (2) prevent or alleviate inundation of landward facilities by wind drift; (3) enhance beach recreation; (4) provide views of the sea for shorefront

Figure 2.3 Bulldozed dune (or sand dike) at Trancas Beach, California, showing narrow, steep-sided dune occupying location on the beach where a dune would not be expected under natural conditions.

residents; and (5) enhance environmental quality. Reshaping may be accomplished on a periodic basis to re-create or eliminate landforms that are difficult to manage as permanent features because they are out of equilibrium with natural processes. Reshaping may also be done on a one-time basis to re-initiate a landform or process that continues to function naturally. Reshaping operations can result in temporary forms, such as excavation trenches, disposal mounds and stockpiles for use as a sediment supply throughout the season (Miossec 1993). Sometimes the rationale for reshaping and the function of the resulting landform are not intuitively apparent, especially considering the number of potential detrimental impacts or alternative means of accomplishing goals (Figure 1.2).

Increasing levels of protection

Types of protection projects

Grading may be carried out along entire barrier islands (McNinch and Wells 1992) but is more likely to be conducted at the municipal level because many municipalities have their own equipment, and co-ordination with other communities is not required. Grading can be conducted on shingle beaches as well as sand beaches (Packham *et al.* 1995).

Grading is often selected as the means of providing shore protection because other options are too costly, are prohibited or take too much time. For example, the restriction of static structures as legally acceptable alternatives in some states in the USA increases the need to examine other local options (Wells and McNinch 1991; McNinch and Wells 1992). Waiting for natural events to create protective dunes is risky in some locations because the onshore winds that fill sand fences are contemporaneous with the waves and surge levels that are the targets of protection efforts. Narrow beaches in some locations restrict the potential source of wind blown sand, making construction of dunes using bulldozers a favored option.

Grading is one of the most rapid means of achieving desired landform characteristics. Estimates by municipal officials in New Jersey indicate that a bulldozer with a 3.7 m blade can push about 10 m³ of sediment per pass and can push a load of sand into the foredune about once every 5 minutes. One bulldozer of this size can repair a 150 m long section of dune in about a day using sediment scraped from the beach. Serious dune building operations involve use of several bulldozers at one time, and a dune can be created by a municipality in a matter of weeks using beach sediments. One of the largest grading operations occurred in South Carolina, USA following Hurricane Hugo in 1989. Operations began 10 days after the storm and extended along 105 km of beach; at the height of construction, five contractors had over 60 machines on the beach; and over 95 percent of the planned dune rebuilding was accomplished within about 2 weeks; total cost was about $1.5 million (Kana et al. 1991).

Grading for protection purposes normally involves removing sediment from low portions of the beach and placing it in a high protective berm or dune (also called a sand dike). These landforms (Figure 2.3) are often constructed to reduce wave overwash in the winter, and they may be eliminated in the summer to enhance recreational use (Mauriello and Halsey 1987; Armstrong and Flick 1989). Grading following the February 1990 storms in northeastern France involved removing sand from a 50 m wide strip near the low-water line, creating a similar strip near the high-water line (Bryche et al. 1993). Grading has also been employed to close openings in shore-parallel runnels to prevent losses and speed their filling with sand (Palmetto Dunes Resort 1990), and it can be used to replace sediment alongshore in what is essentially a backpassing operation (Rouch and Bellessort 1990). This process can equalize the exposure to hazard along a single jurisdiction. The municipality of Avalon, New Jersey, for example, has a state permit to mechanically backpass sediment as a means of repairing beach and dune scarps and extending the life of a beachfill in a way that is feasible within their town budget

(Mauriello 1991). Use of bulldozers to create dunes following storms is common in other municipalities in New Jersey, using sediment from beach, overwash and aeolian deposits or from other sources that are brought to the site in trucking operations. *Ad hoc* operations may be conducted during non-storm periods, and sediment skimmed from the top of the beach in beach-cleaning operations also may be deposited in the dunes (Nordstrom and Arens 1998).

Use of bulldozers to create dunes was more common in The Netherlands in the past, but is no longer necessary because the dunes have been built to required levels of protection. Bulldozers still may be employed to reshape dunes. Sediment may be bulldozed from peaks in the dune crest and placed in lower areas, and bulldozing to a desired slope (about 1 to 3) to enhance planting is common (Nordstrom and Arens 1998).

Among the most unique protection projects involving grading are the sand seawalls used in Namibia to provide temporary protection to coastal mining operations from wave action and to increase the mining area by mining closer to the waterline. At Oranjemund, a total of 12.7 million m³ of material was used for the construction and maintenance of these sand seawalls during a 5 year period (Smith *et al.* 1994).

Assessment of protection projects

Tye (1983) reports on detrimental effects of Grading at Folly Beach, SC, where grading increased upper beach slope from 1:32 to 1:23 by removing 49 m³ m^{-1} of sand from the lower beach face. This action induced offshore transport by plunging waves, allowing erosional conditions to persist. Post-storm grading of a beach segment that suffered no net erosion during a hurricane resulted in increased erosion of 12.6 m³ m^{-1}. Tye suggests that the quantity of sand redistributed by the grading process should not exceed the volume naturally returned to the beach and that eroded beaches that do not recover naturally should be graded (Tye 1983).

McNinch and Wells (1992) conducted a study of 1 km long graded and ungraded beaches at Topsail Beach, North Carolina, USA. The borrow zone was confined to a narrow strip along the lower foreshore and was graded in a way that involved little change in slope. Their study revealed that more sediment was removed from the graded section than the control section during a hurricane, but less was removed during a northeaster. They conclude that grading can be beneficial when conducted over long periods and under conditions less severe than hurricanes. The borrow zone during the period of most active grading filled naturally to its pre-graded profile within 48 hours, indicating that grading did not appear to inhibit onshore transport and allowed for rapid replacement of the artificially removed sediment. It was difficult to achieve a dune of optimal design size because there was limited space available to place the fill and there were limited volumes available from the beach zone. They suggest that effective design of a grading project requires: (1) identification of both the erosion problems and the limitations of grading as a solution; (2) proper siting of the borrow zone and the artificial dune ridge; and (3) removing sediment from the borrow zone at a rate that approximates the rate that natural processes can replace it. Keys to success of their project are: (1) the long period of time that grading is conducted; (2) the small quantity moved at any one time (a 0.15 to 0.20 m veneer moved by a single piece of equipment); and (3) locating the borrow zone low on the profile, where it is inundated twice each day (McNinch and Wells 1992). The implications of grading studies are that the process works best where it provides a way of mimicking natural recovery at a somewhat more rapid pace without causing a dramatic alteration of sediment budgets or landforms.

Preventing or alleviating hazards

Removal of sand washed onto roads occurs following major storms (Bush 1991). The sediment that is removed is often returned to the beach using trucks, but it can be pushed into vacant lots, reducing the volume of sediment in the beach system and altering the shape and sediment composition of remnant natural enclaves. Inundation of coastal properties by wind blown sand is a major problem in many communities on windward coasts (Sherman and Nordstrom 1994), and substantial costs can be incurred for maintenance of roads, parking areas, walls, beach access routes and storm water drains (City of Stirling 1984). The need to grade to prevent inundation is not restricted to windward coasts; sediment that accumulates on roads and against boardwalks in New Jersey, on the east coast of the USA, is routinely removed in community-level operations. Sand blown inland onto private yards is often eliminated by residents in *ad hoc* operations using brooms, shovels and wheelbarrows (Figure 2.4). In some cases, the sand blown against buildings is removed from the beach system (Gusmão *et al.* 1993).

Removal of sediment reduces the hazard of wind blown sand at threatened locations but often results in accelerated transport from the mounds created where the sediments are placed back on the beach. Gaps in the foredune are often closed to reduce the potential for aeolian transport and reduce the problem of besanding of buildings and infrastructure, but often the removal of sand that is inundating houses contributes to future besanding by maintaining a gap in the dune and an unvegetated surface. Sand that is blown under boardwalks but is prevented from being transported through to the other side by buildings or bulkheads may build up to a height that brings it in

Figure 2.4 Manual removal of sand blown onto private lot at Ocean City during winter storm season of 1995–6. Photo by S.M. Arens.

contact with the wooden cross-members and results in timber rot. Locations such as these in New Jersey, USA are graded periodically, and the sediment is moved back to the beach. These operations create a trench under the boardwalk and isolated piles of sand on the backbeach, where the sediment is dumped. In many cases, sand drift can be eliminated or greatly reduced by constructing dunes on the backbeach to function as traps to wind blown sand, although the desire of residents and visitors for views of the sea and ready access to the beach prevent implementation of this alternative.

Blowouts may be mechanically smoothed to prevent wind erosion (Angus and Elliott 1992; Nordstrom and Arens 1998). Dunes that have been scarped as a result of wave erosion may be reshaped to create access ramps and reduce the hazard of slope failure (Nordstrom and Arens 1998). On coasts of high relief, portions of overhanging cliffs have been intentionally broken off to keep cattle from falling (Binderup 1997), and blasting of rock cliffs has been conducted (without success) to reduce the hazards of rocks falling on users of the beach below (Williams and Davies 1980).

Enhancing beach recreation

There is considerable variety in the way beaches and dunes are reshaped to facilitate recreation. The most frequent alterations include: (1) creating flat beaches as recreation platforms; (2) building platforms for beach structures; (3) cleaning beaches of litter; and (4) eliminating barriers to access. Other alterations that are interesting but have only limited local or short-term value include dumping dredged materials offshore to enhance surfing (Wiegel 1993b; Mesa 1996) and creating mounds on the backbeach to provide observation platforms for lifeguards. There are many other ways that landforms are created, destroyed or altered by local managers to enhance beach recreation that are unreported because they are conducted using equipment already available within municipalities and require no documentation for approval for outside funding.

Creating flat beaches

The emphasis on beach recreation often takes precedence over protection in the summer, when the resident and tourist population increases dramatically. The need for wide, flat beaches as platforms for active recreation or for buildings related to recreational uses has led to intentional elimination of dunes in many locations (Cencini and Varani 1989; Paskoff 1992; Nordstrom and Arens 1998). The reshaping can be considered permanent or can be a seasonal adjustment that involves transfer of sediment from the dune to the beach in the summer and transfer of sediment from the beach to the dune at the end of the tourist season (Mauriello 1989).

Providing platforms for beach structures

Grading operations are also conducted to create raised, flat, artificial berms on the backbeach to facilitate construction and use of pavilions (mostly restaurants) and cabanas and to protect them against wave uprush. These sediment platforms are common in The Netherlands (Figure 2.5), where they are called bankets. They are often built several meters above the elevation of the beach, making them considerably higher than could occur there as a result of normal annual wave conditions. These features may be considered functional equivalents of either storm berms or dunes depending on their height above the backbeach (Nordstrom and Arens 1998). They may occur on beaches that are too narrow for them to form in that location or occur at times when storm wave uprush would have been insufficient to create a storm berm.

Cleaning beaches of litter

The call for clean beaches as recreation platforms in urban resorts has resulted in beach-cleaning operations that have removed the litter that provides microhabitat and favors formation of incipient dunes. Raking in developed areas often affects the entire surface of the backbeach. Beach raking at Fisher Island, Florida is done almost daily to clear the beach of sea-grass (Bodge and Olsen 1992). Expo Beach, Okinawa is cleaned 250 times each year (Shiraishi et al. 1994). The beach at Atlantic City is raked to clear debris at a

Figure 2.5 Recreational use structures on the beach at Zandvoort, The Netherlands. The larger structures on the backbeach are built on an artificially constructed banket above the surface of the beach seaward of it.

frequency of about once per week landward of the dune and daily on the seaward side (Nordstrom and Jackson 1998). Kelp removal can be a daily operation on the beaches in southern California (Hotten 1988). The volume of sand removed with the wrack from the Ocean Beach, California littoral cell is from 20 percent to 75 percent of the total load and has been estimated at 7600 to 11 400 m^3 a^{-1} (Hotten 1988).

Beach cleaning eliminates microtopography and loosens the surface by breaking up surface crusts (salt and algae); it breaks up lag elements (shells or gravel); and it eliminates vegetation and litter, resulting in enhanced aeolian transport (Doody 1989; Nordstrom and Arens 1998). Removal of wrack eliminates biomass and nutrients and gives the beach an unnatural appearance, while the heavy equipment disturbs eggs in the beach (Hotten 1988). The lack of ecological value of raked beaches and their clean processed look has resulted in use of the expression "white bread beaches" by environmentalists.

Eliminating barriers to access

Bulldozers are used in The Netherlands to reshape high dunes that have been scarped as a result of wave erosion to create access ramps and reduce the hazard of slope failure (Nordstrom and Arens 1998). Grading has also been conducted to fill ponds in low-lying areas on the backbeach (Mauriello 1991),

Figure 2.6 Graded dune at Platje de Pals, Spain, showing difference between flat bare sand surface of graded area in foreground and topographically diverse, vegetated dune in background.

and to eliminate scarps that form on nourished beaches by dragging large pipes across them (Wiegel 1992b).

Providing views of the sea

Grading of dunes to lower levels to provide a view of the sea from boardwalks or residences (Figure 2.6) is a common practice. Grading for views is unnecessary in intensively used recreation areas where beach cleaning prevents formation of dunes. Dune grading is often conducted illegally in locations where the dune is considered important for shore protection and covered by land use regulations but where local enforcement of these regulations is not possible.

Illegal grading commonly occurs on a lot-by-lot basis, resulting in a conspicuous, unvegetated, flat-bottomed notch in the dune crest and an irregular dune with an incomplete vegetation cover that diminishes the protective value of the dune (Cortright 1987; Nordstrom 1988b). These low areas increase the likelihood for sediment transport to occur from the beach by wave or aeolian processes. Sediment may be moved to the landward side of the dune or pushed out onto the beach in an often misguided or futile attempt to reduce the likelihood that future sand inundation will occur. Grading on a windward coast often leads to rapid besanding of properties during strong onshore winds.

The state of Oregon reversed its policy by allowing grading in one of their shorefront communities to determine whether grading could be accommodated while achieving shore protection benefits. Their initiative is interesting in that it produced general guidelines and a management plan to serve as a model for other local jurisdictions. Requirements were that: (1) a plan was developed for the entire reach; (2) the minimum dune height remained 1.2 m above the 100 year flood elevation; (3) priorities were set for low dune areas that needed to be built up; (4) the dune was widened seaward and the foreslope smoothed to dissipate storm wave uprush; (5) a new zone of aeolian accretion was established seaward of the previous dune crest utilizing sand fences and beach grass; (6) vegetation was used to stabilize bare areas of the dune to prevent deflation of surface sediments and lower local wind velocities; (7) no sand was removed from the beach foredune system; (8) standards and timing for redistribution of sand were prescribed (Cortright 1987; Marra 1993). Subsequent grading created a lower, wider dune with a flat top that differed considerably in appearance from its previous hummocky, more natural-looking form, but the wider, more continuous, stabilized dune was considered to provide enhanced protection while restoring homeowner views (Marra 1993). Grading at this location was considered a viable option because the dunes were high; the shoreline was not eroding; and the community supported a plan that included a strong protection component. Geomorphic constraints make the option inadvisable in most coastal communities in the USA, particularly on the Atlantic Coast and Gulf of Mexico, where the hazard potential from storm surge places residents at greater risk (Nordstrom 1988b).

Enhancing environmental quality

Reshaping landforms to enhance environmental quality includes using bulldozers to create environments that encourage tern breeding (Randall and Doody 1995), breaching barriers to maintain salinities in lagoons (Orford *et al.* 1988) or promote runs of salt-water fish into or out of coastal ponds (Tiffney and Andrews 1989), altering the contours of nourished dunes to make them blend in with the adjoining areas and create naturalistic contours (van Bohemen and Meesters 1992), and breaching foredunes to create new blowouts (Figure 2.7) and slufters (tidally flooded overwash areas). These approaches, that focus on human design and reconstruction of ecosystems using small amounts of effort to control systems in which the main energy inputs still come from natural sources, are called environmental/ecological engineering and are described in greater detail in van Bohemen (1996). Alterations to create blowouts and slufters to increase mobility are considered feasible on the coast of The Netherlands, but only where no economic

Figure 2.7 Reactivated blowout at Terschelling, The Netherlands.

investments require protection and where there is still adequate protection against low-frequency storms (van Bohemen 1996). These kinds of activity, that increase landform mobility, are still considered experimental and only feasible in isolated locations where the shoreline is considered stable. Most alterations affecting landform mobility are designed to increase stability or the likelihood that accretion will occur, as identified in the following section.

Altering landform mobility

Beaches

Most alterations to beach mobility are designed to add sediment through nourishment operations or keep sediment in place using protection structures. These alterations are discussed in the next two chapters. Attention below is devoted to altering beach mobility by changing the locations or characteristics of inlets.

Closing inlets

Inlets may be closed as part of an operation that opens a new one (Møller 1990). One example is Turtle Gut Inlet in New Jersey, that was closed to improve the effectiveness of the jetties and navigation channel at Cape May Inlet. Actions taken during military operations to reduce navigation efficiency (usually by sinking ships in navigation channels) also can lead to inlet closure.

Construction of dunes has been accomplished to eliminate incipient inlets and prevent new inlets from forming (Ehlers and Kunz 1993).

The most common occurrences of inlet closure appear to be at sites of new inlets opened during storms. Closure of inlets is usually done to retain shore-parallel access routes, reduce flooding in backbays, and prevent impact of possible changes in salinity or predation by ocean species on bay shellfish (Sorensen and Schmeltz 1982). Closure following storms is most readily accomplished in a few days or weeks after opening because newly formed inlets are subject to rapid growth. The breach that formed updrift of Moriches Inlet, New York just after it was created was 91 m wide and 0.6 m deep; 4 days later it was 213 m wide, averaging 1.0 m below mean low water; 9 months later it was 884 m wide (Sorensen and Schmeltz 1982). Little Pikes Inlet that formed at Westhampton, New York during the storm of December 1992 widened at an average rate of 13.1 m day^{-1} (with a peak of 35.4 m day^{-1}) and grew from an initial width of 30.5 m to 1.52 km in 8 months; it was closed using 1.14 million m³ of sand and 550 m of steel sheeting (Terchunian and Merkert 1995).

Closure of inlets can eliminate the seaward-directed transport of sediment by ebb currents; cause a net erosion of the delta front; create longshore bars at the seaward edges of the former ebb deltas; smooth relic tidal shoals; and cause sedimentation of the former channels (Louters *et al.* 1991). The complex mix of processes and great variety of landforms at inlets is greatly reduced, and the elimination of inlet-related cycles of sedimentation on adjacent shorelines makes changes there more predictable and landforms more stable.

Stabilizing inlets

Inlets are stabilized to maintain predictable navigation channels or shoreline positions. The inlets that are targets of stabilization may be existing inlets or inlets that form during storms at locations where local interests would like to have them, such as Ocean City, Maryland. Stabilization usually involves use of structural controls, such as jetties that can be used to maintain both the channel and the adjacent shoreline or shore protection structures that cause the inlet to migrate over a shorter distance than under natural conditions. Dredging activities that are conducted to maintain channels at predictable depths and locations can also help stabilize inlets by reducing cycles of erosion and accretion related to channel fluctuations. Inlet stabilization by structures is discussed at greater length in chapter 4.

Increasing inlet mobility

Changing the location of dominant channels at inlets can create profound alterations to the adjacent shoreline. Inlets can be displaced updrift under natural conditions as a result of breaches in the ebb tidal delta or breaches in the updrift barrier island during storms. These breaches may result in creation of a more efficient channel than the former main channel; the former main channel then fills; and the location of inlet-related erosion and deposition along the shoreline changes (FitzGerald *et al.* 1978). Similar changes can be initiated unintentionally or intentionally by human action.

A 1978 dredging project at Townsend Inlet, New Jersey demonstrated unintentional alteration of a tidal channel, accompanied by unanticipated erosion and accretion. In 1978, the main ebb tide channel at the inlet was located close to the shoreline downdrift of the inlet throat, where it impinged on shorefront development. That year, a new channel was dredged through the ebb tidal delta north of the former channel to create a shorter connection to the sea and reduce dredging cost. The resulting changes were predictable following existing natural models of ebb delta breaching (FitzGerald *et al.* 1978) and included deposition on the inlet throat beach and accelerated erosion of portions of the downdrift shoreline that had formerly been in a position to receive sediment delivered offshore of the main inlet channel. That dredging project demonstrated the capability for human activity to dramatically alter the periodicity of the natural cycle and to enhance the capability of the natural system to change the locations of erosion and deposition (Nordstrom 1987b).

Kana (1989) shows how the periodicity of the erosion/accretion cycle can be successfully altered for the express purpose of causing locally eroding areas to revert to depositional conditions. The project involved cutting a new channel updrift of Captain Sams Inlet on Seabrook Island, South Carolina, and filling the old channel, moving approximately 133 000 m³ of sediment at a cost of US$300 000. The amount of sand made available to downdrift locations due to closure of the previous inlet was over 1.5 million m³. Results demonstrate the feasibility of incorporating changes at inlets into environmentally compatible, soft engineering solutions, but the dramatic and rapid changes in locations of erosion and accretion reveal the potential for increasing hazards if predictions of new shoreline changes are not accurate.

Creating inlets

New inlets may be created by dredging to improve navigation or water quality. The initiation of new inlet cutting is perhaps easier than commonly considered, having been accomplished in the past with manual or animal labor and land-based equipment rather than ship-borne dredging plants (Kana 1989; Dean 1991b; Galvin 1995). Artificial inlet location may be rendered easier if the new location corresponds to the location of a former channel (Lloyd 1980).

Cutting of artificial inlets has been common in Florida, where resulting inlets include Government Cut, Bakers Haulover Inlet, Port Canaveral, South Lake Worth and Port of Palm Beach Entrance (Wiegel 1992b; Bodge 1994a; Browder *et al.* 1996). Some inlets, such as Government Cut and Bakers Haulover Inlet, were located where natural inlets previously existed. Changes to the sediment budget and to the location of shoreline adjacent to new inlets can be dramatic. The Port Canaveral Inlet resulted in loss of about 6.1 million m³ of sand removed from the littoral system due to impoundment at the updrift jetty and loss of about 6 million m³ of sand dredged and disposed outside the littoral system (Bodge 1994a).

Modification of inlet channels that periodically open and close naturally as a result of seasonal differences in wave activity or runoff is common, particularly on the coast of California (Webb *et al.* 1991), but modifications have occurred in other areas as a means of controlling flooding (Orford *et al.* 1988). These actions may simply alter the timing of the natural cycles of inlet opening and closure rather than creating a new inlet where it would not have occurred under natural conditions or establishing a new equilibrium condition for a pre-existing inlet.

Dunes

Trapping sand to build dunes

Dunes may be artificially created and stabilized to prevent sand drift and inundation of human facilities, provide a predictable barrier against wave overwash and flooding and provide a barrier to salt spray that can help maintain the existing biological inventory. The most common techniques used to trap sand are employment of sand fences (discussed in chapter 4) and planting of vegetation. Accretion in planted areas is slower than accretion using fences, but it has the advantage of creating a dune with greater natural and protective values. Growth of vegetation and fungi throughout the dune will contribute to its resistance to erosion. Beach grass that is commonly used to stabilize dunes (*Ammophila arenaria* or *A. breviligulata*) does not thrive in artificially stabilized areas where burial by sand cannot confer its selective advantage on this species (van der Putten *et al.* 1993). The establishment of successional species depends on the richness of the soil micorrhizal fungi and other soil nutrients (Koske and Gemma 1997) and on the physical soil structure resulting from a healthy pioneer growth of *Ammophila* in wind blown sand (Maun 1993).

The conversion of dunes to stable features is more widespread than is commonly perceived, particularly in the USA, and includes seashore parks managed by local and state governments (Gares and Nordstrom 1988) and National Seashores such as Cape Hatteras, and Fire Island (Godfrey and Godfrey 1973; Psuty 1989). These human-altered dunes are usually linear and may bear little resemblance to the ones that existed prior to the management projects, although the human-altered dunes are often perceived to be the current standard by which park management plans are evaluated.

Stabilizing dunes

Dunes are most commonly stabilized using vegetation plantings (including the aforestation programs identified earlier), vegetation cuttings and inorganic materials. Foredunes built in large-scale government projects are usually planted. State and federal projects in the USA are initially planted, but subsequent maintenance of vegetation is left to local managers, and the success of maintaining this cover varies considerably with local commitment. Dunes created using bulldozers, usually in small-scale emergency projects, are rarely planted immediately because they are considered short-term, sacrificial features.

Examples of planting methods and successes are presented in Skarregaard (1989), Angus and Elliott (1992) and Avis (1995). Suggestions include: (1) using plant material originating from mobile foredunes, rather than stable foredunes (van der Putten and Kloosterman 1991); (2) planting *Ammophila* in accreting areas where plants are adapted, rather than in areas undergoing deflation (van der Meulen and Salman 1996); and (3) restricting use of vegetation cuttings where this may lead to eutrophication and introduction of alien vascular plant species (Koehler *et al.* 1996), although cuttings can contribute to natural seed germination by trapping wind blown seeds (Tippets and Jorgensen 1991).

Monocultures can occur because the dune is planted using a single species that is perceived to have the greatest utility for stabilization, but species diversity may be encouraged for aesthetic values and to ensure that die-out of one species does not jeopardize the chance to maintain a vegetation cover (Mauriello 1989). Considerable variety is found in the vegetation in the portions of the foredune found on private properties in the USA, because both exotic and natural vegetation is used to landscape yards and gardens. Some of the exotics used on private properties have colonized portions of the dune outside private lots adjacent to where they are planted (Nordstrom and Arens 1998).

Thatching is often introduced to the surface of dunes to stabilize bare surfaces temporarily to provide time for vegetation to grow. A variant on this practice, termed "brushing" is practiced in Western Australia using branches from local coastal shrubs or garden or verge prunings and sowing seeds of dune colonizing plants (Western Australia Department of Planning and Urban Development 1993). Other materials that have been used for stabilization include: straw that is strewn on the surface or punched into the ground (Tippets and Jorgensen 1991; van der Putten and Kloosterman 1991), tires (Western Australia Department of Planning and Urban Development 1993), biodegradable and non-biodegradable matting (Demos 1991; Angus and Elliott 1992), and bitumen spray (Ritchie and Gimingham 1989). Some of the great variety of methods used to stabilize surfaces is provided in Angus and Elliott (1992) who point out that often these efforts at stabilization are aimed at preventing threats that are more imagined than real, resulting in unnecessary environmental losses.

Remobilizing dunes

Dunes may be remobilized (destabilized) to change sediment budgets, stimulate growth conditions for desired species, or reinstate natural processes or appearance. Re-activation of stabilized dune fields to reinstate the beach sediment budget and reduce rates of beach erosion was conducted at the Waenhuiskrans dune field in South Africa (Swart and Reyneke 1988). Activities involved burning and cutting plants, followed by a plowing of the top 0.5 m of soil to break up the root structure and reinitiate the deflation phase of the former mobilization–stabilization cycle. A principal difference in the recent human-induced remobilization phase was leaving a vegetated strip of dunes adjacent to the village of Waenhuiskrans as a buffer against sand drift (Swart and Reyneke 1988).

Mowing, spraying, cutting, pulling or scraping vegetation from the surface have been employed in other areas to reinitiate sedimentary processes for the purpose of eventually causing changes to the vegetation and morphology of the dunes (Klomp 1989; Rothwell 1985; Anderson and Romeril 1992). Aeolian activity on the Dutch foredunes that are not critical for protection has been artificially enhanced in places by creating gaps in the crest to allow the dune profile to move landward (Figure 2.7). Mechanical removal of the vegetation on the foredune has also been practiced to allow the crest to migrate inland and create a more gentle foredune slope while retaining original beach width and dune volume. This kind of feature, termed a rolling foredune, was initiated on Terschelling (Arens and Wiersma 1994; van der Wal 1996).

Small-sized (1 to 5 m²) and medium-sized (50 to 500 m²) formerly stabilized blowouts at Midden Heerenduin near Haarlem in The Netherlands were reactivated by removing soil material as well as vegetation. The soil had to be removed because sand mixed with organic material (top 30 to 50 cm) is not easily moved by wind. Medium-sized blowouts gradually increased in depth at a rate of about 0.05 to 0.10 m a^{-1} in the flat terrain, with larger increases occurring in the undulating terrain. A slow increase in the blowout area of 5 percent a^{-1} occurred in the medium sized blowouts. Deposition in the surrounding area was 4.0 to 5.7 times the area of the blowout, with most sand deposited <30 m away from the blowouts. Most of the small patches that were reactivated stabilized spontaneously (van Boxel *et al.* 1997).

Dune mobility can be increased for the purpose of reinstating the former natural appearance of a human-altered landscape and burying cultural features that are too expensive to remove, as was achieved in the Devesa of Saler, Valencia (Sanjaume 1988; Sanjaume and Pardo 1992). These actions to reshape dunes to enhance aeolian activity are normally taken on a site-specific basis and are often experimental in nature. They are not taken along many developed coasts because they are perceived to increase the degree of hazard. They are usually not taken in natural areas either, largely due to the prevailing conservative approach to dune management (Nordstrom and Lotstein 1989).

Changing the surface cover

The changes to the surface vegetation cover of dunes mentioned above are intentional attempts to alter the mobility and configuration of dunes, but unintentional or indirect changes can have similar results. The human actions include setting fires, mowing, weeding and introducing exotic vegetation.

Setting of fires by aborigines is believed to have been responsible for initiation of sand drift in Australia, although the frequency of fire there increased since the advent of European settlement (Chapman 1989). Some of these fires may have been deliberately set to clear vegetation (Chapman 1989). Prescribed fire has been used on dunes in other areas to burn shrub to make dunes more suitable for grazing livestock (Corona *et al.* 1988), and fire has been suggested as a potential means of controlling alien species, such as *Pinus mugo* in Danish dune heath (Vestergaard and Alstrup 1996).

Mowing, like low-intensity grazing, can be used to manage dune plant communities (and eventually alter the morphology and evolution of the dune), although the long-term effects may be less beneficial than grazing (Hewett 1985). Yearly mowing has been accomplished to prevent succession to willow scrub on damp and moist slacks in nature reserves in The Netherlands (Westhoff 1985). Weed control may also be used to modify the exposure of target species to wind and soil moisture (Barron and Dalton 1996).

One of the principal means of changing the morphology and mobility of landforms by changing vegetation is through accidental or intentional introduction of exotic vegetation. Plant invasions can result in decreased biodiversity, loss of authenticity of vegetation, interference with successional processes, undesirable appearance of landscape, and reduction of value of land for conservation and recreation (Andersen 1995b).

Some of the important exotics that owe their introduction or spread to human actions include: (1) Japanese sedge (*Carex cobomugi*) introduced to coastal dunes in the USA from Asia (Pronio 1989); (2) pine trees (e.g., *Pinus nigra* ssp. *laricio* and *P. contorta*) used to stabilize dune systems throughout the world (Doody 1989; Sturgess 1992); (3) Bitou bush (*Chrysanthemoides monilifera*) inadvertently introduced to Australia from Africa (Chapman 1989); (4) *Acacia cyclops* introduced to Africa from Australia (Hellström and Lubke 1993); (5) the Australian *Casuarina equisetifolia* used to stabilize mobile dunes in Mexico (Espejel 1993); and (6) European beach grass (*Ammophila arenaria*), introduced to stabilize dunes on the Pacific coast of the USA (Cooper 1958; Pinto *et al.* 1972), South Africa (McLachlan and Burns 1992) and Australia (Barr and McKenzie 1976).

The effect of European beach grass introduced to the Pacific coast of the United States (Figure 1.1) is especially significant because it profoundly altered the morphology as well as the biota of the littoral zone along hundreds of kilometers of coast (Cooper 1958; Pinto *et al.* 1972). The invasion was so complete that no dunes dominated by native species remain in the state of Washington, so it is difficult to compare the dynamics of the exotic cover with a natural dune (Canning 1993).

Natural terrestrial coastal vegetation is often considered useless to developers, particularly where natural vegetation creates a surface cover that looks bare (Espejel 1993). Native plant species are prevented from recolonizing many shorelines because of the human preference for using lawn grass and exotic shrubs and trees for landscaping. Many landowners in New Jersey, USA have replaced all vegetation with gravel, using weed killer to maintain an unvegetated surface, and these locations are now barren (Nordstrom 1994a). Exotic species are not always preferred by property owners. The shapes of dunes created on the Washington coast where the native *Ammophila breviligulata* is planted differ in structural and aesthetic characteristics from where the introduced *A. arenaria* is planted (Seabloom and Wiedemann 1994). Use of *A. breviligulata* creates a lower dune that may be more susceptible to breaching during storms, but more appreciated by residents because of visual access to the beach and a lower incidence of illegal dune grading (Seabloom and Wiedemann 1994).

Exotics that were formerly introduced are now being eliminated in some areas to allow for re-establishment of indigenous species (Chapman 1989; Avis 1995). These programs may eliminate many of the problems of exotics, but the species composition of sites that are stabilized using indigenous species may not correspond closely to the natural distribution of those species (Avis 1995).

Actions outside the boundaries of coastal landforms

Indirect effects of actions taken outside the boundaries of beaches and dunes can have greater impact on these landforms than many actions taken within the landforms. Examples of external actions include: (1) reducing sediment supply to coasts due to damming and mining sand from rivers; (2) altering plant succession in dunes (and eventually the dynamics and morphology of the landform) due to changes in nutrient levels or acidity from pollution in precipitation (Doing 1989; Westhoff 1989; van Boxel *et al.* 1997); (3) increasing water pollution that degrades reefs, leading to increase in erosion rates on beaches landward of them (Thieler and Danforth 1994); and (4) changing basin area and tidal prism due to land reclamation and channel closures that can lead to alterations in inlet channels, ebb tide deltas and size and position of barrier islands (FitzGerald *et al.* 1984; Niemeyer 1994; Eitner 1996). Examples of effects of dams and stream mining are included to provide perspective. Other activities are not discussed in detail because of the difficulty of establishing direct process–response relationships.

Dams and water regulation activities

Construction of dams has changed accreting shorelines near river mouths to eroding shorelines in many locations throughout the world (Toyoshima 1982; Marqués and Julià 1987; Awosika *et al.* 1993; Innocenti and Pranzini 1993), although it is often difficult to distinguish between the effects of the dams on the sediment budget from effects of quarrying, land reclamation, aforestation and agricultural use that may occur at the same time.

Construction of dams and relocation of river channels or conversion of rivers from multi- to single-channel systems affect sediment distributions at deltas, resulting in greatly increased rates of coastal erosion at adjacent shorelines (Postalache and Diaconeasa 1991; Aibulatov 1993; Dal Cin and Simeoni 1994) and a smoothening of the coast in plan view (Postma 1989). The sequence of events, as described in McDowell *et al.* (1993), includes: (1) reduction in sediment supply due to reduction in stream discharge; (2) shrinkage of river-mouth bedforms with alteration in nearshore wave patterns; (3) remobilization of shelf sediments with transport onshore and alongshore; (4) onset of erosion around river mouth; and (5) extension of erosion alongshore.

The construction of dams on the upper Nile River cut off almost all water discharge and delivery of sediments to the coast and (coupled with reduced rainfall) resulted in maximum shoreline recession rates of 200 m a^{-1}, although rates of 5 to 10 m a^{-1} are more common (Frihy and Komar 1996). The water capacity of dams in north Africa became 20 percent lower in 20 years because of sediment trapping (Paskoff 1987). Damming has reduced the contribution of beach-forming materials by over 90 percent in some areas (Dzhaoshvili and Papashvili 1993). Solid sediment discharge at the mouth of the Rhône River in the late 1980s was only 10 percent of what it was in the mid nineteenth century (Paskoff 1987). Water regulation activities in southern Spain have reduced stream flow to the point where many rivers are dry for 7 to 9 months a^{-1} (McDowell *et al.* 1993). Dams, withdrawal of water, changes in the geometry of river beds and mining caused sediment flows on rivers on the Marche coast of Italy to be 30 to 70 percent less than flows before 1966 (Dal Cin and Simeoni 1994). About 11 km of beaches are being eroded north and south of the outlet of the Ombrone River, Italy, with values as high as 10 m a^{-1} (Innocenti and Pranzini 1993). Suggestions for sediment bypassing around dams have been made (Wasyl *et al.* 1991), but they have not been employed, and the problem of sand starvation is still critical in many areas.

Mining streams

Gravel and sand are being mined from almost all the principal river-beds in Italy (Postma 1989), and stream mining probably occurs in countless other locations throughout the world. Paradoxically, mining of streams is often conducted to supply materials to construct the coastal resorts that eventually will need the sediment to maintain the beaches (McDowell *et al.* 1993). Alternatively, mining is conducted as part of nourishment operations to supply beach material that may eventually have reached the coast or enhanced the likelihood that other riverborne materials would have reached the coast. Channel mining of rivers on the Russian Black Sea coast has caused a deepening of lower streams by 2 to 5 m and a reduction of gradient 1.5 times in some sections, resulting in deposition of up to 40 percent of the solid load, comprised mainly of beach-quality material (Grechischev *et al.* 1993). Estimates of the removal of bed material from four rivers in the Emilia-Romagna region of Italy between 1957 and 1972 are 9.2 million m³ (Postma 1989).

Summary conclusions

This overview of the degree, scale and rationale for alterations to beaches and dunes to suit human needs indicates that these alterations are widespread and varied, and their effects occur along a spectrum from the immediate to the long term. Many alterations are highly site-specific, but the impacts on shoreline evolution can be widespread if: (1) individual personal actions are conducted by many participants over a long shoreline reach (e.g., residents removing sand inundating properties); (2) communities adopt a program to maintain a specific landform configuration (grading a protective dune or flat beach for recreation); or (3) a one-time human action enhances the efficacy of natural processes to establish a new equilibrium condition that affects adjacent portions of the coast (remobilizing dunes through devegetation or altering channels at inlets, instituting new erosion–deposition cycles).

Outcomes are a combination of both intended and unintended effects. The rationale for intended human alteration of dunes includes hazard management, recreation demands, and enhancement of environmental quality. The range of choice in type and extent of action is defined by human use criteria, although usually only a small narrowly focussed set of criteria forms the basis of the action. Unintended effects are often the result of accelerated changes by natural processes that have been unwittingly enhanced by human actions and commonly involve changes in sediment availability and mobility. Activities conducted outside the boundaries of the shoreline also affect beaches and dunes, primarily by influencing coastal sediment budgets and pollutants, indicating that the chain of events leading to shoreline evolution caused by human action can be characterized by complex interactions over large spatial scales and with long lag times.

3

Replenishing landforms

Introduction

This chapter reviews the ways that sediment lost from beaches and dunes is replaced using material brought from outside the nourished area. The discussion is largely confined to design considerations, overall shape and volume of nourishment projects, as built, along with initial implications for biota. The distinction between nourished beaches and natural beaches and their resource values is further clarified in chapters 5 and 8.

There is no lack of experience with beach nourishment projects, and there are numerous case studies that evaluate these projects. A total of 138 articles on beach nourishment, artificial beaches and sand bypassing were published in the journal *Shore and Beach* alone between 1933 and 1992 (Wiegel 1992c). There are monographs and edited volumes on the topic (Stauble and Kraus 1993a; National Research Council 1995; Tait 1996), summaries and reviews (e.g., Stauble and Nelson 1985; Houston 1996c), and detailed evaluations of some of the larger projects (e.g., Wiegel 1992b; Stauble and Kraus 1993b). The nourishment option is now used nearly routinely throughout the world (Swart 1991; Davison *et al.* 1992), and nourishment has become cost-effective (relative to hard structures) due to improvements in dredging technology and the implementation of nourishment programs over longer reaches of shoreline (van Oorschot and van Raalte 1991).

A distinction can be made in use of the terms "nourishing" (or replenishing) and "restoring." The word restoring is often a misnomer when applied to nourishment operations because beach and dune environments are not restored simply by placing a volume of sediment on a coast. Accordingly, the word replenishing is used in the title of the chapter rather than the frequently encountered word "restoring," and the word "nourishing" or "nourishment" is used to refer to the operation itself. The sediment that is replaced is referred to as "fill."

Nourishing beaches

Uses of fill and cost constraints

The most often cited rationale for beach nourishment is as a soft engineering alternative for protecting human facilities from wave attack in locations where the original beach has been lost, but there are many other uses for nourishment. These uses include: (1) designing and creating new beaches where none existed in order to fulfill a specific human use (Walker 1981b; Bodge and Olsen 1992; Shiraishi *et al.* 1994); (2) eliminating detrimental effects of shore protection structures by burying them (Yesin and Kos'yan 1993; Nordstrom 1994b); (3) providing protection to seawalls and dunes (Smith *et al.* 1993; National Research Council 1995); (4) protecting natural areas (Klomp 1989; Roelse 1990); (5) widening the beach to accommodate new construction (Rouch and Bellessort 1990); and (6) counteracting the effects of contemporary sea level rise (Stive *et al.* 1991). A human-created beach in an urban area is generally well received by the public, and nourishment operations have played a considerable role in determining or enhancing the reputation of many coastal locations as valued tourist destinations (Rouch and Bellessort 1990; Flick 1993; Leidersdorf *et al.* 1993; Dornhelm 1995; Wiegel 1995).

There are distinctly different viewpoints on the value of beach nourishment, depending on whether the observer has primary interest in damage reduction, recreation or retention (or enhancement) of property values (Camfield 1993). Considerable debate has occurred on some of the differences in viewpoints (e.g., Houston 1991; Pilkey 1990, 1992), especially regarding the cost effectiveness of nourishment. The benefit–cost aspects of nourishment and data supporting the value of a beach are presented in Bell and Leeworthy (1985), Dean (1988a), Smith and Piggot (1989), Townend and Fleming (1991), Camfield (1993), National Research Council (1995), San Diego Association of Governments (1995), Stronge (1995), Beachler and Mann (1996), Finkl (1996a) and Mann (1996).

Cost for a m³ of sand can range up to \$30 (Irish *et al.* 1996), and projects range from highly cost effective (Schwartz *et al.* 1991) to excessive, at least in terms of initial estimates (Pilkey 1992). Much of the economic success of beach nourishment operations is evaluated by basing fill longevity against prenourishment predictions using calculation models for design of beach fills (e.g., Dean 1991a; Larson and Kraus 1991; Houston 1996b). Economics *per se* is not treated in detail in this book, but the economics of beach nourishment does have a profound effect on whether the option is favorably received relative to other engineering solutions (Purnell 1996). Once nourishment has

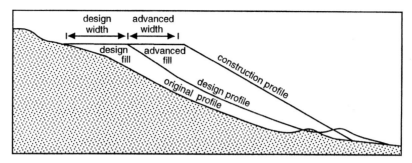

Figure 3.1 Relationship between pre-nourished profile, design profile and advance fill. Modified from National Research Council (1995).

been selected, economics has an effect on the morphology and sedimentary characteristics of landforms through selection of the size, location, method of borrow, transport and placement of materials, likelihood of follow-up operations and, inevitably, the evolution of many segments of coast.

Design considerations

Conceptually, a nourished beach may be envisioned in terms of a design width, a design profile shape, a design fill volume (calculated to provide the necessary protection) and an advanced width and fill volume (sufficient to account for beach loss between renourishment projects) (Figure 3.1). Logistical constraints may prevent the artificial creation of the ideal profile for either the advance fill or design fill, and the shape of the beach immediately after nourishment may represent a construction profile created according to the sand placement capability of the available construction equipment (National Research Council 1995). In this case, the fill volumes will still be predicated on the calculated fill volumes derived from the assumed slopes (Coastal Engineering Research Center 1984; Kana 1993). Federal design policy in the USA requires that the design cross-section is optimized to return maximum net benefits and the design beach is the added width of beach that returns maximum net benefits; the advanced fill volumes are designed to achieve the lowest annual cost for the renourishment program (National Research Council 1995). The overfill required to achieve the optimum profile can be considerable. For example, a surcharge of 40 percent on the design volume has been suggested as an approximation of the required overfill (Verhagen 1992, 1996).

Some operations may treat the construction profile as the design profile to attempt a more permanent change in the characteristics of the beach. The restoration project at De Haan, Belgium was an attempt to create a new beach profile configuration from the inactive backbeach to offshore near closure depth and includes a recreation platform (as backbeach), profile nourishment to over 2 m below mean lower low water and a feeder berm offshore of this location, resulting in placement of about 750 m³ m⁻¹ over a cross-shore distance of about 700 m (Helewaut and Malherbe 1993).

Use of a design profile is the normal placement practice in the USA (National Research Council 1995). Nourishment success is often based on the ability to predict the slope of the design profile – a task that is difficult on high-energy beaches with a relatively high tidal range (Kana 1993). A design profile may be useful for computing the volumes required for a nourishment project in comparison with existing beach profiles, and a construction profile may simplify completion of the project, but the use of these profiles can cause problems in the perception of the success of a nourishment project. Lay people may assume that fill retention and morphology are predictable and that the design profile will be the same as the construction profile. The reworking of the construction profile, that is treated as an anomalous temporary feature in design studies (National Research Council 1995), can have great psychological impact as a template for evaluating future coastal evolution and assessing the success of the operation. One example is the Indialantic/Melbourne Beach, Florida nourishment project, where a scarp in the fill occurred the day after completion; a rapid cut back of the fill berm occurred in the first two weeks (although wave heights were <1.0 m); and 56 percent of the material placed on the subaerial berm was removed in the first 2 months (Stauble and Holem 1991).

Conspicuous losses from the upper beach have led to criticisms of the effectiveness of beach nourishment operations. These criticisms may be invalid because use of the shoreline as a single measurement parameter does not adequately identify the volumetric changes to the profile, only the shape change to the upper profile; the fill materials may remain in the nearshore and still be effective in dissipating wave energy (Houston 1991; Stauble and Kraus 1993b; Pope *et al.* 1994). The concept of failure of a beach nourishment project should not be tied to loss of sand from the subaerial portion of the beach, although this term has been used to describe such losses (Bruun and Willekes 1992). Placing fill sediment in the water to avoid the negative impact of losses from the upper beach does not necessarily overcome negative publicity. A design beach is approached only slowly where offshore nourishment is employed, and the public may perceive these projects as a waste of money as well (National Research Council 1995).

As a result of these perception problems, Kana (1993) has suggested that the design of beaches should be based on profile volume rather than shape. Stive *et al.* (1991) recommend using the expression "shore nourishment" instead of "beach nourishment" to emphasize that an important part of the sediment resides beneath low water and that it is necessary to nourish the complete active profile. Indeed, the design methodology for beach nourishment projects should provide sufficient volume to nourish the entire profile from the berm or dune to the seaward limit of the active profile to avoid underestimating fill requirements (National Research Council 1995). Acceptance of this broader conceptualization of the nourishment process accomplishes two important objectives: (1) it expands the range of choice available to managers and dredging contractors so they can employ the most cost-effective alternatives to beach nourishment operations, regardless of where it is placed; and (2) it increases the acceptability of projects that do not create an initially wide upper beach. The latter consideration supports the concept that serious "errors" cannot be made in the design of a nourishment project (providing the sediment characteristics of the fill are similar to the pre-nourishment sediments) because the needed volume is related to the losses out of the shoreline reach; if the beforehand estimate was incorrect, only the estimated lifetime of the project has to be adjusted (van de Graaff *et al.* 1991).

Beach dimensions and volumes

The width of a nourished beach is often a compromise between the amount of sediment required for protection or recreation for the given reach length and the amount of money available to obtain and transport the materials. The result is often a narrower beach than the optimum, but overly wide beaches may be created in locations that are close to opportunistic sources (such as dredged channels) or in locations that are designated as stockpile or feeder beaches. The width of the beach resulting from the combined nourishment/disposal operation at Perdido Key in 1990 created a 122 m wide beach that almost doubled the width of the barrier island in places (Gibson and Looney 1994). Beach widths in Santa Monica Bay increased by 46 to 152 m in combined disposal/nourishment operations (Leidersdorf *et al.* 1994). A width of 30 m is considered adequate for storm protection on embayments of the Black Sea, while 40 m is more appropriate for more exposed portions of that coast (Yesin and Kos'yan 1993). Other widths reported for restored beaches include 60 to 90 m (Zenkovich and Schwartz 1987; 1988), 90 to 100 m (Byrnes *et al.* 1993; Dornhelm 1995), and 140 m (Boldyrev *et al.* 1996).

The length of shoreline protected by nourishment projects is presently small relative to the amount of natural coast. There were 56 large federal beach nourishment projects conducted in the USA during the period 1950 to 1993, extending along 364 km of coast, leaving 28 600 km of eroding coast and 3439 km of critically eroding coast unprotected by federal projects (Houston 1995). Artificial beaches represent 3.79 km of the 132 km French Riviera coast (Anthony 1994). These lengths are a bit misleading because they only refer to design lengths. Longshore transport extends the impact of nourished sediments for greater distances alongshore. The relative lengths of nourished beaches is bound to increase as erosion proceeds in developed communities that have access to funds for nourishment.

The lengths of nourishment sites may be short, but the volumes of sediment may be considerable. Over 22.3 million m³ of fill were placed on the islands of the German North Sea coast in 47 operations prior to 1988 (Kelletat 1993). Nearly 50 nourishment projects were conducted in The Netherlands by 1990, involving about 60 million m³, with about 7 million m³ being added to the coast each year since 1991 (Roelse 1990; de Ruig 1995). More than 90 beaches in the United States have been nourished in more than 200 separate operations (Pilkey and Clayton 1987). The total amount of sand placed in 56 Corps of Engineers projects between 1950 and 1993 was 144 million m³ (Sudar et al. 1995). Over 100 million m³ has been placed on the shoreline of southern California (Flick 1993), with over 24 million m³ placed on the shoreline of Santa Monica Bay (Leidersdorf et al. 1994) and 26 million m³ placed on the Silver Strand shoreline near San Diego (Flick et al. 1991). A total of 29.8 million m³ has been placed on the New Jersey beaches (USACOE 1990).

Relatively large-scale operations at specific sites (including multiple operations) include >13 million m³ placed on Victoria Beach, Lagos (Awosika and Ibe 1993); 10.6 million m³ placed at Ocean City, New Jersey in several operations (Weggel et al. 1995); 6.8 million m³ placed at Long Beach, California (Clayton 1989); 5.4 million m³ emplaced at Ocean City, Maryland (Stauble and Cialone 1996); >7.0 million m³ emplaced at Jupiter Island, Florida (Hamilton et al. 1996); and >8 million m³ emplaced at Zeebrugge, Belgium in anticipation of erosion due to construction of breakwaters (De Wolf et al. 1993).

Smaller projects may be undertaken to provide protection to localized problem areas or to provide a temporary solution to larger problems. Many of these projects are conducted using trucks. Volumes reported for some of these trucking operations are 23 000 m³ (Bodge and Olsen 1992), 45 600 m³ (Oertel et al. 1996), and 130 000 m³ (Philip and Whaite 1980). Trucking projects as large as 1 million m³ (Flick 1993) and 680 000 m³ (Wiegel 1993b) have been reported.

Estimates of volumes expected in the future are large. Van Dolah et al. (1993) report that about 12 million m³ of sand would be required to restore and maintain a 15 m wide dry sand beach along the coast of South Carolina (approximately 300 km long) for 10 years. Up to 22.8 million m³ is projected for initial beach building for the San Diego region (San Diego Association of Governments 1995).

Sources of fill

Fill materials can include sand from offshore (National Research Council 1995), backbays (Zawadzka 1996), transgressing dunes (Philip and Whaite 1980), desert sand (Fanos *et al.* 1995), upland sedimentary deposits and crushed rock, including waste from road and tunnel construction or from building sites (Rouch and Bellessort 1990; Dzhaoshvili and Papashvili 1993; Wiegel 1993a; Anthony and Cohen 1995; Pacini *et al.* 1997). Sources of fill may change through time due to depletion of sources, changing perceptions of environmental values, changes in technology, or simply the availability of opportunistic sources. Much of the fill emplaced in New Jersey before the 1970s was obtained from the backbays because it could be readily obtained using inexpensive equipment. This material was found to be less suitable than offshore sources because of perceived environmental degradation in the back-bays and the large proportion of fine-grained materials. Recent small-scale operations have used upland sedimentary deposits brought to the site in trucks. Large-scale operations have used sediment from the nearshore and from inlets (US Congress 1976; Weggel *et al.* 1995), making use of dredging plants that can maintain operations in ocean environments. Offshore deposits provide up to 95 percent of the sand in the USA at present (National Research Council 1995). Offshore sources are commonly used in other countries. Sand in The Netherlands and Denmark, for example, is excavated at depths greater than the -20 m contour (Laustrup 1993; de Ruig 1995).

Maintenance dredging of navigable inlets provides a considerable volume of sediment that can be made available to nourish beaches (National Research Council 1995). Douglass and Hinesley (1993) estimated that 38 million m³ of sediment was available from the Mobile Alabama Ship Channel this century. The US Army Corps of Engineers maintains and improves about 40 000 km of navigable channels and dredges an average volume of 230 million $m^3 a^{-1}$ (Hales 1995), although not all of this sediment is located near coastal areas where beach nourishment is possible.

Administrative constraints often make sediment dredged at inlets unavail-able for deposit on nearby beaches (National Research Council 1995). Past inlet maintenance dredging projects in New Jersey discharged the sediment in open water because of the small quantities involved ($<152 000$ $m^3 a^{-1}$), the lower cost of casting sediment to the side of the dredge, and potential time delays caused by weather and co-ordination problems (Mauriello 1991). Increasing recognition of the resource value of even small quantities of sediment has caused many authorities to take the position that dredging projects must be designed with an upland (beachfill) disposal option and encourage nourishment when a sufficient volume of the sediment is uncontaminated sand (Mauriello 1991; Looney and Gibson 1993).

Local sources may be considered finite, and the availability of nearby sand for the life of a project (50 years for federal projects in the USA) is one of the biggest uncertainties in a given operation (National Research Council 1995). There appear to be insufficient sources of materials for nourishment projects in England (Purnell 1996), and sand sources for the southern Florida Atlantic coast may not be sufficient for the next 50 years and beyond (Wiegel 1992b). Most offshore borrow sources within Broward County, Florida have been depleted, making judicious use of sediment deposited at inlets critical (Lin *et al.* 1996). Leidersdorf *et al.* (1994) point to the importance of utilizing all possible sources of beach nourishment material and taking a pro-active role in seeking sand from sources of opportunity. Fill for beaches in southern California, for example, came from dredging of harbors, marinas and residential lagoons, construction and expansion of sewage treatment, power generation facilities, naval facilities and river beds (Flick 1993; Leidersdorf *et al.* 1993; Wiegel 1994). Some states have already taken the initiative to identify sources of sand for a long-term commitment to beach nourishment (van Dolah *et al.* 1993).

The availability of sources is a site-specific issue. Large quantities are available in some areas. An estimated 1.5 billion m^3 is available on Ship Shoal, off the Louisiana coast (van Heerden and de Rouen 1997). The potential yield of sand along the Alabama, USA coast is 535 million m^3 in water depths of 9 to 26 m within 3 to 25 km of the shoreline in a region where an estimated maximum 0.8 to 1.9 million m^3 of sand would be required for each of the 6 restoration areas (Davies *et al.* 1993). Projects in Florida have nearly depleted economically recoverable sand reserves in state waters (National Research Council 1995), but an estimated 30 to 50 billion tons is available to the Florida beaches on the Great Bahama Banks (Wiegel 1987). An estimated 264 million m^3 of sediment is available near Silver Strand State Beach near San Diego, and 146 million m^3 of sediment is available off Mission Beach California (San Diego Association of Governments 1995). The sand supply at the bottom of the North Sea available for replenishing the Dutch coast is estimated to last for hundreds of years (de Ruig 1995).

Nourishment methods and beach characteristics

Figure 3.2 identifies the zones where artificial nourishment is usually applied to the beach and dune. Smith and Jackson (1990) identify the engi-

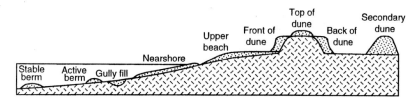

Figure 3.2 Common locations for emplacement of fill used in shore nourishment projects. Modified from van de Graaff *et al.* (1991); Hillen and Roelse (1995); van der Wal (1998).

neering requirements and the physical, social and political impacts of placement in many of these different locations. There is no common internationally accepted terminology for these locations (e.g., nourishment seaward of the upper beach may be termed nearshore, offshore, or shoreface nourishment), and the locations depicted on Figure 3.2 are generalizations. I have used the term nearshore to refer to undifferentiated deposits in shallow water and the terms gully fill and berm to refer to specific types of features created in the nearshore zone or farther offshore. The term stable berm refers to a feature that is located in water that is deep enough that wave action cannot significantly alter its form and volume; active berm refers to a deposit that can be readily reworked by waves and currents and incorporated into the littoral transport system. This distinction is easier to make conceptually than in reality because there is still too little information to predict the conditions under which an offshore deposit will be reworked.

The methods used for nourishment projects vary considerably (van Oorschot and van Raalte 1991), resulting in great differences in the shape, location and grain-size characteristics of the landforms that are created. Some operations are combined nearshore and upper beach nourishment projects (Mulder *et al.* 1994). The traditional approach in the USA uses a simple profile geometry for design beaches using site-specific profile surveys, with a flat berm of desired width, often at the elevation of the natural berm and with an average foreshore slope out to estimated closure depth (Figure 3.1), although some designs have employed a composite slope on the upper and lower beaches (as depicted in Figure 3.2) or an arbitrary slope (Kana 1993).

Temporary landforms

The initial landforms may be viewed at several different temporal and spatial scales. The most conspicuous (and often grotesque) individual landforms, although of short duration, are those constructed to facilitate emplacement of fill or achieve a design profile (Figure 3.3). The load contained in each dredge (or truck) and delivered to the nourishment site is often placed in a

Figure 3.3 Modifications to the beach during beachfill operation at Sylt, Germany, 1985.

small mound that is reworked by wave, swash or aeolian processes (depending on its location). Deposits in the swash zone soon lose their hummocky shape, but deposits placed farther offshore may retain the hummocky appearance because of the limited amount of reworking seaward of the zone of breaking waves.

Sand delivered by truck to the backshore may retain its hummocky appearance for a considerable amount of time before being reworked by winds or storm waves. The isolated deposits may be reworked by earth-moving equipment to form a new, larger landform that is usually more linear than the original fill deposits. Other landforms designed as temporary features include rehandling pits in the nearshore (van Oorschot and van Raalte 1991) and storage areas used to place sand for desalinization prior to placement on the landward side of the dune (van de Graaff *et al.* 1991). Zawadzka (1996) identifies increases in heights of breakpoint bars and changes in their shape, in addition to creation of bars and troughs in locations where they had not occurred, although these forms were temporary. There are likely to be small-scale subaqueous landforms resulting from the interaction of waves and currents with both disposal mounds and dredging equipment that have not been reported or studied because of the difficulty of monitoring them and the perception that only landforms that have long-term survivability are of interest.

Concentration on the protective and recreational aspects of nourishment and the prevailing perception that fill materials will inevitably be reworked into an equilibrium profile approximating the pre-nourishment profile may contribute to the lack of interest in identifying and monitoring temporary nourishment-related landforms. Addition of ecological and aesthetic values to the rationale for conducting nourishment operations and commitment to more frequent nourishment projects at a specific site may result in more frequent occurrence of "temporary" landforms in the water, greater long-term survivability of nourishment-related landforms on the upper beach and more interest in the resulting geomorphological and ecological variability in the coastal landscape.

Upper beach nourishment

Nourishment of the upper beach has been the most common practice in the past. This location is considered appropriate for emergency operations because it places sand close to the coastal facilities that require protection, and it enhances the accuracy of determining payment volumes (Grosskopf and Behnke 1993). Bruun and Willekes (1992) call this method of placement the conservative American approach to nourishment. The method exchanges stability for the time-limited advantages of placement ease and recreational use of the beach (Bruun 1993). It has been termed a public relations disaster because of the conspicuous and rapid removal of sand by natural processes (Smith and Jackson 1990).

Sand is normally transported to the nourishment area by pipeline as a sand and water slurry or (in small-scale operations) by direct dumping by truck. The deposited material is usually reworked by bulldozers to the design profile. An open fill area created using piped sediment may cause high rates of loss and a limited segregation of fines, but it can also produce a slope profile close to the final equilibrium profile. An alternative is to emplace a dike (or bund) to contain the fill material, causing fines to segregate and accumulate against it, creating a higher beach (van Oorschot and van Raalte 1991). Berm height is often made considerably greater than the height of the natural berm in order to achieve protection against wave runup. For example, the artificial berm at Dead Neck Beach on Nantucket Sound, Massachusetts was 1.8 m higher than the backbeach elevation (Wood *et al.* 1996).

The result of subaerial nourishment is a widened, over-steepened upper beach with a shape and composition out of equilibrium with natural processes. Inspection of landforms created in large-scale projects conducted by the US Army Corps of Engineers reveals a post-construction profile of consistent width and elevation for much of the length of the project (Figure 3.4).

Figure 3.4 Characteristics of the beach and seawall at Cape May, New Jersey, following a federal beach nourishment project.

The recently nourished beach retreats rapidly as the fill is reworked by waves and swash to create a gentler upper slope. A prominent vertical scarp is often conspicuous at the upper limit of wave reworking at high tide (Smith and Jackson 1990). Fill sand that is only slightly smaller than native sand can result in a much narrower beach than one using the same size or coarser sand (National Research Council 1995). Even when beach-compatible materials are used, the nourished beach may be physically different from the non-nourished beach in terms of sand compaction, shear resistance, moisture content and grain shape (National Research Council 1995), and sediments on the back-beach may have a conspicuously different appearance from sediments on the eroding foreshore.

Subsequent human actions also play a profound role in the way a nourished beach evolves. Figure 3.5 illustrates the major differences between the characteristics of a nourished beach managed for recreation, involving grading and beach cleaning (Figure 3.5B$_1$) and a nourished beach that is allowed to evolve under natural processes (Figure 3.5B$_2$). Even a cursory glance at this figure reveals the reason why beach "nourishment" is not the same as beach "restoration" in locations where the beach is maintained as an artifact rather than a naturally functioning landform.

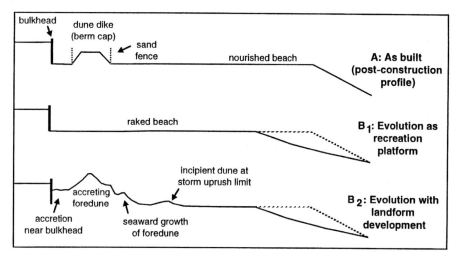

Figure 3.5 Comparison of the post-construction profile of a nourished beach with a nourished beach that evolves as a managed recreation platform and one that evolves naturally.

Veneer beach fills

Veneer fills have been used on the upper beach where beach-quality sand is not available in sufficient quantity for an economically viable operation. This type of fill is often limited to shorelines where seasonal and storm-induced profile changes and rates of erosion are sufficiently well known that the design thickness of the surface layer can be accurately predicted (National Research Council 1995). Examples of fills with veneers include Key West, Florida, where coral rock was used for the lower layer (National Research Council 1995), Ostia, Italy, where the lower layer is mixed sand and gravel and the upper layer is a 1.0 m thick sand deposit (Ferrante *et al.* 1992), and the French Riviera, where materials used for land reclamation are topped by an overlying layer of fine homogeneous gravel or sand quarried from inland sites (Anthony 1994).

The nourishment site in Corpus Christi Bay, Texas (Kieslich and Brunt 1989) provides an example of a veneer of sand over a bottom layer of fine-grained materials that is feasible to use because the beach is in a low-energy environment. The Key West and Ostia sites are on exposed coasts, but the former is sheltered from maximum wave energies by a reef and the latter is sheltered by a sill built to create a perched beach. There is not enough experience with veneer beach fills at present to say that they will be a viable long-term alternative in other exposed areas (National Research Council 1995).

Aeolian transport on nourished beaches

Nourishment of the upper beach can have profound effects on aeolian transport as a result of: (1) an increase in source width for entrainment of sediments; (2) changes in grain-size characteristics that affect the likelihood of entrainment and transport; (3) changes in moisture conditions due to creation of a new surface level; (4) changes in the shape of the beach or dune profile; and (5) changes in the likelihood of marine erosion of the foredune that influences transfer of sand within the beach–dune system (van der Wal 1998).

Van der Wal's (1998) results of grain-size analyses and wind-tunnel experiments of samples from nourished beaches and adjacent unnourished beaches reveal that rates of aeolian transport are highly dependent on both grain-size distribution and the amount of shell fragments. Fill sediment that was moderately to moderately well sorted with a negative skewness caused by shell fragments revealed low rates of sand transport, whereas sand samples from nourished sites where sediments were well or very well sorted exhibited high transport rates. Her results indicate that median grain size is a poor predictor of the rate of aeolian sand transport when coarse shells are present.

The great initial width of most nourished beaches and the presence of fine material lead to increased likelihood of transport by wind soon after placement (Draga 1983; van der Wal 1998), although fill materials are often characterized by a bimodal or platykurtic distribution, and removal of the fine sediment can leave a surface of coarser shell or gravel that resists aeolian transport (Davis 1991; Psuty and Moreira 1992; van der Wal 1998). Field observations by van der Wal (1998) indicate that shell pavements can form within weeks, and they can remain in areas that are not periodically flooded by the sea, such as the tops of dune nourishment projects placed on the upper backbeach (Figure 3.2). These shell pavements were not encountered on the natural beaches van der Wal monitored.

The reworking of the seaward portion of the fill by waves and the landward portion of the fill by aeolian transport results in considerable cross-shore variability in sediment grain-size characteristics. These spatial differences can greatly exceed the variability on the natural beaches in the area (van der Wal 1998). Namikas (1992) reports a mean grain size of 0.99 mm on the surface of the nourished backbeach at Sandy Hook, New Jersey, where the fill body below the surface had a grain size of 0.85 mm and the foreshore had a mean grain size of 0.41 mm. Sediments on the foreshore are usually reworked and well sorted (Psuty and Moreira 1992). Sediments at Namikas's site were best sorted on the foreshore, more poorly sorted within the inactive fill area and most poorly sorted on the deflated backbeach. The foreshore of the fill area was nearly identical in size and sorting to foreshore sediments taken from the natural beach updrift of the nourished area. Davis (1991), in contrast, noted a difference in swash samples at nourished and unnourished sites, but the difference was substantially less than the difference between the swash zone and the backbeach at the nourished site.

It appears that aeolian transport from nourished sites may be better controlled if greater attention is paid to the grain-size distribution and the spatial variability of sediment characteristics across the profile. Initially high rates of aeolian transport may be anticipated, and sand fences can be placed to reduce wind hazard (Helewaut and Malherbe 1993). The presence of a surface lag on the beach does not affect its evolution under the influence of wave swash, but it will affect the likelihood of deflation and formation of dunes landward of the beach. Formation of a lag surface in dunes that are artificially nourished can greatly restrict their evolution as described in chapter 5.

Subaqueous nourishment

Subaqueous nourishment using materials from offshore sediment sources can cost considerably less than deposition by pumping onto the subaerial beach, making this option the standard, least cost alternative for sites with viable sources (McLouth *et al.* 1994). Jackson and Tomlinson (1990) report a cost savings of 50 percent using subaqueous nourishment at Kirra Beach, Australia. Mulder *et al.* (1994) estimate a similar amount of savings in the cost per unit value over nourishment of the upper beach, but they point out that approximately twice the amount of sediment would be required for a subaqueous nourishment to arrive at the same volume of sand on the upper beach after 5 years. Subaqueous nourishment is becoming more common as the economics of the operations improve and the advantages of optimizing fill geometry are better recognized (Bruun 1993), and it may be the only alternative where access for machinery can only be gained from the sea side (Zenkovich and Schwartz 1987).

There is considerable variety in the subaqueaous placement of fill materials, and the forms depicted in Figure 3.2 are only approximations of a few of the many possible alternatives. Sediments may be placed in mounds, ridges, isolated depressions and linear gullys that create a new topography; sediments may be spread over the pre-existing surface to retain the original topography; and sediments may be placed on existing landforms to enhance their size or decrease their topographic variability. Figure 3.2 does not distinguish between types of features created in the nearshore zone that are close enough to the beach to be reworked by waves and swash during frequently occurring storms because these features have such short durations. Sediment deposited farther offshore as berms or as gully fills may survive as distinctive landforms and interact with waves to cause changes in the beach onshore of them.

Nourishment of the nearshore close to the water–beach contact (Figure 3.2) often occurs using the rainbow or jet method of discharging sediment by spraying it in shallow water as a sand and water slurry; shallow draft barges also may be used to dump sediment at this location (van Oorschot and van Raalte 1991). The rationale for nearshore nourishment is that maintenance of the nearshore profile reduces the likelihood of erosion of the beach and dunes landward of it, and the placement occurs at the location where sediment placed on the upper beach would have gone anyway (van de Graaff *et al.* 1991; Mulder *et al.* 1994). The rainbow method can be used to replenish sediment while causing minimal alteration to the shape of the previous topography.

Berms

The term "berm" (or "mound") is used to refer to submerged deposits of dredged material that can be placed as symmetrical piles or elongated ridges to function either as a feeder for nearshore nourishment or as a stable environment for wave attenuation or habitat creation. A hopper dredge is often used to create these features. Benefits of berms include: (1) enhancing fisheries; (2) blocking offshore sediment losses; (3) reducing wave energies; (4) augmenting sand budgets; (5) forming a foundation for offshore structures; and (6) serving as stockpiles for later rehandling (Hands and Allison 1991; Hands and Resio 1994). Nearshore and offshore nourishment have advantages over upper beach nourishment in that losses through time are not visible, and the aesthetics of the beach are not spoiled by fine- or coarse-grained material on the backbeach. A disadvantage of offshore nourishment is that the subtle visual impact of accretion on the backbeach may convey the perception that the operation was not effective (van de Graaff *et al.* 1991). In some cases, it is necessary to wait a year or two until a storm of appropriate size can mobilize the deposit (Zenkovich and Schwartz 1987).

Early experience with berms was often disappointing because these features were placed in water that was too deep for deposits to move landward or to produce any benefits to adjacent shores (Laubscher *et al.* 1990; McLellan 1990; Hands and Allison 1991). Underwater mounds can focus wave energy leading to great local variations in longshore transport potential (Hughes and Brundrit 1995). To be effective as a form of shore protection, a berm must be considered an engineered structure requiring a verifiable design, construction methodology and periodic maintenance, rather than a simple modification of an open water disposal operation (McLellan 1990). Methods of categorizing active and stable depths for berms, based on wave and sediment parameters, evaluation of prototype designs and design guidance, are provided by Hands and Allison (1991) McLellan and Kraus (1991) and Allison and Pollock (1993). Results of limited field examinations of nearshore berms reveal successes in filtering wave energies, increasing sand quantities on beaches to the lee and displacing the beach profile offshore (Allison and Pollock 1993; Hoekstra *et al.* 1996), but there is still considerable research to be done to determine depth of placement and distance of placement from the shoreline (Allison and Pollock 1993).

Berms can function as an alternative to beach disposal if beach disposal conflicts with nesting activity (McLouth *et al.* 1994). Some berms can be justified economically on the basis of shortened haul distance for dredged material, but increased future use of the option as a shore protection measure depends increasingly on being able to demonstrate the economic benefits of berms to eroding shorelines (Richardson 1994). Most berms are constructed near maintained inlet channels, where they simplify the problem of finding suitable disposal areas for dredged sediment (McLellan and Kraus 1991). Some of the benefits identified for berms are theoretical rather than empirically documented and are not the designed intent of the disposal operation. None of the stable berms that have potential value for fishery benefits were designed for that purpose, and there is no quantitative evidence of fisheries-resource enhancement in the scientific literature (Clarke and Kasul 1994). The berm created southwest of Dauphin Island Alabama is being monitored as a national demonstration project to identify the effects on fish as well as stability of fine-grained material and effects on incident waves (Hands and Allison 1991).

The berm created southwest of Dauphin Island is one of the largest underwater features constructed in the USA; it was created in 14 m water depths using about 14 million m³ of sand and mud; it has design crest dimensions of 300 m by 2750 m and is 6 m high (Hands and Allison 1991; Allison and Pollock 1993). One example of a stockpile was the nearshore disposal site for material dredged from St. Marys Inlet, where 825 770 m³ of material was redredged and placed on a beach disposal area in Florida, a few kilometers to the south (Pope *et al.* 1994).

An example of an intermediate-scale berm is the one created at Newport Beach, California, where 969 760 m³ of material were placed in water depths ranging from 1.5 to 9.1 m (Mesa 1996). Small-scale operations creating berms in the USA include Cocoa Beach, Florida (Bodge 1994b) and Silver Strand State Park near San Diego (Andrassy 1991). The berm at Cocoa Beach was created using 250 800 m³ of sediment placed in two operations; it had a relief of about 1.65 m and had a 122 m wide base that was located between 8.2 and 5.5 m

MLW. The material that was placed in depths shallower than about 6.8 m began to migrate rapidly shoreward within days to weeks, whereas material placed in depths greater than about 7.6 m moved comparatively little (Bodge 1994b). The berm at Silver Strand State Park was created using 112 480 m³ of sediment; it had an average relief of 2.13 m and a 183 m wide base that was located between 4.6 m and 8.5 m MLLW. This berm resulted in significant accretion on the beach (Andrassy 1991).

Some offshore structures are designed to replicate natural forms but change the dimensions of these forms and their impact on waves. Sediment may be used to reduce the depth between bars, reduce the topographic variability of the bars or increase the size of the bars. Use of offshore nourishment to enhance bars in the outer surf zone may lead to a cross-shore fixation or slight reversal in the direction of formerly offshore-migrating bars (Hoekstra *et al.* 1994; Kroon *et al.* 1994). Design options for the nourishment at Terschelling, The Netherlands included placing sediment seaward of the outer bar, between bars or a combination of both locations (Mulder *et al.* 1994). The nourishment between bars involved placing 2 million m³ of sediment for a distance of 4.4 km alongshore in a trough that had a depth of 5 to 7 metres (Kroon *et al.* 1994; Mulder *et al.* 1994).

Improvements in dredging technology and computational techniques have enhanced the likelihood for success of future operations designed to create landforms to achieve specific functions, and their use does not require unusual circumstances or expensive, risky placement techniques for success (Hands and Allison 1991; Hands and Resio 1994; Foster *et al.* 1996). The relatively shallow draft and rapid placement using split-hull hopper dredges that can operate safely in shallow water in active littoral systems, and modern electronic positioning to ensure accurate placement for a well-defined submerged feature, greatly enhance the success of the berms that now can be constructed (McLellan and Kraus 1991).

Borrow areas

This book concentrates on the characteristics of dynamic beaches and dunes that are affected by breaking waves and winds rather than features in deep water or features landward of the dynamic portion of coastal dunes. Thus, interest in borrow areas is confined to sites in the dunes and on the active beach or at sites in the water that are sufficiently close to the beach to affect sediment exchanges or waves and currents that rework the beach.

Subaqueous borrow areas can: (1) act as sediment traps, thus changing the local sediment budgets; (2) alter wave refraction patterns, thus changing locations of accretion and scour on adjacent beaches; and (3) change the local environment for biota by creating greater water depths, local differences in water circulation, temperature and surface sediment characteristics. The effects will be similar to those of other mining operations (e.g., Hesp and Hilton 1996). Most borrow areas are outside the zones that are the focus of the nourishment operations, but borrow areas may be close enough to the nourished beach and of sufficient size to noticeably affect wave refraction patterns, contributing to erosional hot spots on the beach (Hamilton *et al.* 1996).

The most commonly occurring projects resulting in borrow areas right on the beach are the scraping projects identified in chapter 2. Most projects that use sediment dredged from subaqueous sources that are close to the nourished beach are conducted at inlets and are often conducted to mitigate critical erosion on the downdrift side of that same inlet (Cialone and Stauble 1998). Inlet shoals provide suitable sediments because these deposits are "sand bridges" between adjacent beaches, and size distributions of sediments are compatible with adjacent native beach sediments, but the mining operation is a perturbation to the equilibrium of the entire inlet system that may alter the location of accretion and erosion on adjacent beaches (Cialone and Stauble 1998). An evaluation of the effects of 8 shoal mining projects ranging in volume from 170 000 m³ to 6 235 000 m³ conducted by Cialone and Stauble (1998) indicates that there is so much site specificity associated with the projects and so little systematically gathered data that it is difficult to determine impacts to navigation, the inlet-adjacent shoreline, the ebb shoal equilibrium and the borrow area.

Nourishment operations that make use of sediment dredged in normal channel-maintenance operations may be distinguished from those that borrow sediments from shoals on the ebb tide delta. Channel dredging is an attempt to maintain the configuration of the topography of the inlet while removing sediments, while shoal dredging alters both the topography and the local sediment budget. Shoal dredging creates potential feedback mechanisms that can affect the entire inlet system. Dredging of new channels, as opposed to maintaining existing channels, has the potential of creating even more profound changes in inlet dynamics and shoreline change as described in chapter 2.

Considerations for transport alongshore

It is likely that most nourishment operations are intended to minimize longshore transport out of the nourished area due to the need to provide site-specific protection and the high cost of providing fill materials. Some nourishment sites may be designed specifically as feeder locations, relying on longshore transport to provide sediment to downdrift locations, for example using capes as feeder beaches for bays (Yesin and Kos'yan 1993). Much of the sand that moves out of the project area can be accounted for in adjacent areas, which are benefitted by the additional sediment (Beachler and Mann 1996). Most nourishment operations designed to account for longshore transport involve sediment bypassing or backpassing operations.

Bypassing

Bypass operations represent artificial transfers of sediment designed to make up for human actions that prevent natural sediment transfers, and many of the nourishment projects identified in previous sections are essentially bypass operations. The State of Florida, USA has implemented a program to maximize use of sediment that would otherwise be lost at inlets due to dredging operations. The state recently charged that all dredging of beach-quality sand (or an equivalent quantity found elsewhere) should be placed on downdrift beaches and that a quantity of sand equal to the natural annual longshore transport be placed downdrift (McLouth *et al.* 1994). The result may be a *de-facto* bypass operation using material from updrift or a *de-jura* bypass operation using material from an offshore site (Lin *et al.* 1996).

A variety of different kinds of bypass operations have been conducted, including mobile dredges, fixed pumps with movable dredges (e.g., mounted on movable booms) and fixed and movable jet pumps (National Research Council 1995). Fixed systems may be relatively inexpensive to run, such as the Indian River Inlet, Delaware system where costs are a little over $2 per m³ (Bosma and Dalrymple 1996). Some operations may be designed to bypass sediment from either side of the inlet, where there are seasonal reversals in the direction of littoral drift (Patterson *et al.* 1991).

Walker (1991) and Wiegel (1994) present reviews of bypass operations on the Pacific coast with emphasis on the many operations in southern California, beginning in the 1930s. Guidelines for design and evaluation of these and other systems are included in USACOE (1991), and recent case studies are provided in Clausner *et al.* (1991) and Coughlan and Robinson (1990), who present an interesting example of use of jet pump technology. The majority of bypass operations are conducted by conventional dredges (Clausner *et al.* 1991) that may yield potentially better bypassing results than a fixed or semi-fixed dedicated plant (McLouth *et al.* 1994).

Backpassing

Small-scale backpassing projects involving beach scraping can be conducted with equipment available at the municipal level (Mauriello 1991). Large-scale backpassing, that requires a major investment, has the greatest likelihood of being implemented if it is designed as a component of a beach nourishment or bypassing operation conducted updrift of it (Yesin and Kos'yan 1993). This kind of recycling plan is seen as an effective technique for enhancing the benefits of the original borrow material while curtailing the need for locating and using new borrow sites (Oertel *et al.* 1996).

Scraping that moves sediment from beach to dune following storms is essentially a shore-normal backpassing operation that accelerates the pace of the beach-to-dune component of the beach–dune cycle. The emergency restoration plan for the center of Pawleys Island, South Carolina, following Hurricane Hugo, using sand dredged from accreted spits at both ends of the barrier island (Kana *et al.* 1991), is both a bypassing and backpassing operation.

Backpassing on a large scale is less common than bypassing, in part because bypassing is an attempt to overcome a deficit caused by a human alteration (creating an issue of accountability), whereas backpassing is often an attempt to make use of an opportunistic source of sediment. It is likely that the number and scale of backpassing operations will increase in the future as ready supplies of sediment outside nourishment areas become exhausted and as nourishment operations become more specifically designed to approximate natural sediment budgets in drift cells. The number of examples of backpassing currently found in the literature and the amount of space given to backpassing in this book is minimal compared to its great future potential.

Nourishment frequency

Frequency of nourishment affects the size of the initial landforms, the likelihood that certain types of landforms will be created, the ability of these landforms to survive and evolve and the periodicity of cycles of growth and destruction. Considering the importance of these factors in creating and maintaining desired components of the coastal landscape, it is unfortunate that nourishment frequency is usually treated in terms of economic viability, rather than geomorphic or ecological desirability. Traditionally, there has been a preference for capital-intensive large-scale operations that attract large initial funding and have minimal maintenance costs, although there has been greater acceptance recently of economic evaluations that provide better balance between initial and subsequent costs (Townend and Fleming 1991).

Beach nourishment is a perturbation to the existing beach in plan and profile (Dette *et al.* 1994; Work and Dean 1995). Initial creation of desired landforms that have been lost due to prior human action may only be possible with the major perturbations associated with large-scale nourishment operations that create a wide platform for natural processes to occur. The initial loss of pre-existing natural environments can occur because administrative

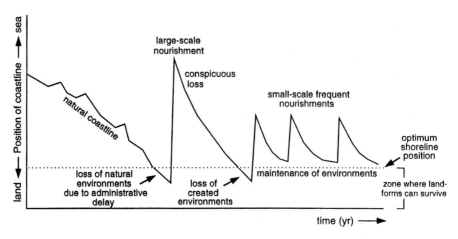

Figure 3.6 Potential for loss of natural and human-created landforms due to inability to implement nourishment projects in a timely manner. Modified from Verhagen (1991, 1992) and Dette *et al.* (1994).

constraints prevent the first nourishment from taking place before erosion has become critical (Figure 3.6). The nourished beach is wide enough to provide the opportunity for new landforms to be created (Figure 3.5B$_2$), but management policies and actions must be in place to allow for this to happen. Loss of these created environments can occur if administrative constraints once again prevent nourishment from occurring in a timely fashion (Figure 3.6).

Shoreline perturbations should be minimized when the object is preservation of existing landforms, arguing for small operations that are conducted at frequent intervals. The optimum nourishment project for retaining sediment in the nearshore system is one that disturbs the dynamic equilibrium profile of the underwater beach as little as possible to minimize mobilization of sand and convection out of the fill region by waves and currents (Dette *et al.* 1994). These considerations argue for small, less costly, frequent operations that have the added advantage of minimizing the negative impact of highly visible losses associated with large-scale operations with long repetition intervals (Figure 3.6) (Bruun and Willekes 1992; Kunz 1993a; Dette *et al.* 1994). A short time period for the first renourishment also can allow for uncertainties in alongshore erosion rates, enable correction of erosional hot spots (local shoreline re-entrants within a longer nourished shoreline) and avoid overbuilding in accreting areas (National Research Council 1995).

Renourishment rates of every 2 to 6 years have been suggested to achieve a beach-volume equilibrium (Kunz 1993a). Isolated erosional hot spots may require more frequent small scale nourishment operations to maintain beach dimensions similar to adjacent areas (Hamilton *et al.* 1996). Frequencies as high as every year have been suggested for small-scale (*c.* 30 000 m³ a⁻¹) truck-ing operations (Townend and Fleming 1991).

Beach planform

Much of the long-term erosion of a nourished beach is caused by increasing gradients of littoral drift along the length of the nourished area. These gradients include the original one that existed prior to nourishment and the one caused by the anomalous planform bulge created by the fill that causes end losses and spreading of the fill as sand is transported alongshore to the adjacent shoreline (National Research Council 1995). Unlike profile equi-libration that may occur within a few years, alongshore equilibration may occur only after decades. Calculation of losses of beach fill that are based on historical erosion rates fails to recognize that the bulge can cause spreading losses, but present US federal guidelines recommend a renourishment factor to account for these losses (National Research Council 1995).

The ends of the nourishment project can be extended at an angle to taper smoothly with the existing shoreline, creating a beachfill transition to elimi-nate the discontinuity and reduce erosion and end loss (Hanson and Kraus 1993). The longer the length of fill, the less the percentage due to end loss (Smith and Jackson 1990). On a long beach, the spreading out of sand has the effect of creating a longer project (Figure 3.7); diffusion losses will be less; and losses will decrease greatly with time (Beachler and Mann 1996; Dean 1997). Where sand comes under the influence of an inlet, the sand may be lost from the beach system. The longevity of beach fill can be improved by setting the end of the fill back from the end of the inlet (so the fill material is not lost so readily) or by building a terminal structure (Figure 3.7) to hold the sand in place (Dean 1997). The extension of fill into adjacent areas allows for the potential for new or larger landforms to be created at those areas, providing controls are in place to allow this to occur.

Other planform irregularities associated with nourishment projects include erosional hot spots (Figure 3.7) and individualistic or experimental shore-perpendicular landforms created using spoil, such as sand humps and sand groins that are gradually shaped by natural processes (Kunz 1993a; Dette *et al.* 1994; Niemeyer *et al.* 1996). Among the more interesting of these features are the "ground groins" that are a series of fan-like deposits created in the Gagra resort area on the Black Sea using fill brought by trucks (Zenkovich and Schwartz 1987).

The behavior of beach nourishment in front of a seawall may differ from beach nourishment on a long beach composed of compatible sand, in that the centroid of the planform anomaly in front of a seawall migrates with initially

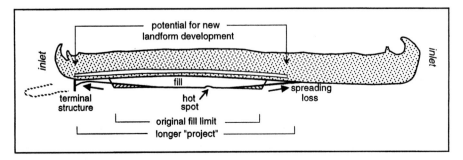

Figure 3.7 Plan view of nourished beach on a barrier island, comparing original fill area and longer length of nourished area caused by spreading losses.

increasing speed in the downcoast direction, whereas the centroid of the anomaly on a sand beach remains nearly fixed under both normal and oblique wave attack (Dean and Yoo 1994). The implications are that planform migrational characteristics should be considered in evaluating the performance and economic viability of projects fronting seawalls, and that wave direction may have greater significance on seawalled coasts than on sandy coasts (Dean and Yoo 1994). Wave direction is also important where retention structures are used, because reorientation of the fill at the structures to the direction of wave incidence could facilitate bypass around the tips of the structures to downdrift areas (Dean 1997). This bypassing effect could actually decrease some of the sand starvation effect of the structure, but it would be difficult to design a structure to bypass the optimum amount of sand.

Ecological considerations

Environmental effects of nourishment are often considered temporary (Roelse 1990), but little is presently known about the ecological impacts (Gibson and Looney 1994; Rakocinski *et al.* 1996; Gibson *et al.* 1997), and it is too early to make informed statements about the long-term implications of large-scale projects. It is reasonable to suggest that the adverse aspects of nourishment on biota are inversely related to the need for nourishment, in that areas requiring sand for shore protection are usually the ones most lacking in natural beach and dune habitat. The implication is that nourishment will provide net ecological benefits in critically eroding areas. This statement must be tempered by the consideration that nourished beaches in developed locations are often raked to create a more desirable recreation platform (Figure $3.5B_1$), providing a greater constraint on value of subaerial habitat than considerations of grain-size characteristics, size of fill or method of placement.

Potentially negative consequences of beach nourishment include disturbance of indigenous biota, alteration of foraging patterns of species that feed on them and disruptions to species that use the beach or adjacent areas for nesting, nursing and breeding (National Research Council 1995). Physical alterations to adjacent areas include: (1) burial of bottom habitat in the surf zone as the beach is widened; (2) increased sedimentation in areas seaward of the surf as the fill material is redistributed; (3) change in the nearshore bathymetry, with associated changes in wave action and beach morphometry (e.g., from dissipative to reflective); (4) elevated turbidity levels; and (5) higher salinity levels in sediments emplaced or in aerosols associated with placement by spraying (Roelse 1990; Nelson 1993; Löffler and Coosen 1995; National Research Council 1995; Rakocinski *et al.* 1996).

Vegetation

Beach nourishment can have a positive impact on threatened plants, such as seabeach amaranth (*Amaranthus pumilus*) (National Research Council 1995), although it can also favor aesthetically unappealing vegetation (Looney and Gibson 1993). The flat subaerial profile that usually characterizes nourished beaches may restrict colonization, as can a low nutrient availability where sediments are primarily quartz sand (Gibson and Looney 1994). Natural colonization in the first few years can include the same species that characterized the former strand line and foredune, but total cover can be considerably less (Looney and Gibson 1993). Primary succession can proceed slowly, although the use of sand fences can significantly increase the density of some species by acting as a trap for wind blown seeds and sand (Looney and Gibson 1993; Gibson and Looney 1994; Gibson *et al.* 1997).

Selection of appropriate elevation of the nourished beach can have a pronounced effect on vegetation. Colonization of nourished beaches can be relatively slow where the post-nourishment profile is low and flat, contributing to continuous salt spray, dessicating effects of the wind and inundation from storm overwash that can uproot germinating seedlings or kill salt-water intolerant species (Looney and Gibson 1993; Gibson and Looney 1994). Alternatively, this inundation can remove coarse lag deposits and provide a mechanism for natural reworking of nourished sediment, and species involved in primary succession may be the result of strandings due to storm overwash and inundation (Looney and Gibson 1993). Dredge spoil placed at an elevation that is higher than the natural dunes can cut off the salt spray that is important in germination and zonation and the primary source of nutrients for foredune species (Looney and Gibson 1993).

Poorly planned deposition of dredge material can result in aesthetically unacceptable vegetation (Looney and Gibson 1993). Natural growth of vegetation on beachfill placed at a greater height than the natural berm at Sandy Hook, New Jersey created a mono-specific stand of *Solidago sempervirens*. This species is native to the area, and it is an expected colonizer of coastal dunes, but the sole presence of *Solidago* (conveying the impression it was planted in preference to *Ammophila*) gave an unnatural appearance to the shore.

Fauna

Temporary loss of infaunal communities through sand burial during nourishment is expected and considered largely unavoidable; the more important issue is often the recovery rate after project completion (National Research Council 1995). Some studies indicate only temporary (weeks to months) alterations in abundance, diversity and species composition of sub-aerial habitats and return of abundance and diversity of benthic fauna within the borrow sites within a year; other studies have documented changes in species composition of benthos that lasted much longer, especially where bottom sediment composition was altered (National Research Council 1995). Macrobenthic assemblages occurring deeper than 3 m occupy a relatively stable environment that may take longer to recover than nearshore sandy beach assemblages. Assemblage structures at Perdido Key, Florida shifted seaward, probably in response to increased silt/clay loading and were evident more than 2 years after nourishment (Rakocinski *et al.* 1996). Supplemental studies are warranted in other geographical regions, and additional data are required to assess effects where extensive intertidal habitats exist and where temporary alterations in the benthos will affect foraging (National Research Council 1995).

Reviews of existing studies indicate that beach nourishment can provide suitable habitat for nesting by turtles and threatened or endangered bird species such as the piping plover (*Charadrius melodus*) (National Research Council 1995). Beach nourishment operations will not categorically improve conditions for these species, and considerable costs may be incurred in either minimizing losses or maximizing gains. Alterations in the compaction, density, shear resistance, color, moisture content and gas exchange of beach sands can influence hatching success of turtles, and scarp formation can increase the number of false crawls, although some effects (e.g., sand compaction and scarp formation) can be reduced or eliminated by grading operations (National Research Council 1995). Some of the effects can be superficially inconspicuous but nonetheless significant. For example, although hatching and emergence successes of sea turtles in native (silicates) and imported (aragonite) fill materials indicate that the imported sand provides suitable substrate for sea turtles, the small differences in temperature regimes changed hatchling sex ratios (Milton *et al.* 1997).

Many of the adverse impacts on fauna may be related to the timing of the operations (Lawrenz-Miller 1991). It may be possible to conduct operations around the activities of the fauna using the beach, but the result may be greatly increased costs. The need to avoid the nesting and hatching season of loggerhead turtles required operations at Folly Beach, South Carolina to be conducted in the most extreme weather conditions in the winter months (Edge *et al.* 1994).

Nourishing dunes

Nourishment methods and sources of sediment

Dune nourishment projects may be conducted as part of beach nourishment projects or as independent operations. Dune nourishment is more common than is generally perceived, especially if the definition is extended to buildup of dunes using sediments scraped from nearby beaches or from overwash sediments. Exchanges of sediment between beach and dune occur frequently under natural conditions, and it is not unreasonable to consider dune replenishment as beach nourishment if the purpose of the project is to replace a volume of sediment rather than restore natural environments. The landforms may be designed to function as dunes, or they may be created simply as a convenient or less costly way of dumping sand.

The sediments used to nourish dunes often come from the same sources as those used to nourish beaches. Much of the sediment placed in the dunes in New Jersey after the March 1962 storm was dredged from backbarrier lagoons, transported through pipes and placed directly within retaining dikes. Emplacement of the slurry on the beach, with reshaping done subsequently using bulldozers, is now a common practice (Nordstrom and Arens 1998). Trucking operations are common in New Jersey, where municipal operations can be as small as <100 m³ or involve tens of thousands of m³. The smallest operations are usually those that recycle sand that inundates yards and roads back to a convenient access point on the beach or dune (Nordstrom and Arens 1998).

Dune-building projects in The Netherlands have been conducted independently of beach nourishment projects in the past, but they are not likely to occur frequently in the future because dunes have been built to Delta height.

Priority is now given to nourishment of the beach, allowing dunes to form by aeolian action – a practice that has considerably greater long-term value as discussed in chapters 5 and 8.

Although the sources for beach and dune nourishment may be the same, the materials may have a different value as dune deposits. For example, fine sand sizes that may result in rapid and unacceptable losses on beaches may provide the optimum substrate for artificial dunes because they facilitate evolution of dune environments through natural aeolian transport.

Landforms created by nourishment to function as dunes may be termed storm berms, dune dikes or sand dikes to highlight their human resource value as a barrier to overwash and to acknowledge their lack of aeolian origin. "Berm caps" (e.g., 1.2 m or less) placed above normal berm design elevations (as portrayed in Figure 3.5A) may be used to minimize the impact of wave overtopping (Bocamazo 1991), and sand dikes (Figure 2.3) are constructed on the California coast to reduce wave overwash in the winter (Armstrong and Flick 1989). The departure of nourished dunes from true dunes is enhanced where the resulting feature is shaped to optimize protection or ease of creation or management. The initial design dune at Revere Beach, Massachusetts had a 1:5 shoreward slope, a 15.2 m wide dune crest and a 1:15 seaward slope (Smith *et al.* 1993). The dune at Meeuwenduinen, Schouwen, The Netherlands (Figure 3.8) shows the flat top and linear appearance that can result from landforms designed primarily for protection.

Locations of nourished dunes

Nourishment projects may be conducted to create dunes where they did not exist or to place sediment on the front, top, and back of existing foredunes (Figure 3.2). All of these locations have been used in shore protection projects in The Netherlands (van Bohemen and Meesters 1992). Existing foredunes have also been bulldozed to the front of the secondary dune when the foredune became too small, although the implications of the 1990 Dutch policy are that foredunes will remain near their present position. Stuifdikes (Figure 3.9) are still created to extend the foredune seaward or to create valued natural environments (such as dune slacks) behind them. Stuifdikes were used in the past to connect small Wadden islands to create larger islands (e.g., Ameland and Texel) and they remain as conspicuous features on these landscapes (Nordstrom and Arens 1998).

Placement of fill material on the back of the dune is appropriate for locations where: (1) the protective value of the dune must be increased; (2) existing land use behind the dune does not preclude use of the space; and (3) the fill

Figure 3.8 Artificially nourished dune at Meeuwenduinen, Schouwen, The Netherlands.

Figure 3.9 Stuifdijk at Terschelling, The Netherlands built across overwash area between segments of existing dune to provide continuous flood protection.

materials are not contaminated with salt (van de Graaff *et al.* 1991). The salt content is an issue because many vegetation species found behind the dune are not salt tolerant. In The Netherlands, it has been found that natural desalinization due to rainfall can take place within two years; sand is initially placed on the seaward side of dunes and subsequently moved to the back side after this time period (van de Graaff *et al.* 1991). A rare case of creating a new dune behind an existing remnant dune occurred at Cape May Point, New Jersey, leaving the foredune to become the sacrificial dune. The new dune was built in this position because the natural dune was more resistant to erosion and its value for protection was perceived of greater use than its value as natural habitat.

Nourishment of the top of the dune is appropriate where land use behind the dune prevents backdune nourishment. This method of nourishment often creates an unnatural appearance, requiring creation of artificial peaks and valleys (van de Graaff *et al.* 1991). Sand from behind the dune area can be used to build a high and continuous dune crest (Philip and Whaite 1880).

Nourishment of the front of the dune is appropriate where: (1) land use behind the dune prevents nourishment of the back of the dune; (2) the width of the pre-existing dune is not great enough to retain its protective value during storm erosion; or (3) a higher dune is not considered desirable because of slope instability or restrictions to recreational use (access and views of the sea). Displacement of the beach–dune contact farther seaward will result in large losses due to storm wave attack, and nourishment of the front of the dune must be considered, in part, as an upper beach nourishment project (van de Graaff *et al.* 1991).

Nourishing uplands

The process of creating an artificial upland is far less common than creating a feature designed to function as a dune or nourishing a beach in front of an eroding cliff. New upland materials may be added as a result of dumping sediment or fill material over the side of an eroding bluff to ease placement. A process more analogous to creating a new eroding upland is one that is popular on the Black Sea coast and involves building of embankments of heterogeneous sediment that are designed to be eroded by waves and separated into beach-forming fractions (Shuisky and Schwartz 1988). Various methods of waste disposal, such as dumping of mine tailings over cliffs (chapter 2) have the effect of nourishing uplands, as does the creation of artificial islands (discussed at the end of chapter 4).

Table 3.1. *Questions to be evaluated when considering beach nourishment as resource replenishment and enhancement*

What are the benefits of nourishment that are unseen to the lay public and politicians?
What is the value of alterations to the subaqueous portion of the profile?
What is the role of the new backbeach?
   Will it be occupied by cultural features or replaced by them?
   If left undeveloped, will it function naturally or as an artifact (graded, raked)?
How does project timing affect changes to the shoreline?
   Will nourishments be conducted before landforms are eliminated or severely truncated?
   How will landforms prior to nourishment compare with landforms at the end of the project?
Will new landforms on the backbeach be considered nuisances or viable natural features?
Will landform variability and dynamism be considered valuable?
What are the opportunities for enhancing relationships between cultural and natural features?
What are the opportunities for enhancing natural processes relative to cultural processes?
How do economic choices constrain optimization of long-term effectiveness of the project?
How are the important non-economic criteria for decision-making (ecological, aesthetic and
   geomorpological) being incorporated?

Summary conclusions and new themes for study

Beach nourishment serves human needs that are perceived to have an economic benefit, and the new beach may be considered an economic resource. The morphology of the nourished beach and associated coastal land-forms is traceable to economic decision-making, and it is not surprising that these landforms look more like artifacts than natural features. A broader conception of nourishment as resource replenishment and enhancement is needed in planning shoreline modifications. The questions in Table 3.1 are designed to stimulate identification of opportunities for re-establishment of naturally functioning landforms that will extend the range of future benefits. Several of the themes identified in this table apply to other human alterations of the coast and will be discussed in subsequent chapters. More specific suggestions for enhancing the resource values of nourished beaches are made in chapter 8, following discussion of the relationship of human uses to natural shoreline characteristics.

4

Effects of structures on landforms and sediment availability

Introduction

This chapter identifies the direct effects of human structures on geo-morphic processes and the resulting effects on coastal sediment budgets and landforms. The structures that are discussed are designed for shore protection, navigation, recreational boating, beach recreation, housing and transportation. Artificial islands and structures outside the coastal zone that affect sediment budgets are also addressed. The effects of modifying beaches and dunes to build these structures and the effects of human actions in redistributing sediment to use these structures or mitigate their effects were discussed in chapters 2 and 3. The characteristics of landforms associated with many of these structures are addressed in chapter 5.

Protection structures

The body of literature on shore protection structures is enormous, and the functions of the many types of these structures are well known (Silvester and Hsu 1993). Impetus for conducting most geomorphic studies of these structures is provided by the need to address the initial erosion problem, invariably related to local deficiency in the sediment budget. Frequently, the effect of these structures is to displace the *locus* of erosion to adjacent unprotected areas, resulting in a need to address subsequent erosion problems in those areas. These kinds of geomorphological investigations are most often couched in an applied investigative mode, examining the suitability of alternative means of protection.

There is ample evidence of accelerated erosion rates, pronounced breaks in shoreline orientation and truncation of the beach profile downdrift of shore perpendicular structures (Figure 4.1) (Everts 1979; Leatherman 1984; Nersesian *et al.* 1992) and reduced beach widths (relative to unprotected

Figure 4.1 Downdrift end of the groin field at Cape May, New Jersey.

segments) where shore-parallel structures have been employed over long periods of time (Hall and Pilkey 1991). Examples abound of failed structures and structures about to fail, in part because structures have been under-designed (Bush 1991; Gusmão *et al.* 1993; Rogers 1993; Knight *et al.* 1997). It is likely that many other "failures" are due to political constraints that provide funding for the structures themselves but exclude funding for beach nourishment that is often included in the original design. Downdrift erosion in these cases is a result of fiscal constraints and not faulty engineering design (Nordstrom 1987b; Nersesian *et al.* 1992).

Many of the undesirable effects of protection structures are so well known that the descriptions are hackneyed, and many studies are characterized by a conspicuous lack of scientific rigor and objectivity in their condemnation. Many writers refer to the inefficiency of coastal structures in providing protection to the coastal landscape without providing examples or documentation. Many papers describing the functioning of groins tend to restate conclusions from other studies without quoting sources, and independent confirmation of negative effects are often lacking (Kraus *et al.* 1994). The purpose of this book is not to repeat the litany of problems associated with shore protection structures, but to discuss their effects on landforms and the geomorphic evolution of the coast. In that context, it will be seen that many structures are more compatible with natural processes than may be initially apparent.

Groins

Kraus *et al.* (1994) provide an overview and critical evaluation of the functional properties attributed to groins. They define groins as shore perpendicular structures emplaced to maintain the beach updrift of them or to control the amount of sand moved alongshore. They deliberately avoid the negative concepts of "building a beach" or "trapping littoral drift" that imply removing sand from the littoral drift system. They point out that groins can be used to advantage when they are located where: (1) sediment transport diverges from a nodal region; (2) there is no source of sand (such as downdrift of a breakwater or jetty); (3) passage of sand downdrift is undesirable; (4) the longevity of beach nourishment can be increased; (5) an entire reach is to be stabilized; and (6) currents are especially strong at inlets (Kraus *et al.* 1994).

Effects of groins

Lack of guidance on functional design of groins and examples of poor performance caused by poor design or failure to fully fund projects have turned groins into a cliché representing beach destruction (Kraus *et al.* 1994). Greatly accelerated shoreline retreat rates are reported downdrift of some of these structures. The erosion zone downdrift of the Cape May groin field was about 2500 m long, with a retreat rate of about 6 m a^{-1}, when Everts (1979) conducted his evaluation. Downdrift erosion occurred at an average of 6.5 m a^{-1} over a 14-year period at Westhampton Beach, New York, USA (Nersesian *et al.* 1992), and erosion occurred at an average of 11 m a^{-1} for a number of years on the west coast of Denmark (Laustrup 1993). A short-term retreat of 20 m in 2 years was reported at a site in northern Portugal (Granja and Carvalho 1995).

Groins trap sediment moved alongshore, but they also affect wave refraction and breaking and surf-zone circulation, producing additional rip currents with conspicuous rips close to the structures, and they redirect sediment offshore to form lobate deposition zones downdrift of the tips of the structures (Sherman *et al.* 1990; Bauer *et al.* 1991; Short 1992; McDowell *et al.* 1993). Groins can interrupt existing beach-bar systems (Short 1992) and create rhythmic features offshore of them (Gayes 1991), resulting in a more complex topography. Groins also result in differences in sediment characteristics on the updrift and downdrift sides (Orme 1980).

Groins have been found to be effective in diminishing longshore currents near inlets and halting migration of tidal channels (Kunz 1990, 1993a, 1996; De Wolf *et al.* 1993; Pluijm *et al.* 1994). The effect of tidal scour requires construction into deep water, resulting in structures that can extend to a depth of almost 20 m (Kunz 1990). Groins also are considered effective as terminal structures at the ends of drift systems that can be used in combination with programs that return sediments to updrift areas (Yesin and Kos'yan 1993; National Research Council 1995).

Alternative designs and uses

Some of the adverse effects of groins can be overcome by building permeable groins. Over 900 pile groins have been built along the German coastline and about 950 along the Polish coastline (Trampenau *et al.* 1996). These groins can have the effect of reducing the longshore current and eliminating the effect of the local structurally induced rip current, creating a more linear shoreline than occurs with an impermeable groin and creating an underwater terrace that can help reduce the erosion potential of waves crossing it (Trampenau *et al.* 1996).

T-groins are favored in some areas to reduce scour and redirect rip currents (Marcomini and López 1993: McDowell *et al.* 1993; Yüksek *et al.* 1995; Oertel *et al.* 1996). They may be designed and built as T-groins, or terminal hooks can be added later as identified by Marqués and Julià (1987). T-groins can help reduce unwanted sedimentation offshore of nourished areas where sea-grass beds would be threatened (Bodge and Olsen 1992).

Groins were common features of early nourishment projects but became less common as a result of attitudes against structural solutions (Truitt *et al.* 1993). This trend may be reversing because it appears that fixed structures in conjunction with nourishment projects can have a positive effect on fill retention (Leonard *et al.* 1990; National Research Council 1995). Either new groins, extension of existing groins or notching of groins to allow for sediment movement are now often included in beach nourishment plans (Rouch and Bellessort 1990; Bocamazo 1991; Leidersdorf *et al.* 1993; Ulrich 1993; Olsen 1996; Denison 1998). These structures may improve fill longevity to the point where their benefits outweigh negative downdrift impacts, and potential negative impacts can be overcome by placing a small percentage of the fill material downdrift (Truitt *et al.* 1993).

Groins have considerable recreational value for fishing because they create new habitat and provide access to deep water, and combined pier/groin structures have been built to enhance this value (Laubscher *et al.* 1990; Prestedge 1992). Other creative use of groins include using a submerged groin to extend headlands to retain a pocket beach (Prestedge 1992), building submerged groins that are actually shoals composed of unconsolidated materials (Pacini *et al.* 1997), and building groins to change the orientation of nourished shorelines to correspond to a natural shoreline or to cultural features, such as promenades (Rouch and Bellessort 1990). Groins connected to breakwaters are favored in some areas (Rouch and Bellessort 1990; Bull *et al.* 1998), and groins along with breakwaters have been employed as components of new artificial beach/harbor projects to hold the sediment in place and provide a sheltered beach area (Sato and Tanaka 1980; Gómez-Pina and Ramírez 1994). In these locations, it is often difficult to distinguish between the function of the structure as a breakwater or as a groin.

New groins or increases in the size of existing groins have been suggested (Sorensen 1990; Günbak *et al.* 1992; Kaplin *et al.* 1993; Yüksek *et al.* 1995), but groins also have been banned in many areas (Meyer-Arendt 1991) or strongly discouraged in management policies (Kraus *et al.* 1994). Some instances of removal of groins have been reported (Fernandez-Rañada 1989; Meyer-Arendt 1991; McDowell *et al.* 1993), but there is little documentation of the results of this action on beach change. The state of Florida, USA requires removal of non-functioning groins and groins that have adversely impacted the beach prior to beach nourishment projects, and 20 groins were removed in one project alone (Beachler and Higgins 1992). Dismantling of groins and breakwaters is planned in other areas (Peshkov 1993) and is suggested for nourished areas if the nourishment project is abandoned (National Research Council 1995). Shortening or lowering of groins to allow for some bypass of sediment is more common than removal and is better documented (Nersesian *et al.* 1992; Peshkov 1993; Kraus *et al.* 1994; Granja 1996).

Effects on aeolian transport and dunes

Groins form effective traps for sand blown alongshore, forming small, usually temporary shore perpendicular dunes on the beach and more permanent dunes at niches formed by the contact between the structure and the landward portion of the backbeach (Figure 4.2). The increase in beach width on the updrift sides provides a wider source of sand for aeolian transport and protects the new dunes from storm wave erosion. The protection provided by the structures can allow the dune and its vegetation to occupy a more seaward location and last longer than it does along the adjacent unprotected shoreline (Nordstrom *et al.* 1999). Groins can create deposits of sufficient size to allow massive dunes with multiple accretion ridges to form through natural aeolian accretion (Nersesian *et al.* 1992). In contrast, dunes downdrift of groins may be eliminated by erosion. Where they survive, they may be steeper than updrift dunes, because of truncation of the seaward portion of the dune by wave erosion (Nordstrom *et al.* 1986).

Shore parallel structures

Types of structures

Bulkheads, seawalls and revetments are shore-parallel structures built to protect landward development after the natural protection has been

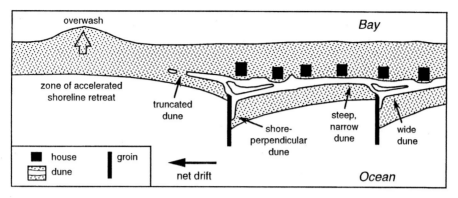

Figure 4.2 Plan view of the beach and dune at the downdrift end of a groin field on a developed barrier island where dunes are used as shore protection.

lost or reduced by erosion or by encroachment of human facilities. Shore-parallel structures are a popular method of protection because they are often easy to build (Paskoff 1992), and they can be built above mean high water, simplifying the process of obtaining a permit. Bulkheads are vertical retaining walls that are not intended to provide protection against direct attack by breaking waves. They are usually built landward of an existing beach, although they may be built farther seaward on sheltered coasts and thus be considered primary protection. Wooden bulkheads are favored in some areas because the materials are cheaper than rock transported from distant quarries (Rogers 1993). Seawalls and revetments are usually built where the protective effects of the beach have been eliminated. Revetments differ from seawalls in that they have more gentle slopes, designed to dissipate wave runup. These structures, like bulkheads, can also serve to retain materials landward of them and can help support bluffs and prevent landslides on high-relief coasts (Hearon *et al.* 1996). The distinction between the three basic types of shore parallel structures is often blurred, and they are often treated as one generic type under the heading of seawalls. That convention is followed here. Reviews of the state of the art in field, lab and modeling studies of seawalls are provided by Dean (1986), Kraus (1987, 1988) and Kraus and McDougal (1996), and long-term field observations have been conducted in the vicinity of Virginia Beach, Virginia (Basco 1990; Basco *et al.* 1992; Hazelton *et al.* 1994) and Monterey Bay, California (Griggs *et al.* 1991a; Plant and Griggs 1992).

Effects of seawalls on beach change

Seawalls, like groins, are a cliché; in this case they are identified as the undesirable end point in the conversion of the coast from a natural system to one that has no geomorphic features (Figure 4.3). Part of the problem is that the seawall option is implemented on a shoreline where erosion has

Figure 4.3 Seawall at Sea Bright, New Jersey, 1988 prior to a large-scale federal beach nourishment project.

proceeded to the point where there are few natural geomorphic features left (Pope 1997). Some walls are even built seaward of the existing mean high-water line (Wiegel 1992b). Criticisms of shore-parallel walls refer to scour and accelerated erosion seaward of these structures, leading to a reduction of beach width. Many of these criticisms have been based on speculative conceptual arguments on the distribution of wave energy and have lacked field or laboratory evidence (Weggel 1988; Tait and Griggs 1990; Hazelton *et al.* 1994), although recent field studies of sediment transport and wave reflection indicate that suspended sediment concentrations in the water column can be increased in front of a seawall and made available for longshore transport that is enhanced by the presence of the structure (Miles *et al.* 1996).

Pilkey and Wright (1988) identify three different kinds of losses of natural shoreline resulting from construction of shore-parallel walls, including: (1) loss of the beach due to placement of the wall on it; (2) passive erosion (representing the narrowing of the profile in front of the structure as the natural regional shoreline migrates landward; and (3) active erosion (representing the loss due to intensification of surf-zone processes). The updrift side of a wall extending out onto the beach can also function as a groin and impound sand, at least until the wall fails (Kraus 1988).

Hydrodynamic and morphologic response of a beach to a seawall depends on the position of the wall on the beach profile relative to breaking waves and swash (Weggel 1988; Plant and Griggs 1992; Hearon *et al.* 1996). Breaking waves can create vertical movements that greatly exceed those that could occur in that location in the absence of a seawall. The water and entrained sediment can then be blown landward. FitzGerald *et al.* (1994) report water fountains as high as 15 m during the Halloween storm of 1991 in Massachusetts, resulting in 0.3 m deep deposits landward of the structure. The water heights are not necessarily greater than can occur under natural conditions at coastal headlands (e.g., the 30 m heights reported by Fletcher *et al.* 1995), but they are greater than would occur on natural low-relief coasts.

Visual observations indicate that a seawall affects swash velocity, duration and elevation; beach groundwater elevation; and beach slope variability (Plant and Griggs 1992). Interactions between waves and the seawall and fronting rip-rap monitored by Plant and Griggs (1992) during peak energy events revealed that the structure resulted in higher backwash velocities and longer backwash durations, along with greater turbulence, as wave reflection truncated the uprush portion of the swash cycle. Swash uprush distances were less in front of the seawall where the structure truncated the uppermost portion of the cycle but were greater on the beach flanking the seawall where waves reflecting off the structure increase local water depths. They also noted local differences in elevation of the beach water table, rate of rise of the water table and deflection of the water table effluent line due to the relative impermeability of the buried rip-rap at the toe of the structure. Changes in beach profile variability caused by the seawall included: (1) increased numbers of storm-related cycles of erosion and recovery at the seawall and downdrift relative to the updrift beach; (2) greater amplitude of cycles at the seawall; and (3) offshore displacement of the patterns of profile variability. Beach features caused by the structure included rills formed by streams of backwash flowing through rocks of the fronting revetment, and scarps at the intersection of the seawall and downcoast beach berm (Plant and Griggs 1992). The scour that occurs under water is rarely revealed in the beach surface after it is reworked by swash, although small erosion troughs may sometimes be visible at the base of rip-rap structures (Nelson 1991).

A longer-term study by Griggs *et al.* (1991a) revealed that the berm was cut back sooner in front of seawalls relative to adjacent beaches during the transition from the beach shape characteristic of the summer to the beach shape characteristic of the winter. Once the berm had retreated landward of the position of the seawall, there was no significant difference between beach profiles fronting the wall and on adjacent beaches. Scour was observed up to 150 m downcoast of the structure and appeared to be controlled by return–wall

orientation, the angle of wave approach and wave period. The berm rebuilding phase after the winter storm season appeared to be little affected by the
seawall (Griggs *et al.* 1991a).

Shore-parallel walls replace mobile coastal materials with a cohesive structure, and they are considered most appropriate where the primary problem is
one of storm-induced damages rather than of chronic erosion (Pope 1997).
This kind of structure would be located well landward of the active beach,
where it would not influence coastal processes except during high water
(Weggel 1988), thus minimizing adverse impacts on the surrounding beaches
or adjacent properties (Hearon *et al.* 1996). Engineers do not consider passive
erosion in evaluating whether seawalls cause erosion because the purpose of
the structure is designed specifically to prevent this further erosion (Kraus
and McDougal 1996), but this erosion is critical from the standpoint of maintaining natural sediment budgets and long-term evolution of the coast (Kraus
and McDougal 1996) as well as maintaining the viability of landforms and
habitats in adjacent shoreline reaches. As a result, Basco *et al.* (1997) have suggested that stakeholders who construct a wall could be required to mitigate
detrimental effects by adding a quantity of sand to the beach that represents a
calculated annual loss.

Edge effects on shorelines adjacent to seawalls still remain problematic;
locally important effects include accelerated erosion rates due to elimination
of the sediment source landward of the structure and scour from ebb channels
created by drainage of seawater trapped behind the wall during storms
(Lennon 1991). The reduction of beach width in locations where no new sediment is introduced into the system is now accepted by the research community as a detrimental geomorphic effect. This does not mean that erosion is
actually accelerated in front of seawalls or that seawalls inhibit recovery of the
beach after the winter storm season (Kraus 1988; Kraus and Pilkey 1988; Basco
1990; Nelson 1991; Basco *et al.* 1992; 1997; Plant and Griggs 1992; Hazelton *et
al.* 1994).

The effect of shore-parallel walls in the deeper water seaward of them is
poorly understood. Bars may pass in front of shore-parallel structures that
lack an intertidal beach (Short 1992), indicating that large-scale effects do not
extend far seaward of them.

Effects of seawalls on aeolian transport and dunes

The size, shape, type of construction materials and position of shore-
parallel structures on the beach profile result in differences in characteristics
and evolution of dunes. Size affects the degree to which a structure rises into
the boundary layer and acts as a barrier against wind processes or migration of

aeolian deposits; shape and type of construction materials affect the ability of sediment or dune forms to move across the surface of the structure. Bulkheads and seawalls built to low elevations can be buried by wind blown sand. Observations of bulkheads in New Jersey indicates that sand passes over these structures by first forming a ramp in front of them by aeolian deposition that then acts as a transport surface that facilitates movement of sand inland (Nordstrom and Arens 1998). Sediment ramps can be a conspicuous feature following major storms (FitzGerald et al. 1994) and may also form from wave action.

Sand that is not intentionally removed from the landward side of shore-parallel structures forms dunes that can remain contiguous to feeder dunes on the beach or become detached from them if the seaward portion is lost through wave erosion or is removed by earth moving equipment. The shore-parallel structures can then protect the dune remnants landward of them from subsequent wave erosion, resulting in perched dunes that can survive longer than they could occur under natural conditions (Nordstrom et al. 1999). The absence of dunes behind low seawalls is often more a function of human actions to eliminate them than the incompatibility of structures with aeolian processes.

In some cases, shore-parallel structures have been intentionally buried under an artificial dune to remain inconspicuous until the beach erodes during an extreme storm (Kraus and McDougal 1996). This kind of structure provides a reasonable compromise between the need to provide protection and the need to provide natural values. Most shore-parallel structures are built as a last resort and can only develop a dune cover when beach dimensions are increased through post-storm accretion, artificial nourishment or dune construction programs.

Breakwaters, headlands, sills and reefs

These structures are designed to alter the effects of waves to reduce rates of shoreline change. They are generally built parallel to the shore and may include: (1) headland breakwaters or artificial headlands that are constructed at the shoreline or near the shoreline, where accretion landward of them is expected to form a tombolo; (2) detached breakwaters (emergent and submerged) that are not connected to the shore by a sand-retaining structure; and (3) submerged sills that reduce the rate of offshore transport, creating a perched beach or allowing an artificial beach to be retained (Chasten et al. 1994). There is great variety in design and use of these structures, and the distinction between them may be blurred. Breakwaters may be constructed on land in anticipation of their eventual need (Silvester and Hsu 1993). In this case, they could conceivably evolve from seawalls to headlands to detached breakwaters to reefs as erosion and sea level rise proceed. Breakwaters may be emplaced during or after damaging storms and become covered by post-storm accretion (and by beach fill), only to be exhumed when needed (Wiegel 1992a). In some cases, groins are used with breakwaters, either behind offshore breakwaters or at gaps in breakwaters (Sato and Tanaka 1980). T-shaped groins are a form of headland breakwater (Tsuchiya *et al.* 1992), and breakwaters may be built as spurs on groins (e.g., Weggel and Farrell 1990). Reefs have been constructed at the ends of groins to form partially enclosed cells that can reduce losses that would occur if the reef had been built as an open-ended structure (Bruno *et al.* 1996).

Headlands

For design purposes, the effect of artificial headlands is different from breakwaters in that headlands are spaced farther apart than offshore breakwaters; the dimensions of the beach in proportion to structures are greater; and the structures are designed to reorient the beach line to create crenulate-shaped stable bays (Silvester and Hsu 1993). These bays are often asymmetrical on a beach with a dominant direction of longshore transport, but they may be symmetrical where wave direction varies seasonally (Saito *et al.* 1996).

Headlands are less common than breakwaters, perhaps because their design requires somewhat more creativity (and potential problems) than a more traditional structure and the crenulate nature of the resulting shoreline may be considered undesirable for traditional beach recreation. There is evidence that headlands are becoming more common as design problems are worked out and their effectiveness is demonstrated (Saito *et al.* 1996). They have been suggested to help stabilize erosional hot spots in nourished beaches (Hamilton *et al.* 1996), a use that seems suited to the site-specific nature of their design and function. Widely spaced headlands have been seen as a means of stabilizing a beach in a way that preserves scenic beauty better than visually intrusive breakwaters (Saito *et al.* 1996).

Breakwaters

These offshore structures (Figure 1.2) are designed to reduce the amount of wave energy reaching the shoreline by dissipating, reflecting or diffracting oncoming waves, thereby enhancing deposition and creating a beach environment where uses and protection strategies compatible with low wave energies may be implemented (Chasten *et al.* 1994). The alteration of the energy of the wave environment may also result in a change in slope, grain size and temporal variability of the beach, in that sheltered beaches are usually steeper, have coarser and more poorly sorted surface sediments and undergo less change in profile.

Breakwaters are a common feature on the Mediterranean coast, especially in Italy, where there is great variety in designs. They are uncommon as a means of protecting exposed tidal shorelines due to high construction costs (Marcomini and López 1993), but increased benefits from use of breakwaters with beachfills may justify their costs and encourage their future construction (Chasten *et al.* 1994). The value of breakwaters for establishing habitat (Kawaguchi *et al.* 1994) may also contribute to their acceptance and use.

Perched beaches

A perched beach is a nourishment operation, combined with a shore-parallel structure (sill) that helps hold the fill material in place. The sill also creates a more gently sloping beach that reduces both offshore sand losses and longshore transport (Ferrante *et al.* 1992). Perched beaches are effective in low-energy environments such as estuaries (Douglass and Weggel 1987) and sheltered portions of ocean bays (Wiegel 1993a), but they also hold potential in higher-energy environments (de Ruig and Roelse 1992; de Ruig 1995).

Perched beaches may be preferred over emergent coastal structures where tourism, aesthetics and ecological considerations are perceived important to the well-being of the community and where there is a scarcity of suitable marine sand for nourishment (Ferrante *et al.* 1992). The sill at Ostia, Italy is a rubble mound structure located about 150 m seaward of the shoreline, with a toe level at −4.0 to −5.0 below msl and a 15 m wide crest at −1.5 m msl (Ferrante *et al.* 1992). The sill can be identified by a change in the color of the water, and it is marked with buoys, but it does not interfere with beach recreational activities, and it has favored the development of marine fauna and stimulated fishing (Ferrante *et al.* 1992). A different slant on the perched beach concept is the nearshore hardbottom protection structure emplaced in the Holywood/Hallandale beach nourishment project to keep sand placed on the upper beach from covering the nearshore hardbottom seaward of it (Beachler and Higgins 1992). The use of artificial reefs containing nourished beaches that function as zonal protection works (Sawaragi 1988) is another variant of the perched beach concept as is the use of emergent groins in association with submerged barriers (Lamberti and Mancinelli 1996). In many cases, the distinction between perched beaches and artificial reefs is blurred.

Reefs

Reefs are submerged structures that are designed to control rather than resist wave activity; they may be simpler and less expensive to build than emergent breakwaters and are less susceptible to damage (Ahrens and Cox

1990). They may be a favorable management option when a structural solution is desired but visibility from the beach would interfere with aesthetics (Pacini *et al.* 1997) and where emergent breakwaters would result in reduced water quality in the protected area (Tomasicchio 1996). These structures hold great promise for future deployment, but there are still unresolved questions about the optimum crest height of the structure and distance offshore to avoid ponding of water trapped behind the structure resulting in unwanted rip currents and increases in longshore current velocity on the lee side that could lead to scour (Browder *et al.* 1996; Lamberti and Mancinelli 1996; Nobuoka *et al.* 1996; Dean *et al.* 1997).

Dikes

Dikes are structures designed to prevent inundation of inland resources. They may be of virtually any size and may be composed of natural or cultural materials. Dune dikes (essentially artificially created dunes) may be designed and placed in a location where they can be reworked by waves. Other dikes are designed to resist waves and raised water levels and are usually protected from erosion by beach nourishment or an armored surface that functions as a revetment. These latter structures can be managed as purely static features without vegetation cover or with a non-dynamic vegetated veneer (mowed grass), or they can be integrated into the landscape by allowing aeolian processes and natural vegetation to shape the surface veneer. Most dikes are designed to protect cultural features, but dikes have been used to protect natural areas, altering the processes and the resulting morphology and vegetation. The construction of a sand barrier across the Boschplaat in The Netherlands, between 1931 and 1936, for example, converted an almost bare overwashed sand flat to a salt marsh that is now a nature reserve (Beeftink 1985). In other cases, dikes fronting natural environments have prevented overwash from creating intermittently wetted environments (slufters), resulting in loss of vegetational diversity.

Dikes are designed to prevent inundation by water, but they need not prevent aeolian transport or prevent dunes from forming. Paved surfaces of dikes in The Netherlands allow sand to be transported inland to form new dunes on the landward side of the structure. The physical separation of the dune from the beach does not preclude aeolian transport because the dike is an efficient transport surface. Artificial dunes may be used to cover dikes composed of cultural materials, such as on the geotubes at Atlantic City, New Jersey (Figure 1.9) and on the asphalt dikes at Zeeuws-Vlaanderen and De Brouwersdam in The Netherlands (Nordstrom and Arens 1998). The linear nature of the artificial structure may reveal the human origin, but the surface environment has more natural value than its paved alternative.

Alternative designs and construction materials

Alternative designs and construction materials are often employed in place of traditional protection structures to keep costs low or overcome shortage of natural rock in some areas (Pilarczyk 1996). Numerous proprietary designs now exist for erosion control structures of unique geometry or construction materials (usually made of concrete or sand-filled bags) built in the form of fences, walls, mazes or flexible elements designed to trap sand in shallow water. Many of these structures function in a way that is similar to traditional methods, but they are often emplaced without first obtaining objective laboratory or field evaluation and may not live up to pre-installation estimates of their performance (Chasten *et al.* 1994; National Research Council 1995). There are also many types of structures built by residents requiring little or no capital cost that have interesting designs but limited useful lives (Mimura and Nunn 1998).

Little definitive information exists on the performance of non-traditional approaches, including those that have failed to achieve their claimed potential (National Research Council 1995). The success of a non-traditional alternative is usually a function of (1) the stability of the units under storm conditions and over the prescribed economic lifetime (considering weight and durability of units or anchoring); (2) how it affects adjacent areas as well as how it performs within the area to be protected; (3) how the cost compares to traditional alternatives; and (4) whether the structure can be easily repaired or removed when damaged (Pope 1997).

Hard structures

Hard structures can be built in modules that facilitate handling during construction and readjustment or removal in the future (Carbognin *et al.* 1989; Oertel *et al.* 1996; Olsen 1996). Linked concrete mat armor consists of concrete blocks that are joined by flexible linkages to form a continuous, articulated cover. The modular installation achieves great stability using lightweight armor units, and it can accommodate changes in the subgrade without allowing major losses of fill, but it has a relatively high cost and is difficult to repair if there is significant damage (Gadd and Leidersdorf 1990; Leidersdorf *et al.* 1990).

Other applications of shoreline armoring include: (1) using gunite or bitumen to armor surface slopes (Figure 4.4); (2) using concrete or gabions on

Figure 4.4 Canon Beach, Oregon, showing variety of measures used to stabilize
and protect eroding coastal cliff.

or within the seaward slopes of dunes for toe protection (Technische Adviescommissie voor de Waterkeringen 1995); and (3) using old car bodies as armor units (Kuhn and Shepard 1980). Foreshore designs usually consist of wave-damper baffles. These structures hinder beach use, and they may work effectively for a short time but are readily reworked by storm waves.

Rubble structures may be designed to have a dynamic response to wave attack and adjust and deform. The advantages are that stone size may be smaller than required for traditional armor, and placement does not require special care (Ahrens and Heimbaugh 1989; Medina 1992). These structures differ from many proprietary designs used on the beach in that they are intended to be exposed to high-energy conditions. The volume of rock may be large, but the cost of obtaining and transporting small units may be relatively low and may require less specialized and smaller equipment for installation and repair (Viggósson 1990).

Geotextiles

Geotextile bags filled with sand can be used as armor units, and geo-textile fabric can be used to create membranes that function as patches in natural landforms, as surface armor, or as internal armor that provides backup protection during severe storms. Geotextile bags filled with sediment have provided slope protection on nine artificial islands in the Beaufort Sea, where they appeared to be best suited for use in shallow and intermediate water depths where wave heights seldom exceed 3 m (Leidersdorf et al. 1990). Their advantages were seen to be moderate cost, ease of performing repairs and relative simplicity of installation. Problems included susceptibility to ice damage and the environmental nuisance of litter from destroyed bags (Leidersdorf et al. 1990). The temporary nature of geotextile structures can be an advantage as well as a problem in that they can be constructed in experimental forms and later strengthened to provide a more permanent function if the design performs well (Pilarczyk 1996).

An example of use of geotextiles to patch a natural landform is the artificial bar segment used to close off a 300 m wide gap in the longshore bar at Sylt, Germany (Ahrendt and Köster 1996). Anchored geonets (netting that is stretched over and pulled down on the ground surface) have been suggested as a way of increasing the resistance of the surface of sandy landforms without having to use rock armor in order to provide protection in locations where regulations prevent the use of structures (Gray et al. 1996). Armor layers have been constructed within dunes (d'Angremond et al. 1992; Dette and Raudkivi 1994). In some cases, the membrane is open on the landward side, with the top and bottom ends of the fabric held in place with anchor tubes.

Designs that mimic nature

Designs for artificial protection that mimic natural processes are appealing because they appear to be environmentally compatible in function if not in materials. One of the more interesting examples is artificial seaweed, a method that was used as early as the 1960s (Rogers 1987; Pilarczyk 1996). Problems with artificial seaweed include ineffective anchoring and abrasion. Effects are difficult to monitor in the field, but preliminary conclusions are that this technique is ineffective in controlling shoreline erosion by waves on an exposed coast (Rogers 1987) but effective in preventing localized scour at offshore structures (platforms and pilings) where wave and current forces are not as strong (Pilarczyk 1996).

Pumps, drains and fluidizers

The hydraulics of flow through the beach can be altered to increase or decrease the likelihood of sediment entrainment or deposition and increase the capability of the waves and swash to do their work. Drainage systems can be implemented to lower the beach water table and enhance deposition of swash uprush while diminishing erosion on the backwash. Most drainage installations employ buried pipes and electronic pumps, but drainage using buried strip drains or artificially constructed permeable layers that require no subsequent operating costs is possible (Davis *et al.* 1992; Curtis *et al.* 1996; Katoh and Yanagishima 1996). The drainage can be toward the seaward side (Katoh and Yanagishima 1996) or toward the backbeach, although the water is eventually discharged on the seaward side (Curtis *et al.* 1996). Drainage systems are structural solutions that appear to work with, rather than against, natural physical processes. They also have little to no impact on the aesthetic attraction of the protected beach (Turner and Leatherman 1997).

A fluidizer is a pipe with holes placed along its length, through which water flows as high velocity jets, increasing the pore pressure and liquifying the sediment substrate (Patterson *et al.* 1991). The sediment is then more susceptible to movement by waves, currents, gravity or pumping. Fluidizers have great potential for maintaining channels (Parks 1991) and for bypass operations (Patterson *et al.* 1991), but their specific effects on the upper beach and dune relative to traditional bypass operation are unknown.

Drainage systems have a problem in common with most other methods of shore protection, namely that accretion at the site of the project is associated with scour at some other location because the pumps and drains do not increase the regional sediment budget. It is also too soon to draw any conclusions on their effect on beach infauna to determine whether they have adverse environmental effects. Turner and Leatherman (1997) conclude that the early promise of the beach dewatering concept is yet to be demonstrated at the prototype scale, and new dewatering systems should be regarded as experimental. Their conclusion is based as much on the lack of basic knowledge of the time-varying saturation of the beach and the modification of sediment transport induced by infiltration and seepage as it is based on the lack of data from installation sites (Turner and Leatherman 1997). There are also potential adverse environmental impacts on fresh-water supply, dune vegetation, beach meiofauna and water quality in the discharge area that may be associated with enhanced drainage (Curtis *et al.* 1996).

Inlet stabilization measures

Jetties

Jetties are designed to constrain and direct the flow of water at inlets and to provide a barrier to longshore transport to prevent shoaling of the channel. The effects of jetties are well documented and include: (1) impoundment of sediment updrift; (2) increase of erosion rate on the downdrift shoreline; (3) migration of the pre-existing channel (usually toward and parallel to the updrift jetty); (4) loss of pre-existing ebb distributary channels; (5) erosion of the pre-existing ebb delta; (6) transport of sediments moved by ebb flows to deeper water; and (7) creation of a new ebb tidal delta and associated bars farther seaward (Kieslich 1981; Dean 1988b; Hansen and Knowles 1988; Pope 1991). Jetties also can induce lower nearshore gradients and reduce breaker heights toward the jetties (Short 1992).

Change on the shorelines near jettied inlets is unidirectional over the long term, with accretion updrift and erosion downdrift (Figure 4.5B) instead of bidirectional (erosion–accretion cycles) as occurred before jetty construction (Figure 4.5A). Beaches in the inlet throat usually are removed by tidal scour where jetties are constructed parallel to each other and are closely spaced. Sediments removed from the inlet throat are not replaced because the seaward extension of the jetties blocks sediment from the ocean beaches and the ebb tide delta. The total elimination of throat beaches from some jettied inlets removes the bare sand areas that provide valuable scenic attractions for people and habitat for shore birds. These losses are compensated by human recreational-resource opportunities added by jetties, including provision of a more predictable shoreline for development, an increase in the variety of wildlife in and around the structures and provision of recreational uses (boating and fishing from the jetty) that are complementary to those provided at unjettied inlets (Nordstrom 1987b).

A. Natural inlet (rapid and unpredictable change, cyclic development)

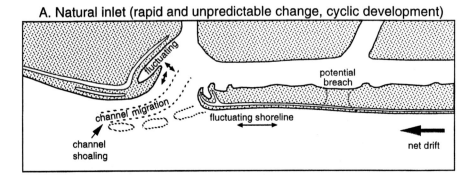

B. Jettied inlet (predictable change, unidirectional development)

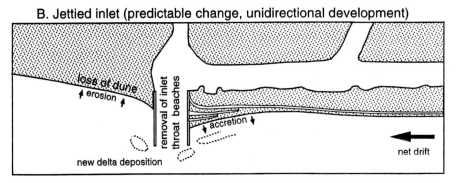

C. Unjettied inlet with dredging & structures (stable or unidirectional)

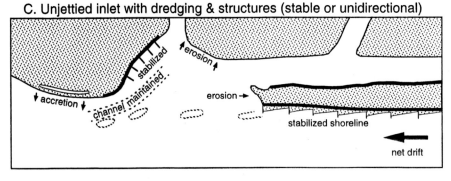

Figure 4.5 Comparison of a natural inlet, jettied inlet and unjettied inlet with dredging and protection structures on a barrier island. Modified from Nordstrom (1987b, 1988a).

The complexity of inlet processes and topographic variability of the inlet shoreline is greatly reduced by closely spaced perpendicular jetties (Nordstrom 1987b), but currents within shoreward converging (arrowhead-shaped) jetties can result in considerable local shoreline mobility causing erosion problems that require frequent site-specific management solutions (Morang 1992). Where jetties are spaced far apart, as at Barnegat Inlet and Absecon Inlet in New Jersey, beaches in the inlet throat are not always eliminated, although the mobility of the shoreline is diminished because the jetties dampen ocean wave energies (Nordstrom 1987b).

Accretion updrift of jetties with erosion on the downdrift shoreline are well-documented effects of jetty construction (Coastal Engineering Research Center 1984), although sediment bypassing to downdrift shorelines can occur in the presence of jetties (Stauble and Cialone 1996). Examples of average rates of shoreline retreat downdrift of jetties include 3 m a^{-1} (Everts, 1979), 6 m a^{-1} (Tye 1983) and 25–30 m a^{-1} (Awosika and Ibe 1993). Sand starvation downdrift of jetties can create rates of erosion that are too great for dune building to keep pace with landward displacement of the beach profile, converting the characteristic shoreline topography from dunes to overwash flat and altering vegetation assemblages (Roman and Nordstrom 1988). This short-distance effect of sand starvation is a readily observed effect, but there is a long distance, relatively inconspicuous effect caused by the materials-deficit (Bruun 1995), and the impact of controlled inlets in many areas may be even greater than previously suspected. The sediment starvation is often conceived as a direct reflection of the net sediment transport rate, but account must be made for interruptions to the gross transport rate; downdrift effects should be related to reduction in rates of accretion not just rates of erosion (Bodge 1994a). It is reasonable to consider the effects of jetties to extend at least as far as the influence of the inlet prior to jetties – a distance extending as far updrift as 5.4 km and as far downdrift as 18.4 km at the Colorado River, Texas inlet (Heilman and Edge 1996).

The construction of jetties may increase the amount of shoreline to be protected, even in areas where there are no buildings (Nordstrom 1987b). Erosion downdrift of jetties often results in construction of seawalls and groins to mitigate the effects of the sediment deficit and to ensure that a new inlet will not be created to jeopardize the maintenance of the controlled channel. These adverse effects of sand starvation may be prevented by making provisions for sediment bypass. Bypass operations are often included in designs for converting natural inlets to jettied inlets, but bypass is often not accomplished due to cost constraints.

Dredging activities do not cease following construction of jetties. Annual dredging in the jettied East Pass, Florida inlet, for example is about 74 000 m^3 annually (Morang *et al.* 1996). The annual amount of material removed through maintenance dredging may be reduced to less than it was prior to jetty construction, but the frequency at which dredging is required remains high (Nordstrom 1987b). Maintenance histories for inlets, including jetties are often complex and characterized by increasing scale of human involvement. The history of maintenance of the St. Marys River entrance channel, can be separated into 7 epochs, representing periods of jetty construction (1881 to

1904) and modification (1924 to 1953) and numerous actions taken to deepen, widen and lengthen the channel; these activities resulted in increases in volumes of materials removed in maintenance dredging operations, from 16 700 $m^3 a^{-1}$ to 616 200 $m^3 a^{-1}$ (Pope *et al.* 1994).

Jetties can have a positive effect on formation of dunes by creating wider beaches updrift of them that increase the likelihood that dunes will form and survive. Foredunes in the accretion zone caused by jetties can form multiple ridges as the accretion moves the location of sand trapping seaward (Figure 4.5B). The accretion zone is also likely to attract the interest of developers. Accordingly, provisions must be in place to prevent development of these accreting areas if they are intended to evolve naturally.

Stabilization caused by other structures

An indication of the changes that can occur at unjettied inlets under conditions of development is provided by Nordstrom (1988a) for the shoreline of southern New Jersey and depicted in Figure 4.5C. Shorelines at these locations have been less mobile than prior to intensive development (before about 1935). Part of the reason for the reduction in rate of shoreline change updrift of inlets is the lack of new permanent breaches in the barrier islands as a result of protection provided by bulkheads, seawalls, nourishment projects and protective dunes that are higher and more continuous than at comparable locations at natural inlets. Continued downdrift migration of inlets has been impeded by seawalls, bulkheads and groins that stabilize the shoreline downdrift of the inlet throats. The beaches in the inlet throats have been completely eliminated at times at two of the New Jersey inlets and truncated at the bulkhead line at one of the inlets. Dredging projects at two of the inlets reduced the periodicity of erosion/deposition cycles associated with breaching of the ebb tidal delta and the shoreline updrift. The human activities have kept the navigation channels from fluctuating as widely as they would under natural conditions, thus reducing the periodicity of erosion/deposition cycles associated with breaching of the ebb tidal delta and updrift shoreline. Episodes of erosion and accretion continued to occur downdrift of the inlet throats, but rates of change have been less since the implementation of protection structures and channel dredging (Nordstrom, 1988a).

Marinas and harbors

Marinas and port facilities built out from the shore (Figure 4.6) act as artificial headlands that break up the orientation of the shoreline, change refraction and diffraction patterns, trap sediment, deflect sediment offshore

Figure 4.6 Castelldefels, Spain, showing marina built into the sea that provides a trap for sand moved alongshore and shelter for natural and artificially nourished pocket beaches.

and starve adjacent downdrift beaches (McDowell *et al.* 1993; Moutsouris 1995). Beaches have been eliminated in the process of creating marinas, but new beaches may be intentionally created adjacent to these same marinas or within the marinas as a recreational amenity. Beaches created within the marinas are low-energy beaches and are potentially complementary to the outside exposed beaches in terms of recreational benefits and habitat, providing they are managed to allow for these complementary uses. Beaches may form just inside the protective breakwaters, although these features are undesirable because they interfere with navigation and are inaccessible for recreation. They are likely to be removed, such as at the one at Dana Point, described by Wiegel (1993b).

Marinas are occasionally built as complexes, including harbors, reclamation fill and new artificial beaches (Anthony 1997). Positive results of harbor creation and management are possible where beach management is an integral component of a plan of action that includes creation of new beaches using the dredged sediment (Gómez-Pina and Ramírez 1994). Marinas that are located where there are large numbers of residents and shorefront visitors are likely to be managed for beach-related activities, but provisions also have to be made to protect or augment natural values to ensure that maximization for recreational use does not preclude development of natural features.

Pilings and elevated piers

Inventories of human structures and historic photographs indicate that elevated piers were common in the USA at the beginning of the twentieth century (Meyer-Arendt 1992; Leidersdorf *et al.* 1994), although these structures were often built for recreational fishing and were too small to last through major erosional events. Piers designed to have long useful lives are often built to larger dimensions and may have multiple use activities. Piers in intensively developed resorts, such as Atlantic City, New Jersey (Figure 1.9) contain clothing stores and gift shops in addition to amusement rides.

The effects of a pier on waves and winds depend on length, width, pile diameter, pile spacing, number of piles per bent, bent spacing and pile bracing. Horizontal members are important in their effect on wind flow and as a barrier to waves and swash during storms. Effects may be considered in terms of the local impact of isolated pilings and the cumulative regional effects of the entire structure.

Scour at pilings

There is little information gathered in the field on scour around piles or pile groups, and the scope of existing studies does not extend to effects on adjacent areas (Miller *et al.* 1983; Nicholls *et al.* 1995). Visual inspection of subaerial beach surfaces after storms rarely provides insight to the effect of waves and currents, because scour holes are filled by swash associated with falling water levels, but trenching can reveal heavy mineral lag layers that can be used to identify scour depths (Nicholls *et al.* 1995). Pile diameter, pile shape and orientation to the flow are key parameters affecting the dimensions of the scour hole. Nicholls *et al.* (1995) found scour radii of 400 to 420 mm with depths of 170 mm to 300 mm caused by 200 mm diameter piles on New York and New Jersey beaches following the December 1992 northeaster. They found these values to be consistent with a maximum scour hole radius 2 to 3 times the pile diameter and a depth 1 to 1.5 times the pile diameter reported in previous studies.

Miller *et al.* (1983) identified depths under active scour conditions at the Coastal Engineering Research Center (CERC) Pier using lead line soundings and detailed underwater surveys. This 561 m long structure is supported on 108 piles spaced at 4.6 m intervals on center across the pier and at 12.2 m intervals along the pier. The piles are 0.8 and 0.9 m in diameter and have 1.2 to 1.4 m diameter collars. Miller *et al.* (1983) found that maximum scour around individual piles under moderate wave conditions was 1.0 m below the depth of the nearby bottom; scour radii extended about 3.7 m from the center of the

1.2 m diameter pilings (with abrasion collars). The shapes of the local scour holes were asymmetrical, being steeper on the seaward side and deeper on the landward side.

An increase in the structural strength of piling foundations may increase susceptibility to scour. Miller *et al.* (1983) found that scour holes coalesced across a bent that had a pile spacing of 4 diameters. Bracing that provides extra strength to pilings provides an additional obstruction to flow if not aligned parallel to the flow, while collars around pilings that provide lateral support act as additional obstructions to flow and increase the effective diameter, enhancing scour (Nicholls *et al.* 1993).

Effects of piers on beach characteristics

Noble (1978) and Simison *et al.* (1978) used field inspections and aerial photographs to identify the effects of 20 piers in southern California that extended from landward of the mean high water level through the nearshore zone. The piers varied from 191 m to 762 m in length and from 4.6 m to 91 m in width, and they had pile diameters from 0.30 m to 0.76 m wide with considerable differences in pile spacing and bent spacing. They concluded that pile supported piers had no appreciable impact on the adjacent shoreline. In support, they cited previous field studies that revealed no significant effect on accretion or erosion of adjacent shorelines and theoretical studies indicating that reflection and eddy losses were of minor importance and the ratio of transmitted wave height to incident wave height should approach unity when pile spacing is greater than four times pile diameter (Simison *et al.* 1978). Part of the reason for their conclusions could be the broad spatial scale of their investigation that masked local effects.

Results of studies by Miller *et al.* (1983) and Weggel and Sorensen (1991) contrast with these findings. Weggel and Sorensen (1991) noted that five pile-supported piers at Atlantic City, New Jersey measurably influenced local long-shore transport rates and beach planform. They found that there is often a tombolo-shaped bulge that develops at or downdrift of piers and that the pier with the highest pile density had the most observable effect on the beach.

The more detailed field study conducted by Miller *et al.* (1983) revealed significant effects in both the shore normal and alongshore directions associated with the CERC Pier. Profile measurements along the pier showed greater depth, steeper slope and greater vertical variation than measurements of natural profiles located away from the pier. Scour, resulting from the interaction of the pier with waves and currents, produced a long, shallow permanent trough under the pier. The trough usually had one steep and one gently sloping side, and the depth, width and symmetry changed with changing

wave conditions. The trough was 100 to 200 m wide and extended from the
−3 or −4 m contour seaward past the end of the pier. Depth differences up to
3 m were greatest within a large scour hole just inshore of the seaward end of
the pier. This scour hole deepened and expanded downdrift during storms
but acted as a sink for fine sediments during calmer conditions. Depth contours were generally straight at a distance from the pier and curved landward
in the vicinity of the pier. There was a strong rip current under the pier during
storms, and cross-shore bottom currents were up to 0.3 m s⁻¹. Miller *et al.*
(1983) conclude that the pier acts as a permeable groin. During long periods of
unidirectional longshore transport, sand is trapped on the updrift side with
concurrent erosion on the downdrift side, greatly altering shoreline orientation. These effects were also noted at a nearby fishing pier.

Effects of piers on aeolian transport and dunes

Piers are designed to be elevated above the effects of wave action but
not above the elevation of land access. As a result they have increasing impact
on the ground surface with distance inland, where they can act as traps for
aeolian transport on the upper beach. Observations of piers in New Jersey
indicates that they provide a sufficient barrier to aeolian transport to create
dune spurs perpendicular to the beach. These dunes can build up faster than
shore-parallel dunes because they trap sand moved alongshore (where the
width of the beach as a sand source is unrestricted). The resulting deposits
survive longer than the incipient dunes that form on the adjacent beach
because the pilings prevent use of beach cleaning equipment that would
remove them.

Promenades and boardwalks

Seafront promenades and boardwalks provide longshore access for
pedestrians. These two types of structures have similar utility functions, but
they can be differentiated by their effect on the beach. I use the term promenade to describe structures that are built on the beach or dune surface or
placed on fill and protected by a shore-parallel wall on the seaward side
(Figure 1.5) and boardwalks to describe structures that are elevated above the
surface (Figure 4.7).

Promenades

These structures are more common than boardwalks. They appear to
be especially popular in developed European coastal resorts and may be considered aesthetically indispensable in places (Miossec 1993). Promenades may

Figure 4.7 Boardwalk at Rockaway, New York, where the absence of important cultural features landward of the structure has allowed the dunes to survive.

be built onto the beach (Fernandez-Rañada 1989), where, unlike boardwalks, they result in a net loss of active beach. The value of infrastructure that justified construction of the promenade is likely to justify expenditures for beach nourishment. Thus construction and use of promenades need not result in elimination of beaches, but alterations to the beach to maximize recreational value of promenades may preclude evolution of coastal landforms by natural processes.

Construction of low promenades can lead to pressure to eliminate or limit the height of protective dunes seaward of them in order to retain views of the sea. The absence of a dune can then lead to besanding of cultural features to the lee. Observations at Manasquan, New Jersey, where the promenade is built at the elevation of the backbeach indicates that inundation by blowing sand is common, resulting in aeolian landforms that have great variety in size, shape and orientation due to interaction with cultural features. These landforms survive the winter but are removed early in the summer tourist season. Aeolian landforms at raised promenades are confined to the beach until they form ramps that facilitate transport inland. At that point, the ramps are frequently removed to prevent besanding of buildings and support infrastructure.

Boardwalks

Boardwalks are common features on many segments of shoreline on the northeast coast of the USA. These structures were originally designed to protect the plush interiors of hotels from sand on visitors shoes, but their commercial value was discovered in the late nineteenth century, and they began to feature shops and amusements (Koedel 1983). Boardwalks in the resorts in New Jersey are wooden plank structures built parallel to the shore and elevated above the backbeach on pilings. They are designed for use by pedestrians and occasionally light public transportation and safety vehicles. They are usually located just seaward of commercial establishments, although they may extend to adjacent portions of the shoreline that are zoned for private residences.

Boardwalks provide a recreational function that is similar to promenades, but their impact on beaches and dunes differs from those structures because boardwalks are raised above the ground surface, allowing sand to be transported under them by both wave and aeolian processes. There is often a space between the planks that restricts sediment accumulation on top of the structure. Elevation above the backbeach is usually >1.0 m and <4.0m. Boardwalks are usually lower than the crests of dunes that could form near them, but dunes seaward of these structures are rarely allowed to grow much higher than the surface of the boardwalk, so users can view the sea. The boardwalks themselves are not protected by shore-parallel structures on either their seaward or landward sides because their elevation affords them direct protection, but a bulkhead may be built on the landward side to protect adjacent buildings and infrastructure from wave damage. Where bulkheads are not used, local managers may place a sand fence under the boardwalk to reduce aeolian transport and prevent besanding of developed properties landward. Beaches and dunes may form landward of boardwalks where managers allow natural sediment exchanges to occur (Figure 4.7). Thus the passive effects of boardwalks are relatively benign, but maintenance of the recreational value of these structures results in human actions that eliminate dunes on the seaward side (to retain views) and beaches and dunes on the landward side (to mitigate wave damage or besanding).

Minor beach structures

Recreation-related structures.

A great variety of small or temporary structures are placed on the beach to enhance recreational use (Figure 2.5). These structures include bath houses, restaurants (Fabbri 1989), makeshift commercial stalls (Awosika and Ibe 1993), chairs, beach baskets (chairs that protect against wind) (Draga 1983), lifeguard towers and buildings that contain rescue or first-aid equipment. The buildings designed for recreational use are often grouped together in the literature and referred to as cabanas, bagnos or beach houses, although they may have different function, size and impact on the beach. Some of these structures are simple and some include facilities such as restaurants, coffee shops and video arcades. A rough estimate of the number of entrepreneurs who construct these buildings on the beach in Italy is 50 000 to 70 000 (Fabbri 1989). Umbrellas, tents and beach chairs are often placed seaward of the the buildings and close to the water, where damage from minor storms leads to complaints by the entrepreneur to provide protection structures that deteriorate beach quality for bathing (Fabbri 1989). Maximization of use of these recreation structures may also involve grading operations (chapter 2), and calls for converting them into permanent buildings, resulting in a geomorphic impact that greatly exceeds that of the initial structures.

The surfaces near small structures that remain on the beach after the tourist season may not be maintained as a recreation platform, allowing wind and wave interaction with the structures to create distinctive landforms. Wind usually causes scour beneath elevated buildings and around the sides of buildings, with a small accretion zone on the leeward side. The localized zones of accretion and scour associated with aeolian processes rarely have a local relief greater than 0.5 to 1.0 m, in part because the landforms are soon eliminated by wave action in the winter storm season. The major exceptions occur at the permanent structures on wide beaches (e.g., buildings used to store rescue and first-aid equipment) that survive wave attack and survive beach cleaning operations because equipment cannot operate close to them. Localized zones of accretion at these sites may develop into dunes and develop a vegetation cover.

Drainage structures

Drainage culverts and pipelines on the beach can change locations of accretion and erosion, slow littoral drift (Bandeira *et al.* 1990; Otvos 1993) and form traps for accumulation of wind blown sand (Nordstrom *et al.* 1999). Water from storm drains that empty onto the beach can scour the surface and fluidize the sediments. The effects on the beach of high discharge of storm water runoff from these drains is not well known (Weggel and Sorensen 1991), but there are single-sentence references in the literature to runoff accelerating beach erosion and increasing the likelihood of swash reaching the dune line (City of Stirling 1984; Western Australia Department of Planning and Urban Development 1993). Ends of storm drains that are buried by beach accretion have to be exhumed by earth-moving equipment to function properly, resulting in pits and mounds that remain after the maintenance work has been finished.

Sand fences

Rationale for emplacement

Sand fences are used to trap sand moved by wind in order to: (1) aid in filling gaps in dune ridges; (2) create an entirely new dune ridge; (3) create a sacrificial ridge to protect a more valuable dune behind; and (4) prevent inundation of cultural features landward. The same fences used to trap sand are often used to control access, unintentionally increasing the impact of fences on aeolian transport and dune formation. Sand fences are one of the most important human adjustments affecting the geomorphology of developed coasts because: (1) they are one of the few structures permitted seaward of the dune crest in many jurisdictions; (2) they are relatively inexpensive; (3) they are easy to construct; and (4) their deployment usually occurs at the highly dynamic boundary between the beach and dune.

Fences are usually placed on the seaward side of an existing dune, creating a broader dune platform. In mid latitudes they are most often emplaced at the end of the summer tourist season and at the beginning of the winter storm season, but they may also be constructed at the end of the storm season (in order to repair fences damaged by wave uprush) or at the beginning of summer (to keep people out of the dune zone) (Nordstrom and Arens 1998). The method of emplacement is often according to the whim of managers despite the existence of technical assessments and guidelines for their use (Coastal Engineering Research Center 1984; Hotta *et al.* 1987, 1991).

Fence types

The type of fence used may vary from region to region but is often similar within a region. Many communities in the USA use a standard size fence that is also used to control snow drift. These fences have wooden slats about 38 mm wide by 1.2 m high with a porosity usually reported as 50 percent (Mendelssohn *et al.* 1991). The size and spacing of the slats varies through time due to weathering, and actual porosity is often greater. Field measurements of fences in New Jersey (Figure 1.10) reveal that the slats average about 35 mm wide and are spaced 60 to 70 mm apart (averaging about 65 percent porosity). Most communities in New Jersey use a single or double row in a straight alignment, although zigzag configurations are commonly used.

Fencing materials in The Netherlands are branches or reed stakes, placed close together to form barriers 1 to 2 m high. Sometimes they are placed in a grid, the same way as marram grass is planted (Nordstrom and Arens 1998). Rows of branches or reeds placed parallel to the foredunes are most common, with cross-rows facing into the dune. Short rows may be placed perpendicular or oblique to the foredunes depending on the most active wind direction.

Effect on dune morphology

Sand accumulation efficiency and morphologic changes depend on fence porosity, height, inclination, scale of openings, shape of openings, wind speed and direction, sand characteristics, separation distance between fences, number of fence rows, and placement relative to existing topography (usually pre-existing dunes). As a result of so many variables, no standard method of foredune construction can be directly applied on all beaches (Hotta *et al.* 1991). In general: (1) the upwind dune slope becomes steeper with increase in wind speed; (2) the crest of the dune moves downwind with increase in wind speed; (3) a porosity higher than about 50 percent diminishes the accumulation rate; and (4) fences of higher porosity create higher and longer dunes, but a fence of almost any porosity will be buried after a sufficiently long time (Hotta *et al.* 1987).

Hotta *et al.* (1987) describe flow pattern around fences of different porosities. Flow around a fence with zero porosity reveals a small circulation pattern upwind of the fence and a large circulation pattern downwind. With increasing (but low) porosity, the small circulation cell in front of the fence disappears (at about 20 percent), and the large leeward circulation cell is reduced in size and shifts downwind. With increase in porosity (ranging from 35 percent to 80 percent), the downwind circulation cell disappears (Hotta *et al.* 1987). These differences in porosity result in differences in form of the accretion.

Hotta *et al.* (1991) identify the effects of multiple sand fences. They indicate that: (1) an increase in fence porosity results in a gentler slope on both the upwind and downwind sides and a downwind shift in the crest; (2) a large dune will form with a large accumulation downwind from the front fence if the separation distance is great; (3) a stable dune forms if the front fence is higher than the rear fence; (4) a single fence has almost equivalent trapping capacity as a two-row fence if the separation distance is narrow and the wind speed is low; but (5) a second fence can reduce the tendency for sand trapped at a single fence to be blown off at high wind speeds (Hotta *et al.* 1991).

Dunes of different forms can be achieved by using sand fences in different configurations. Diagrams presented in Hotta *et al.* (1987) reveal that considerable variety in plan and cross-section can occur using fences inclined from the vertical and wing fences placed at different angles relative to the direction of dominant winds. Snyder and Pinet (1981) compared transverse dune profiles taken at paired straight fences (spaced 6 m apart) and at paired zigzag fences (spaced 6 m apart on the outer edges). Their results revealed that dunes at zigzag fences were about 50 percent wider, had more gently sloping dune faces on the seaward side (7 to 9° as opposed to 17 to 21°) and had more symmetrical cross-shore profiles. Longshore profiles at their sites revealed horizontal crests on the dunes created using straight fences and undulating crests on the dunes created using zigzag fences. The latter more closely approximated the shapes of natural dunes in the area. Dunes created 2 years earlier using zigzag fences retained the characteristic crest shapes in the longshore direction and were wider and more gently sloping than the dunes created at the same time using linear fences (Snyder and Pinet 1981).

Sand fences used for crowd control result in many interesting dune shapes. The most common forms other than shore-parallel forms are the shore-normal ridges that are created adjacent to fenced access paths through foredunes. Buildup of these forms is aided by the enhanced sediment flow from the beach through gaps in the foredune crest that are created for or result from pedestrian access.

Accretion rates

Foredunes up to 10 m elevation have been constructed using sand fences (Hotta *et al.* 1991). The rate of growth is greater than under natural conditions and is concentrated in a smaller zone than occurs with natural vegetation. Growth rates are highly variable between regions. Rates of up to 10 to 20 m^3 m^{-1} a^{-1} occur in The Netherlands, aided by prevailing onshore winds (Nordstrom and Arens 1998). Coastal Engineering Research Center (1984) indicates that the trapping capacity of the initial and subsequent lifts of a 1.2 m high fence averages 5.0 to 8.0 m^3 m^{-1} of beach . Annual sediment accumulation measured over 3 years on a transgressing barrier island in Louisiana was 82 to 90 percent less than the average reported for dune-building sites in Massachusetts, North Carolina, Texas and Oregon (Mendelssohn *et al.* 1991).

The sand fencing configurations that accumulate the most sand in the first year may not be the designs that accumulate the most sand in subsequent years. Mendelssohn *et al.* (1991) found that a straight fencing with perpendicular side spurs accumulated the greatest volume of sand initially (although a straight fence without spurs created a taller dune), whereas the straight fence alone accumulated far more sand after 3 years.

Use of sand fences alone to create a dune requires a sufficient amount of sand in the coastal sediment budget. Despite rapid accretion rates, use of fences is often a secondary action taken after the primary structure is created using bulldozers. Fences then help stabilize the bare sand surface and help prevent besanding of property landward of the dune. The accretion trend at the dunes studied by Mendelssohn *et al.* (1991) was eventually reversed due to wave erosion, leading to the conclusion that fencing in conjunction with vegetation plantings can be used to build dunes in a sand-deficient environment, but, inevitably, beach nourishment is required to maintain a healthy, well-vegetated dune on a transgressing barrier island (Mendelssohn *et al.* 1991).

Buildings

There are far fewer studies of the direct effects of buildings than of actions taken to facilitate their initial construction and subsequent use or to protect them from coastal hazards. Buildings (like shore protection structures) replace the backbeach and dune and reduce the source area for wind blown sand (Morton *et al.* 1994), and they alter wave uprush and wind directions and speeds, thereby altering depositional patterns (Nordstrom and McCluskey 1984, 1985). Buildings can reduce the magnitude of wave and wind processes and contribute to deposition by providing barriers, but they can also accelerate flows by constricting them between structures, increasing the likelihood of local scour. Buildings also may provide barriers that serve to detach landforms from their beach or foredune sources. The direct effect of buildings is related to their location on the beach/dune profile and their method of construction (including size, shape, materials, density). The effects are separated here into those that occur from waves and currents and those that occur through interaction with aeolian transport and dune migration. The impact of buildings on aeolian processes, in turn, is separated into effects of houses and effects of high-rise structures

Effects on waves and currents

Buildings designed as permanent structures are rarely built on the beach (unless they are behind protection structures) because common sense and legal restrictions prevent construction in this hazardous environment, but subsequent erosion may result in an unprotected building being located on the beach. Examination of post-storm conditions at communities where houses built on pilings end up on the backbeach indicates that they have little effect on landforms, whereas houses built on the ground create obstructions to flow, creating localized zones of accretion and scour. The dynamic interactions between these structures (and the debris from them) and waves and currents are not possible to assess during a storm because of the difficulty of taking measurements in debris-laden storm waves, but the morphologic changes are conspicuous after the storm.

Gayes (1991) noted that beaches were steeper and nearshore bars were less well developed offshore of human structures than in other locations after Hurricane Hugo. Entire houses can be moved 50 to 100 m inland (FitzGerald et al. 1994; Fletcher et al. 1995). There is considerable litter remaining from damaged structures after large storms (Stauble et al. 1991). Cultural debris can be moved offshore to distances farther seaward than 4 m below mean low water (Gayes 1991), where it can be buried beneath sand (Leadon 1996) or create local, small-scale landforms.

Rubble can remain on the beach to interact with waves and currents years after buildings are abandoned (Meyer-Arendt 1993b), and ruins of houses remain in the water where they affect sedimentation and become safety hazards (Gusmão et al. 1993). Military structures are often even more conspicuous. These structures may initially be built in the dunes but end up on the beach as a result of coastal erosion (Guilcher and Hallégouët 1991). The great strength of these structures allows them to survive for years after being exposed to waves and currents. They are conspicuous features on the shoreline of many parts of Europe (Guilcher 1985; Guilcher and Hallégouët 1991; Miossec 1993) and can be a dominant coastal structure in locations where urban development and shore protection projects are uncommon (Ciavola and Simeoni 1995).

Effects on winds and aeolian transport

Houses

Direct actions taken by residents to make use of their coastal properties (by constructing sand barriers, removing deposits and planting exotic species) may result in greater changes to landforms than the passive effects of houses. The passive effects of houses are best evaluated where there are strict controls on use of the dune and there is limited support infrastructure, reducing complications due to roads, parking areas and human actions. The shorefront communities at Fire Island that were examined by Nordstrom and McCluskey (1984, 1985) allow for this kind of assessment. Nordstrom and McCluskey (1985) compared the effects of changes in wind speed along shorenormal transects at two houses built on pilings. One house was on the crest of the dune (Figure 4.8A); the other house was landward of the dune crest and lower than the crest (Figure 4.8B). Offshore wind speeds were much reduced to the lee of the house on the crest of the dune. Wind speeds were not much

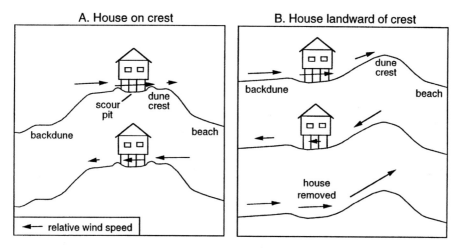

Figure 4.8 Effects of houses and topography on wind speed. Modified from data and topography depicted in Nordstrom and McCluskey (1985).

reduced below the house because of channeling of the airflow under the elevated structure. Wind speed decreased downwind of the house located landward of the dune crest but not as much as at the house on the dune crest. The offshore reduction in wind speed downwind of the house landward of the crest was less because the sampling location was at a higher elevation on the dune, where flow compression contributed to higher wind speed.

During onshore winds (Figure 4.8), there was a sizable reduction in downwind velocity at the house built below the dune crest because of sheltering by both the structure and the dune ridge. The wind velocity under that house was considerably lower than under the house on the dune crest, providing further documentation of the significance of sheltering by the dune.

The house landward of the dune crest was removed during the study, so the effects of topography could be differentiated from the effects of the house. Comparison of wind data taken before and after removal of the house revealed that there was little difference in the average velocities at the center of the house site, but the velocities to the lee of the center location were considerably higher in the absence of the house (Figure 4.8). The field study revealed that there can be a substantial difference in the degree of interference with wind velocity by houses located at different elevations with respect to the dune crest and in different topographic configurations (Nordstrom and McCluskey 1985).

Examination of wind flow and characteristics of buildings and topography within entire developed communities on Fire Island (Nordstrom and

McCluskey 1984) revealed that close spacing of buildings could channel winds, increasing their velocity and reducing the likelihood of survival of vegetation. The bare areas then became subject to increased deflation. Pilings without breakaway walls had a scour pit of bare ground beneath them and a well-vegetated lip of sand around the building on all sides (as depicted in the house on the dune crest in Figure 4.8). The scour was due to high wind velocities beneath the structure, coupled with the absence of stabilizing vegetation there. The lip of sand represents deposition due to reduced wind velocity in the lee of the houses. The scour pits on the dune crest were commonly about 0.5 m below the line of the undisturbed ground surface next to the house; the lip was about 0.5 m above the surface. The pits did not show much evidence of further growth once the typical configuration was reached (Nordstrom and McCluskey 1984).

Houses on pilings landward of the dune crest and at a lower elevation caused greater modification of the dune form than similar houses located on top of the dune. Landward migration of the dune crest does not always bury these buildings because channelization of wind under and around the structures maintains relatively high wind velocities that prevent deposition. The dimensions of the depression created by this process are a function of the elevation of the bottom floor of the structure and the height of the dune crest.

Houses built on the ground, including those with breakaway walls, act as a barrier to the wind and to the movement of sediment from all directions, and they are more complete sediment traps than elevated houses (Nordstrom and McCluskey 1984). Sand accumulation at buildings is reported in other locations (Canning 1993; Gusmão et al. 1993) along with reports of this process raising the height of the foredune (Canning 1993).

Examination of the dune crest at the more intensively developed community of Westhampton Beach, New York (Nordstrom et al. 1986) revealed that the dune was linear on the oceanside but developed a crenulate, barchanoid shape on the landward side at houses (depicted in Figure 4.2). The crenelate shape resulted from the interference of houses with the wind stream and the direct attempts by residents to modify the location of sediment deposition using earth-moving equipment or vertical walls. The deceleration of onshore wind in front of the vertical obstructions caused by houses appears to create an echo dune. Offshore winds (the dominant direction in that location) are accelerated as they blow around the house, and the barchanoid shape contributes to convergence of the wind stream that helps scour the landward face of the bay of the dune. The irregularity in shape of dunes was increased by residents pushing sand to the sides of their property (Nordstrom et al. 1986).

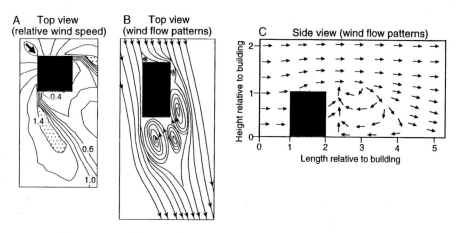

Figure 4.9 Wind flow associated with high-rise structures such as those identified in Figure 1.10. (A) Distribution of mean wind speed during 45° wind; values represent overspeed and underspeed, normalized by free stream velocity (modified from wind-tunnel results in Uematsu *et al.* 1992); (B) Wind flow patterns during oblique offshore wind (modified from Gundlach and Siah 1987); (C) Direction of recirculation using standard computational model (modified from Baskaran and Stathopoulos 1989). The specific patterns and values of wind flow and locations of scour and deposition differ among the studies due to differences in building sizes and configurations, scale effects and boundary conditions. Modified from Nordstrom and Jackson (1998).

High-rise structures

Hardware scale models and computational models of effects of high-rise structures on winds (Figure 4.9) indicate high speeds near the building and flow separation in the lee, with reversals in regional wind direction and pronounced upward flows. These effects may increase rates of aolian transport and alter locations of accretion and deflation on the beach if high-rise buildings are built close to the shore. The resulting sediment losses from the beach may create scour depressions that facilitate storm wave uprush and increase susceptibility of shorefront structures to inundation (Gundlach and Siah 1987; Nordstrom and Jackson 1998).

Nordstrom and Jackson (1998) used results from a wind-tunnel study and a field investigation to examine effects of a 21 m high casino with an 81 m high hotel tower at Atlantic City, New Jersey (from which Figure 1.9 was taken). The building complex was separated from the beach by a 15 m wide boardwalk and an artificial dune with a crest 25 m seaward of the boardwalk and an elevation 0.60 m to 1.90 m above the backbeach. Flow directions in the wind tunnel revealed that winds from several directions were deflected to a more onshore orientation than the regional wind. These alterations in flow did not occur seaward of low-rise (2 to 4 storey) structures located at similar distances from the beach (e.g., foreground of Figure 1.9). Alongshore winds were nearly parallel to the shoreline near the water, but they were deflected onshore at the sides of the high-rise building. Obliquely onshore winds were deflected to a more shore-parallel direction, both in front of and next to the high-rise building, apparently aided by the wall of high-rise structures that extended alongshore (Figure 1.9 background). The obliquely offshore winds were the most dramatically altered of all winds affecting beach change; there was a separation in flow about mid-beach in the lee of the high rise-building, with offshore flow seaward of this zone and onshore and upward flows landward of it, similar to those depicted in Figure 4.9C.

Field observations at the high-rise complex during obliquely offshore winds over <17.0 m s^{-1} revealed that onshore reversals in flow direction and high local wind speeds caused besanding of the boardwalk from the beach side (resulting in considerable discomfort to pedestrians) and created new, highly localized, scour and deposition zones on the backbeach. Winds on the beach just seaward of the boardwalk in the lee of the high-rise tower varied over nearly 180° over time spans of less than one minute. Sand moving along this zone was transported at heights well above 2.0 m. Vertical flows in the lee of the building and collisions of sediment particles on the boardwalk steps, railings and other cultural features created a sand cloud that was visible periodically at elevations of 10 m above the boardwalk. In contrast, sand in transport on the beach near the water, was nearly unidirectional, in the same direction as the regional wind, and the heights of saltating sand grains were <0.5 m (Nordstrom and Jackson 1998).

Scour zones associated with changes in wind flows near high-rise structures at Atlantic City are nearly 1.0 m deep and 40 m wide (Gundlach and Siah 1987; Nordstrom and Jackson 1998). Depositional features can also occur near high-rise structures, creating an additional problem. Reversals of sediment transport direction during offshore winds at condominiums in New Jersey result in inundation of boardwalks and buildings, necessitating removal by earth-moving machinery (Nordstrom 1987a).

Swimming pools

Swimming pools are often located seaward of coastal buildings, where they may provide obstructions to free flow of floodwater and increase turbulence and scour (Nnaji *et al.* 1996; Yazdani *et al.* 1997). Yazdani *et al.* (1997) in their study of representative swimming-pool designs in Florida, found that pool construction materials consisted of concrete, gunite, fiber-glass, timber, masonry and vinyl and that pool builders preferred the rapid on-site construction ease of concrete or pressure-sprayed gunite. Most pools were rectangular, with average dimensions of 5.2 m to 10.4 m and average maximum depth of 1.8 m (Yazdani *et al.* 1997).

Scour at swimming pools may be similar to scour at seawalls in some ways, but it differs in that corner effects are maximized at swimming pools (Nnaji *et al.* 1996). Based on theoretical studies and scour models for non-swimming pool structures, Nnaji *et al.* (1996) conclude that: (1) a round pool is expected to result in half of the scour around a square pool; (2) a small angle of wave attack on a rectangular structure (perpendicular to the wall) will cause less scour than a greater angle (e.g., 45°); (3) placing the smaller side of the pool perpendicular to the flow will result in less scour than placing the longer side perpendicular to the flow; and (4) a smaller length to width ratio causes less scour.

Maintenance of swimming pools seaward of houses restricts the development or migration of dunes into these areas. Special landscaping elements may be introduced for use of the pools, and barriers may be constructed to limit the nuisance of wind blown sand or provide visual buffers for privacy (that also affect shore processes). The presence of pools landward of houses may also result in calls for shore protection well before principal residences are threatened.

Roads and parking lots

Roads and parking lots are impermeable surfaces that provide pathways for overwash (and aeolian transport) during large storms (Hall and Halsey 1991; Fletcher *et al.* 1995). Overwash on shore-perpendicular roads may penetrate all the way to the bayside of barrier islands on these corridors during large storms (Nordstrom and Jackson 1995). Sand transported by waves does not usually extend as far landward as sand transported by winds during moderate-intensity storms. At these times, aeolian transport can continue landward for several blocks (several hundred meters).

Overwash deposits on roads and parking areas are short-lived and are either bulldozed to sides of roads or returned to the beach by bulldozers and trucks. Sediment bulldozed to the sides of roads and parking lots creates raised, linear surfaces that are usually landscaped according to the tastes of local managers. Deposits landward of shorefront roads that are not backed by buildings are often left to develop naturally. Aeolian transport across roads often results in a dune that is physically detached from the beach source but is still supplied by it. Roads and parking lots are often protected against wave erosion by seawalls and against wind blown sand by sand fences or sand barriers (low impermeable structures), greatly restricting the zone over which aeolian processes can occur.

Artificial islands and environments

Offshore islands have been created for petroleum extraction, transportation (including airports and sea ports), flood control, shore protection and recreation. There is an accumulating body of knowledge of methods of constructing these features (van't Hoff *et al.* 1992), and it is only a matter of time before the technology is adequate to create new coast in locations where wave energies and water depths are now considered too difficult or costly. Many of the artificial islands bear little resemblance to natural areas, but many are built with environments that mimic natural features, and there is evidence that artificial landforms, once created, can evolve naturally (Andersen 1995b).

Artificial islands are often associated with the petroleum industry (Noble 1978; Leidersdorf *et al.* 1990). More than 20 artificial islands, causeways and coastal pads have been constructed in the Alaskan Beaufort Sea since 1976; some of these were constructed in exposed locations in water depths up to 15 m (Leidersdorf *et al.* 1990). One example is Northstar Island, an approximately 270 m wide, 6.1 m high oil exploration island with an intended design life of 3 years that was built in an unprotected location in 13.7 m water depth about 25 km offshore. It was created using 800 000 m³ of terrestrial gravel delivered in trucks over a floating ice road (Gadd and Leidersdorf 1990). Some of these offshore islands were protected by sacrificial beaches and subsequently developed spits at their downdrift ends. The rates of erosion and cost of transporting beach materials (gravel) offshore limited application of sacrificial beaches to water depths of 2 m or less (Leidersdorf *et al.* 1990), so geomorphically evolving islands appear to have been considered practical under low-energy conditions.

Efforts to develop artificial islands have occurred over 1500 years in Japan (Nagao and Fujii 1991), where offshore airports now provide one of the greatest incentives for constructing new and large islands. Important airport islands include the Nagasaki Airport and Kansai International Airport. The latter used 180 million m³ of material to create a 511 ha island 5 km seaward of the coast in a water depth of 18 m (Iwagaki 1994). Other offshore airports are planned, including Chubu International Airport, Kobe Airport and New Kitakyushu Airport (Iwagaki 1994). Most offshore islands have been constructed in sheltered bays, but massive works are planned for exposed coasts (Maeda *et al.* 1991; Yamazaki *et al.* 1991).

Figure 4.10 Ocean side of Neeltje Jans, The Netherlands, showing an artificial beach and dune on an artificial island.

An example of a small island designed to serve as a fishing port is the one at Kunnui, Hokkaido, Japan. An offshore location was selected over an onshore location to prevent deposition at the new port entrance and to minimize beach erosion downdrift (Kawaguchi *et al.* 1994). The entrance to the port was built at a depth of 6.5 m and was located over 400 m offshore. It was expected to have an effect onshore similar to a breakwater, and a major consideration was how to balance the need to prevent development of a tombolo with the need to minimize construction costs, in that a longer offshore distance would result in a smaller tombolo but at greater costs (Kawaguchi *et al.* 1994). In many cases, a tombolo would be considered a desirable feature in that it would provide a means of gaining access to the new land, and it could provide a sheltered recreation beach.

Islands and artificial headlands are being built to hold recreational, sport and health complexes and to create artificial bays to stabilize beaches by making creative use of offshore breakwaters and underwater berms in combination with perched beaches (van Oorschot and van Raalte 1991; Grechischev *et al.* 1993). Some of the artificial structures have recreational beaches included in the designs. A prototype of this kind of beach is at Büsum in Germany where an artificial sand beach has been built for recreation on the landward side of a breakwater.

Neeltje Jans in The Netherlands (Figures 4.10 and 4.11) is one of the most

Figure 4.11 Artificially created dune on the bay side of Neeltje Jans, The Netherlands.

impressive artificial islands because of its size (>3 km long), its location in a high energy environment and its resemblance to a natural barrier island. The island was converted in a dredge and fill project from a natural sand bank in the Eastern Scheldt to a permanent artificial barrier island as part of the Delta Project for storm surge protection (Watson and Finkl 1990). The island contains a naturally functioning beach and a well-vegetated dune on the seaward side (Figure 4.10). A shallow inlet/overwash channel (slufter) was also created by leveling the artificial dune (van Bohemen 1996) in order to create a more diverse habitat. Landforms on the landward side (Figure 4.11) are even more interesting in that an artificial dune has been created in a location and with an orientation that would not occur on a natural barrier island, although it is shaped, vegetated and managed to resemble a coastal dune in every other way.

Wine Island, Louisiana provides an interesting case of reconstruction of an eroded barrier island. This feature was once an important resting area and breeding site for several species of shore birds, but it was little more than a shoal in 1990; it was modified by construction of a rock containment dike that was filled with sediment obtained through maintenance of a nearby navigation channel and planted with cordgrass, black mangrove and winter rye grass (Steller *et al.* 1993). The resulting feature is in some ways a large breakwater with subaerial habitat, much as Neeltje Jans is a large dike with subaerial habitat. These two islands reveal the enormous potential for constructing naturally functioning barrier islands where they do not exist or where they have existed but have been virtually eliminated by human or natural processes. The technology exists, and the funding can be made available, if the need exists.

Summary conclusions

Review of field investigations of structures reveals that they have pronounced local effects on erosion rates, shoreline configurations and landforms. Inattention to the actual or potential value of these landforms and to geomorphological and ecological criteria or broader recreational values (e.g., aesthetics) have restricted the value of many protection structures and buildings beyond providing a narrow range of design functions. As a result, many structures and practices have fallen into disrepute. Structures that replace natural features have an immediate and lasting negative effect on the value of shoreline resources, but structures need not have a long-term negative effect if natural processes and sediment transfers are allowed to continue in their presence. The decrease in support and approval for use of protection structures at the shoreline may not be the result of the incompatibility of these structures with natural processes. Technology can protect, increase and even create natural habitat (although a more predictable beach may also attract the kind of development that makes the shore unsuitable for wildlife habitat).

The specific design of a structure greatly affects its impact on coastal processes and landforms. A T-groin, for example, causes different effects on waves and currents and results in a different shoreline configuration from a straight groin, and an elevated house has different effects from a house built on the ground. Buildings and shore protection structures are so varied that they can be chosen for or adapted to work with the natural processes of the coast, but this kind of adaptation and fine-tuning is often proscribed by economic or time constraints. Like the beach nourishment alternative, evaluation of structures should be extended to include the opportunities they provide for maintaining, re-establishing or enhancing naturally functioning landforms that will extend the range of future benefits.

5

Characteristics of human-altered coastal landforms

Introduction

The great variation in types and intensities of human actions affecting landforms identified in chapters 1 to 4 results in great variation in their characteristics. These landform characteristics are evaluated in this chapter in terms of location, dimensions, orientation, topographic variability, sediment characteristics and mobility. The purpose is to place human actions (or non-actions) into perspective as either intrinsic or extrinsic to landform evolution and place landforms in perspective in terms of the degree to which they are natural features or artifacts. The temporal and political dimensions of landform evolution are addressed to show how human actions affect coastal conversions at scales of landforms and landscapes.

Locations of landforms

Beaches

A focus on erosion problems on fully developed shores, where beaches are narrow and the waves break close to coastal structures, conveys the impression that beaches in these areas are farther landward than they would occur under undeveloped conditions. Often, it is more true to say that the landward portion of the beach is farther seaward than its natural counterpart because human encroachment on the backbeach prevents the beach from moving inland at the expense of the developed upland.

Nourishment operations used to restore dimensions of beaches that previously existed and to create beaches where they would not occur under natural conditions extend the beach farther seaward, displacing the location of the breakers and surf farther offshore. Structures built for shore protection and navigation improvement may enhance accretion and displace the zone of wave action seaward in the accretion zone but accelerate the process of landward displacement in downdrift areas. Viewed in the long term, the net result of human action contributes to maintaining beaches within a more restricted zone than under natural conditions but also farther seaward.

Dunes

The location of dunes on the cross-shore profile in developed areas often differs from the location where they would occur under natural conditions because dune position is dictated by direct human actions and by the passive effects of existing human structures rather than the free interplay between vegetation growth, sediment supply and wave erosion. Dunes that form in developed areas that use large setback distances to control the location of fixed structures may mimic natural dunes in many of their characteristics. Dunes that form where human structures are close to the beach can take on many forms, determined by the nature of the structures and the actions of the local managers.

Figure 5.1 portrays some of the most conspicuous dune forms found on a coast where there is great variety of human structures and management, using New Jersey, USA as prototype. Dunes that form on the beach include incipient natural dunes (that are initially trapped by beach litter that contains seeds of pioneer species) and dunes that are trapped by structures. The incipient dunes would grow into foredunes on undeveloped shorelines, but this process is often prevented in developed areas because the litter and pioneer vegetation are eliminated during beach cleaning operations.

Dunes on the beach

Structures on the beach that function as sand traps include shore-perpendicular outfall pipes and groins and shore parallel seawalls and bulkheads (Figure 5.1). Dunes that form at these structures are often shortlived, either because the beach is narrow and the landform is soon eroded by waves or because the landform is removed in beach cleaning operations. The dunes that form in front of shore-parallel structures become ramp-shaped when winds blow oblique to the shoreline (Figure 5.1A) and, while they exist, facilitate transport of sediment inland to dunes landward of these structures (Figure 5.2). Dunes that accumulate on the landward portion of the backbeach in niches created by shore-parallel and shore perpendicular protection structures (Figure 5.1A) or at buildings that extend out on the backbeach may last longer and be better vegetated because they are better protected against wave erosion.

Foredunes that are intentionally created by artificial means usually are built for shore protection (Figure 5.1B). The location of artificially created foredunes is usually closer to the ocean than would occur under natural

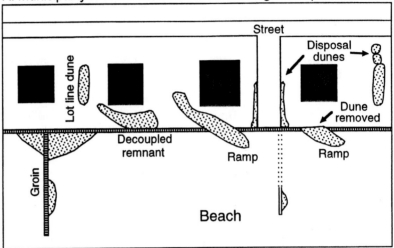

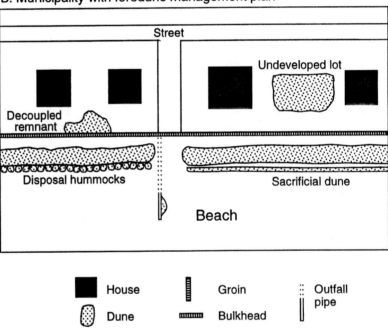

Figure 5.1 Locations of dunes that form in developed communities. The prototype for this figure is the coast of New Jersey, where there is a great variety of structures and human actions to alter the characteristics of dunes and their potential for growth.

Figure 5.2 Dunes on lots landward of bulkhead on Seven Mile Island, New Jersey. Photo by S.M. Arens.

conditions because beach width is restricted by structures on the landward side, and the dunes are built seaward of these structures for extra protection (Figure 1.10). A sacrificial dune may be created even farther seaward than the protective foredune.

The locations of foredunes created for shore protection in The Netherlands are more variable than in many other locations. They have been placed on the landward side of the crest, on the crest and on the seaward side of the crest (van Bohemen and Meesters 1992), and they have been bulldozed to the front of the secondary dune when the foredune became too small. Dune dikes are created to extend the foredune seaward or to create valued natural environments (e.g., dune slacks). Despite these creative dune placements, most new dunes created in locations where dunes already exist are placed in front of the existing dune line, often on the landward portion of the backbeach to create a sacrificial landform (Nordstrom and Arens 1998).

Dunes landward of the beach

Dunes may be located just behind shore-parallel protection structures, in front of buildings, between buildings and in undeveloped lots, either between lots with houses on them (Figure 5.1B) or in locations where development has been restricted alongshore, caused by divergence of the orientation

Figure 5.3 Dune south of Sete, France that has been decoupled from the beach source by shore-parallel road and protective seawall.

of the linear road network from the curving shoreline (Nordstrom *et al.* 1999). These dunes may be contiguous with the foredune (Figure 5.2) or they may be decoupled from it when the portion of the dune on the beach is removed by storm waves or by bulldozers to prevent besanding of inland properties. Dunes are often decoupled from the beach as a result of construction of shore-parallel roads, even in locations where there are no buildings (Figure 5.3). The loss of contiguity of a dune with the beach does not necessarily disconnect it from its sand source because the intervening hard surface enables aeolian transport.

The landforms that are allowed to survive on private properties on the landward side of the beach (Figures 5.2, 5.4) usually are incorporated into the cultural landscape; exotic species are often planted; and some of the exotics colonize portions of the dune. Remnant dunes often exist on undeveloped lots in developed areas (Figure 5.1B). These remnants often show many signs of degradation (Sanjaume and Pardo 1992). Many dunes are created at the side of lots where they are shaped by wind passing around houses. These lot line dunes (Figure 5.1A) may be augmented by residents dumping sand removed from other portions of the property, such as gardens and driveways. Other disposal forms that are not true dunes but may mimic them in size and surface cover include the disposal ridges adjacent to roads that are created during

Figure 5.4 Dune forms on private properties at Lavallette, New Jersey, showing difference in dune characteristics resulting from differences in resident preferences.

beach cleaning operations following inundation of roads by storm wave overwash or wind blown sand (Figure 5.1A). The result of these highly localized activities is a mosaic of cultural, natural and semi-natural forms landward of the bulkhead or dune crest (Nordstrom and Arens 1998).

Height

Beaches

Grading of natural dunes to create a better recreation platform may create a locally higher beach. Nourished beaches are often higher than natural beaches because they are designed to achieve protection goals (Kunz 1993a). Bankets constructed on the backbeach to function as a platform to accommodate recreation structures are higher than the backbeach that would occur as a result of normal annual wave conditions. Higher than normal beaches can occur updrift of shore protection structures. Local heights are increased, although often temporarily, by constructing mounds on the beach using earth-moving equipment to dispose of sediment removed from streets and lots, to build barriers to storm uprush or to construct recreational features, such as lifeguard platforms.

Lower than normal beaches occur in sand-starved locations downdrift of protection structures. Lower beaches also occur where shore-parallel walls landward of the beach prevent the upper beach profile from achieving its full cross-shore extent. Grading of beaches to create wider platforms and raking of beaches to eliminate litter also result in a decrease in beach height.

Dunes

Dunes in human-modified environments may be smaller or larger than natural dunes for a given source width or grain size because of active or passive human modifications. Dunes where management decisions are made at the local level with strong input from property owners tend to be narrow and low because the value of property close to the water and the desire to view the water from shorefront homes cause residents to demand a small dune. Human-modified foredunes are highest where safety is the principal value and government control is strong. Dunes in these locations may be built with the aid of sand fences and vegetation plantings and may be higher than dunes that would form behind narrow beaches under natural conditions. The protective dunes designed for the developed portion of Brigantine Island, New Jersey, for example, would have a top design elevation of 4.6 m, and the lowest elevation of the dune would be as high as the highest elevation in 1885, prior to development (Nordstrom 1994a).

Attempts to achieve adequate protection in The Netherlands have resulted in high crest elevations (Figure 5.5) that are, in many cases, considerably higher than the lowest elevations of foredunes under natural conditions. Sea defences on the central and southern coasts of The Netherlands are now designed to withstand a storm surge level of one in 10 thousand years; other sections of the coast have safety standards related to the value of the real estate and infrastructure of the shore (Koster and Hillen 1995), but the lowest prescribed level of protection is against a 1 in 2000 year event (Ministry of Transport and Public Works 1990). The only exceptions are natural areas in the Wadden Islands where there is no prescribed level of protection. The maximum heights of individual peaks following dune modifications may be higher or lower than pre-existing maximum peaks. Lower heights may result where the sediment in the peaks is used to augment low or narrow portions of the dune. Some dunes in The Netherlands have exceptional heights because of artificial fixation and may have peaks in the crest up to 43 m above the back-beach (Nordstrom and Arens 1998).

Human-modified foredunes are low where recreational values are favored over safety considerations, and they can be intentionally eliminated where the desire to enhance beach recreation dominates management decisions. The

Figure 5.5 Dune at Callantsoog, The Netherlands, showing high dune with limited topographic variability characteristic of dune intended to provide protection against high storm floods.

heights of dunes in many communities in the USA represent a compromise between recreation and protection; the bulk of the dune is built to adequate heights to provide protection against small storms but not interfere with views, and low points (often at the elevation of the backbeach) are allowed at intervals alongshore to favor beach access (Figure 5.1B).

Width

Much of the variation in beach volume or width along developed shorelines is now more a function of landscaping and protection structures on the backshore than differences caused by processes, sediment type or other natural factors (Kana 1993). Construction of buildings, parking areas and other use structures out onto the beach and prevention of onshore migration of the beach profile contribute to narrower beaches under developed conditions than under natural conditions. Beach nourishment may temporarily create a wider beach than could occur in the same location under natural conditions, but this greater width is temporary. Except in selected urban resorts (that are nourished frequently or "overnourished" to enhance recreation) and artificially created shoreline environments (where no beach formerly existed), it is unlikely that the average nourished beach is wider than occurred prior to development.

Dunes in developed areas are usually narrow, because of restricted beach widths and the proximity of structures landward of them. These dunes are often wider than could occur on a natural beach with similar (small) dimensions and sediment characteristics because of active human efforts to build dunes, but they are probably not wider than dunes that would exist in the same region in the absence of any human presence. Wide dunes exist in developed areas where accretion has occurred near shore-perpendicular structures, providing that regulations have prevented new development from encroaching on these accretion zones.

Orientation

Beaches that are directly affected by structures will achieve a new orientation (in plan view), whether this orientation is an intended outcome (e.g., groins, artificial headlands) or is unintentional (buildings that end up in the surf zone of an eroding beach). Protection structures transform longer natural beaches into a new set of smaller drift cells (Byrnes *et al.* 1993), producing more breaks in orientation per unit length of shoreline. As a result, human-altered shorelines can become less linear. Shorelines can become more linear where nourishment operations or attempts to grade existing beaches to enhance recreation bury groins, outfall pipes and natural headlands, but nourishment operations that are localized (Figure 3.7) and erosion hot spots on nourished beaches can create pronounced local breaks in orientation.

Dunes in developed areas are often more linear than natural dunes because a linear feature is less vulnerable to deflation and overwash, and management is simplified. Foredunes are linear whether they are created using sand fences, vegetation plantings, or earth-moving equipment. Artificially-created dunes generally follow the regional trend of the foot of the former dune and have a smooth orientation, even if the local orientation of the contact between the original dune and the backbeach was highly variable, although pronounced shore-perpendicular ridges can result from aeolian deposition at the fences placed for crowd control along access paths.

Topographic variability

Beaches

Beaches in human-altered areas usually have less topographic variability measured in the cross-shore direction. Nourished beaches are often constructed with a simple profile shape to facilitate construction and make it easier to calculate fill volumes (Figures 3.1, 3.4). Recreation beaches are often graded flat to facilitate beach access and use. Beach cleaning eliminates incipient dunes and prevents creation of new dunes or extension of existing dunes seaward. Truncation of the landward portion of the beach by shore-parallel structures results in limited potential for formation of storm berms and incipient foredunes that comprise the landward portion of the backbeach in natural environments.

Topographic variability of the beach in the longshore direction may be increased at shore perpendicular protection structures as a result of formation of local zones of accretion and erosion. Variability may be greatly increased at groins as a result of creation of rip cells, depositional lobes and interruptions to bar systems.

Dunes

In general, dunes have less topographic variability than their natural counterparts when they are formed by sand fences or earth-moving equipment, but they may have greater topographic variability where they form by aeolian processes at isolated or shore perpendicular structures or where their creation using earth-moving equipment is for disposal purposes. Actions taken by municipalities or higher levels of government usually result in a dune with decreased topographic diversity because landforms are designed to a common standard. Actions taken by individual residents, in contrast, may create considerable diversity at the scale of individual lots (10 to 30 m).

Topographic variability of foredunes that are managed to provide protection against flooding and overwash is generally limited, due to efforts to create a continuous barrier of similar elevation. A single narrow ridge is a common characteristic of human-created dunes fronting developed communities (Mauriello 1989). The major exceptions to maintenance of a uniform crestline are at the gaps used for access to the beach. The seaward side of dunes shaped by bulldozers is often initially as steep as it can be piled (about 40°), and it stabilizes at the angle of repose of dry sand. The tops of dunes created or shaped by bulldozers may be flat to retain adequate volumes of sand without overly high crests and to minimize initiation of blowouts. Slopes of dunes created by sand fencing are determined by the angle resulting from trapping, creating slopes that are much gentler (about 20 degrees) and rounder than bulldozed dunes. Many dunes in developed communities are largely unvegetated, and slip faces are common on the landward side of these dunes following strong onshore winds (Nordstrom and Arens 1998).

When a foredune is initially constructed in The Netherlands, the tendency is to create a gentle slope (1:3) by bulldozing the seaward side in order to facilitate planting and to reduce the likelihood of a large cliff forming during erosion events. The slope of these human-created dunes is never steeper than the slopes that are created by accretion at incipient vegetation under natural conditions (Nordstrom and Arens 1998).

Sediments removed from the beach during cleaning operations and dumped at the dune line result in a hummocky surface that reveals the method of mechanical deposition. The piles may be placed seaward (Figures 2.2, 5.1B) or (less typically) landward of the dune crest, either contiguously (creating a linear dune with a hummocky crest) or in isolated hummocks dumped at the closest suitable location. Piles placed on the landward side of the dune crest may be buried by subsequent aeolian activity, and the hummocky origin may be obscured. Disposal dunes placed in front of the foredune (Figure 2.2) may retain their hummocky appearance for months (Nordstrom and Arens 1998).

Management plans for some locations where dunes were stabilized or were graded flat have been changed to allow the dunes to be shaped by natural processes. Examples include: (1) allowing portions of a formerly raked beach to evolve naturally to preserve habitat for endangered species; (2) using earthmoving equipment to simulate natural dunes by creating varied topography and blending the contours with adjoining unaltered areas (Adriaanse and Choosen 1991; van Bohemen and Meesters 1992); (3) allowing bare sand in dunes to evolve naturally after removing human structures (Nordstrom and McCluskey 1984); (4) remobilizing dunes by removing the vegetation and surface soil (van Boxel 1997); or (5) suspending use of sand fences to repair breaches (Gares and Nordstrom 1991). These actions to create a more natural dune, with greater topographic diversity and greater mobility, have been taken in areas devoted to habitat enhancement or recreational uses that do not require maintenance of support facilities. The result is a dramatically different landform from the more typical linear, managed dune, although it may still bear many of the imprints of its human-modified form. Imprints of former human actions in dunes include blowouts that correspond to the location and dimensions of former houses (Nordstrom and McCluskey 1984); linear dune crests between newly formed blowouts (Gares and Nordstrom 1991); and piles of sand and soil removed from artificially created or remobilized blowouts that cannot be removed from the dune environment (van Boxel 1997).

The dramatic changes in topographic diversity resulting from suspension of raking can be seen at Ocean City, New Jersey (Figure 5.6). Spotting of plover

Figure 5.6 Re-establishment of pioneer vegetation and incipient dune at a protected nesting site for piping plovers on the backbeach of a nourished site at Ocean City. The bare-sand zone between the incipient dune and the foredune is due to destruction of vegetation and topography by off-road vehicles.

nests by the New Jersey Division of Fish, Game and Wildlife led to establishment of a small temporary preserve to prevent trampling by recreational use and loss through beach cleaning operations. The site was protected by warning signs and a symbolic fence that allowed the birds (and natural processes) to pass unimpeded but restricted human access. The result of this management practice was colonization of the backbeach by pioneer vegetation and evolution of a dune from an incipient form to one colonized by *Ammophila*, creating a hummocky topography and diversity of sedimentary environments and species. The new dune contrasts greatly with the adjacent mechanically groomed beach (Figure 5.6). The site does not look and function in a fully natural way because the enclave does not extend far across or along the shore. The new dune lacks integrity alongshore and lacks continuity with the foredune landward of it because the beach between these dunes is used as a transport corridor for beach vehicles. The site does not fully represent the kind of environment that could occur if the target environment were fully protected, but it does reveal the potential for re-establishing a naturally functioning system on a nourished beach that otherwise had limited natural value (Nordstrom *et al.* 1999).

Sediment characteristics

New materials are introduced to the beach matrix as a result of nourishment operations using sources outside the immediate beach environment. Sources from backbays, harbors, marinas and residential lagoons can yield sediments that are considerably finer than native materials. Materials dredged from channels may be coarser than native materials. Sources offshore and on ebb and flood deltas at inlets may approximate the mean size of native materials, but they may have different sorting characteristics. Sediments derived from river beds or upland quarries may differ in size, shape, mineralogy and degree of weathering (Pacini *et al.* 1997). Even where sediment is nearly ideal in textural composition, the method of handling the borrow material in its transportation to the fill site will affect its local textural composition (Swart 1991). Most changes from native sediment characteristics are not designed as such but result because the new source of sediments is the most cost effective for a given location.

Beaches

Dramatically different beach materials may be introduced in less traditional nourishment projects or projects that use the beach or nearby environments as disposal areas. Materials include crushed glass (Finkl 1996b), coral fragments dredged with sand (Wiegel 1992b), crushed rock mined from quarries (Rouch and Bellessort 1990; Wiegel 1993a), crushed rock waste from road and tunnel construction (Dzhaoshvili and Papashvili 1993) and from building sites associated with tourism development (Anthony and Cohen 1995). Exotic materials may be used simply because they are available or they can be sought out to provide special value to the new beach environment. Aragonite has been used to enhance recreational values of beaches in Florida (Bodge and Olsen 1992), and this option is likely to be pursued in the future by at least one local government (Beachler and Higgens 1992). Gravel beaches are likely to be favored over sandy beaches in the United Kingdom, potentially leading to a widespread replacement of sand by gravel (Bray *et al.* 1997). Pebbles have been used to prevent pollution from runoff being "absorbed" by fine sands (Yesin and Kos'yan 1993).

Natural reworking of sediment that is artificially produced or delivered to the area may produce landforms that mimic natural landforms in their form and function, but the color or texture may depart dramatically from native materials. At Solvay, Italy (Figure 5.7), the white color of the beach and dune and the light-blue color of the water that result from disposal of processed material lend an alien appearance to the landscape, despite the otherwise

Figure 5.7 The beach and dune at Solvay, Italy, showing beach and dune created by natural processes reworking sediments produced by plants in the river in the background.

natural appearance of the landforms. These nourishment and disposal operations are not widely reported in the literature, but they can have considerable local effect.

Large-caliber rocks from destroyed coastal protection structures remain in the beach matrix and become conspicuous features after the structures cease to have value (Guilcher 1985). In most cases, the problem concerns aesthetics or safety rather than a large-scale change in sediment characteristics. Cultural materials used in construction (e.g., concrete) may readily reveal their human origins, but most finer sediments will be inconspicuous. For example, finer fractions of quarry rock mined hundreds of kilometers from their location on the beach may appear natural as a result of weathering and abrasion (Guilcher 1985). The high energy of waves and swash in ocean environments ensure that most introduced materials will be reworked, and the resulting active portion of the beach will be natural in function if not appearance. The portion of the beach that is not reworked by waves may be unnatural in both function and appearance, and it may diverge even more through time as aeolian transport creates a surface layer of more resistant exotic materials.

Impermeable layers in fill materials and layers that are compacted by vehicles can cause poor drainage and contribute to ponding of water on beaches (Ferrante *et al.* 1992). Nourishment may also introduce shell particles of species not found on the beach (Hotten 1988) or change the temperature and drainage qualities of the beach that change its effects on fauna (Bodge and Olsen 1992). Scraping of the beach exposes sediments lower in the beach matrix and brings the surface closer to groundwater level. Bulldozing of the beach may eventually contribute to diminished rates of transport due to compaction after a few aeolian events have removed material from the loose piles (Nordstrom and Arens 1998).

Dunes

Dunes that form through aeolian processes around obstacles that are introduced by humans, called artificially inseminated dunes by Goldsmith (1989), have sediment characteristics and internal structures found in natural dunes. Dunes created by bulldozing or dumping from trucks or beach cleaning equipment, in contrast, have poorly defined internal stratification, and they may contain sediments that are too coarse or too fine to be aeolian deposits. Landforms emplaced by pumping in a sediment and water slurry reflect sorting by hydraulic processes rather than aeolian processes; sediments may be well sorted, but some fractions may be too coarse to be of aeolian origin. Exotic sediments may have a different color and texture. Sediments used as dune fill materials in New Jersey are often from upland quarries that have a source in glaciofluvial outwash and are of a yellow-brown or rust color resulting from iron oxide weathering. Walkover paths in New Jersey are often built using fine-grained materials to make them resistant to trampling and deflation. Regardless of source, foredune sediments may not be stratified if they are reworked by bulldozers following initial deposition (Nordstrom and Arens 1998).

Sediments used as fill materials in dunes in The Netherlands are usually derived from offshore, below 20 m depth and >10 km from the beach (Nordstrom and Arens 1998). These sediments are often coarse grained and shell rich relative to natural dunes (van Bohemen and Meesters 1992, van der Wal 1998) and may have different leaching characteristics that affect the flow of water and nutrients (Adriaanse and Coosen 1991). Formation of lag surfaces on the windward slopes of nourished dunes in The Netherlands has kept these locations from retaining sand and allowing marram grass to grow (van Bohemen and Meesters 1992; Nordstrom and Arens 1998).

Dunes may have cores of sediment that could not have resulted from aeolian transport but may be capped by sand resembling dune sand (Tippets and Jorgensen 1991). These cores may be designed specifically to resist erosion (d'Angremond et al. 1992; Dette and Raudkivi 1994). Concrete cores have been placed in the foredune in The Netherlands, for example at Schouwen and

Delfland (Technische Adviescommissie voor de Waterkeringen 1995), and clay cores have been used at Cape May Point in New Jersey. Dunes also have been built to function as veneers over other protection structures, such as at Atlantic City, New Jersey, where bulldozed beach sand and vegetation cover tubes filled with sand (Figure 1.9) and at Zeeuws-Vlaanderen and De Brouwersdam in The Netherlands (where artificial dunes cover asphalt dikes). Locally, World War II era bunkers in The Netherlands are covered with sand using bulldozers (Nordstrom and Arens 1998). Low revetments in Denmark are normally covered by sand as a result of nourishment of the backbeach, creating a structure that looks like a narrow dune (Laustrup 1993).

Mobility

Mobility is the key to ensuring that the beach environment will retain ecological values and many human use values because dynamism of the beach is responsible for its physical characteristics, its temporal and spatial diversity and its aesthetic appeal. It is a paradox that stability of the beach becomes the goal once infrastructure is in place to make use of it. Managers of natural areas also succumb to adopting counter-productive static management practices when they direct attention to preserving the inventory of natural features within their management units rather than to preserving the processes that created these features.

Beaches

Most direct human action is designed to reduce the long-term mobility of the beach in the landward direction (to protect infrastructure) and in the longshore direction (to retain beach volume or to prevent excessive sedimentation downdrift), but mobility can be greatly increased in the short term by human actions. High rates of mobility can occur locally as an unwanted byproduct of shore protection operations or navigation improvements (e.g., the effects of creative dredging at Townsend Inlet, described in chapter 2). Shoreline reorientation due to construction of protection structures is rapid but eventually a new equilibrium is achieved and subsequent mobility is greatly reduced. Mobility associated with beach nourishment (both wave- and wind-induced) can greatly exceed natural rates, especially where the new beach is higher and wider than the natural beach and the fill materials are finer than the native materials.

Control of inlets through construction of jetties or channel dredging has decreased natural fluctuations in shorelines at considerable distances from these inlets. The human actions at inlets on the southern New Jersey coast, described in chapter 4 and depicted in Figure 4.5C, resulted in mobility rates updrift of the inlets that were 22.7 percent of rates prior to development, while rates downdrift in the inlet throats were 53.6 percent of undeveloped rates. There was a slight increase in the mobility of the oceanside shoreline downdrift of the inlets (1.15%), but the overall net effect appears to be a conversion from cyclic to unidirectional change in some locations and a general reduction in shoreline mobility in others (Nordstrom 1988a).

Dunes

Aeolian transport rates may be greatly increased during construction phases of building programs, when stabilizing vegetation is removed. González-Yajimovich and Escofet (1991) showed that three times as much sediment was transported landward of a barrier undergoing initial development than in the adjacent area that had initial characteristics similar to the altered area. Once dunes in developed areas are shaped according to human needs, attempts usually are made to protect them in place to retain their utility, and the dunes may become less mobile than their undeveloped counterparts (Nordstrom 1990). This conclusion applies to the dune crest and the landward side of the dune. The practice of constructing new protective foredunes well seaward of the previous foredunes or the bulkhead line or boardwalk (Figures 1.10, 5.1) enhances the mobility of the landform due to wave erosion.

Coarse particles can result in a surface lag that resists deflation and alters the mobility of the dune (Baye 1990), but most restrictions to dune mobility are due to structures. These structures include the fences emplaced to stop or retain wind blown sediment or the buildings and shore protection structures that provide barriers to dune migration.

Distinguishing natural from human-created landforms

Ranking landforms by degree of naturalness

A distinction can be made between landforms that are created intentionally to provide specific utility functions (intended or designed landforms) and landforms that are created by natural accretion as a result of human activity designed to accommodate other utility functions (unintended or opportunistic landforms). Humans can be considered intrinsic agents in evolution of intended landforms and extrinsic agents in evolution of unintended forms. Beaches and dunes created directly by human action at beach nourishment sites are designed as such. Beaches that form downdrift of these sites as a

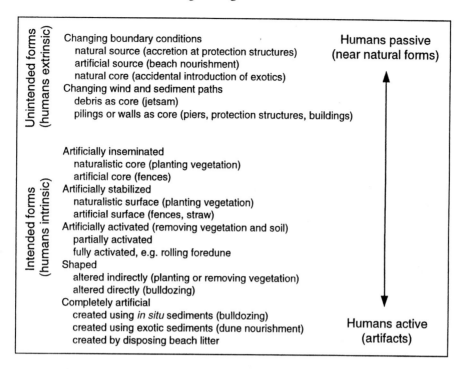

Figure 5.8 Ranking of dune characteristics from most natural forms (top) to least natural (bottom). Modified from Nordstrom and Arens (1998). This figure does not identify forms that result from unintended actions where humans can be considered intrinsic (e.g., dunes that are artificially activated by trampling).

result of additional sediment inputs and dunes that form landward of these sites as a result of increased aeolian activity are opportunistic landforms (Nordstrom and Arens 1998).

Figure 5.8 distinguishes between unintended and intended dune forms, ranked in terms of the degree that humans are active participants in dune creation. Dunes are sedimentary landforms, and the categories are structured to follow a general progression from top to bottom based on degree of manipulation of the sedimentological and geomorphological process. The most natural forms (top of figure) occur where human actions are associated with changes in boundary conditions (beach nourishment, accretion at structures, accidental introduction of exotics) or changes in wind and sediment paths that affect dune development (jetsam on beach, structures) The actions that are intentional include changing the susceptibility of the surface to erosion or deposition (by planting or removing vegetation, trampling, installing fences) and changing the dune volume directly (by disposing litter, bulldozing, nourishing the dune) (Nordstrom and Arens 1998).

Figure 5.8 represents an ordering according to the conditions that create the initial landform. Direct human action may affect the short-term shape or relative position of dunes on the beach profile. Either natural processes or human actions may be dominant in the long term. There is considerable opportunity for aeolian processes to modify the foredunes after they are created, even in locations where exotic materials are dumped and shaped by bulldozers (Figure 2.2). One season may be all that is required to establish an aeolian veneer over a landform that is initially created using earth-moving equipment (Nordstrom and Arens 1998).

The appearance or functions (e.g., recreation, habitat) of the dunes are not evaluated in Figure 5.8, and it is likely that certain types of artificially manipulated dune would have greater perceived natural value than certain types of dune that formed by natural processes. A dune that forms through deposition around exotic species may be perceived to be of less natural value than the same dune that has been subsequently "improved" by removing the exotic species and planting indigenous species to create a specific biological inventory. A dune that is created directly by nourishment and planted with native vegetation (an artifact) may be perceived as having greater natural value than a dune that forms when sand blown across a natural beach is trapped at a shore protection structure (a true dune). The structure of Figure 5.8 could be rearranged to create a function-oriented continuum that would have greater use in interpreting the value of human-altered dunes for management and evaluating the rationale for the human processes leading to landscape conversion. Recognition of the many ways that human alterations contribute to our conception of a dune and its value will help in developing more sophisticated models for foredune evolution that place human alterations in their proper perspective (Nordstrom and Arens 1998).

Defining human-altered landforms

The term "beach" and the term "dune" have been applied to features of both natural and human origin, although the terms may not be literally correct. The backbeach built during a nourishment project is neither created nor shaped by waves; many dunes in developed areas are at least partly created by non-aeolian processes; and many dunes created for shore protection could be termed sand dikes rather than dunes. The terms "sediment mounds," "artificial berms," "embankments" and "dune ridges" have also been used to describe these features (McNinch and Wells 1992; Hotta and Horikawa 1996).

The definition of a dune is often somewhat arbitrary. The landform created by artificial nourishment at Schouwen in The Netherlands (Figure 3.8) was called a beach nourishment project, but the resulting feature is as high as a dune and it provides functions similar to those of dunes because of its location behind the beach (Nordstrom and Arens 1998). In contrast, the raised barriers placed on the landward portion of nourished beaches to provide extra protection against overwash (dune dikes or berm caps) are often called dunes although they may be only 0.6 m above backbeach elevation (Figure 3.5).

Some geomorphologists may argue that new terms should be devised for human-altered landforms, while others may resist using new terms in order to avoid proliferation of jargon or to avoid polarizing opinion about the suitability of human alteration of landforms. Nordstrom and Arens (1998) argue that use of the term "dune" to describe the feature that occupies the position that a dune would occupy on a shoreline subject to only natural processes is considered appropriate if the landform has similar form and functions (natural and human) to the natural feature and if it is modified by aeolian action, even if this modification is restricted to a surface veneer.

Other definitions of dunes in human-altered environments reveal the critical role played by the function of the landform. The 1982 Coastal Primary Sand Dune Protection Act of Norfolk, Virginia, USA defines a primary dune as a mound of unconsolidated sandy soil contiguous to high water that must rise sharply from the adjacent terrain and have one of a list of ten species of plants on at least a part of it (Blair and Rosenberg 1987). This definition retains some of the sedimentological and ecological characteristics of natural dunes. The state of Massachusetts, in contrast, uses a definition that can include gravel barriers – features that cannot be shaped by wind and that have different ecological and protective values from natural dunes. Regulations in New Jersey define a dune as a wind or wave deposited or man-made formation of sand (mound or ridge) that lies parallel to and landward of the beach (New Jersey Department of Environmental Protection 1996). The changing conception of the definition of foredunes from natural features to artifacts is part of the recognition of the role that humans play in the evolution of coastal landforms and the changing role that landforms have in the coastal landscape (Nordstrom and Arens 1998).

The historical dimension

Problems of interpreting the past

Human alterations may be of such large scale or may have occurred for so long in the past that they may be forgotten. Visitors driving the streets on Rockaway spit, New York or Atlantic City, New Jersey can easily forget that these cityscapes are on coastal barriers that owe their origin to wave and wind processes and are still vulnerable to them. Few visitors (or residents) realize that developed portions of coastal cities (e.g., San Francisco) are built on coastal dune fields or that foredunes (and some secondary dunes) have been consciously eliminated or prevented from forming landward of the beach (Nordstrom and Psuty 1983; Sherman in press).

Before development and subsequent human action, neither Miami Beach nor many parts of southern California had the sandy beaches that subsequently gave them their special recreational value. Miami Beach appears to owe its former beach to accretion in the 1910s and 1920s following construction of the inlet at Government Cut and subsequent stabilization with jetties (Wiegel 1992b) and its present beach to a massive nourishment project. The beaches in southern California owe much of their considerable value to massive quantities of sediment added by humans in the past (Woodell and Hollar 1991; Flick 1993; Leiderdorf *et al.* 1994). Artificial beaches on the French Riviera blend in so well with natural beaches that they may not be viewed as an artificial shoreline type by the public (Anthony 1997).

There are now virtually no dune systems in Great Britain that can be considered entirely natural, although most visitors would consider them so (Doody 1989). Mather and Ritchie (1977) note that no area of dunes or machair in the remote highlands and islands of Scotland is free of human influence, with the possible exception of a few minute patches at the foot of cliffs (although human artifacts can accumulate as jetsam in these areas), and there are no natural beach complexes that can be used as ideal models for management of agricultural or recreational beaches (Mather and Ritchie 1977).

Corona *et al.* (1988) report that human intervention has been so profound in the presently undeveloped Doñana National Park in Spain that the ecosystems appropriately may be termed "man made" landscapes. The newly managed landscape in some sites bought by the French Conservatoire de l'Espace Littoral show aspects that are different from the natural ones, especially where local managers have cut the grass. The dunes have been protected from destruction or intrusion of buildings, but they have been turned into meadows that are part anthropic features (Guilcher and Hallégouët 1991).

The conversion of landforms by humans often can go unnoticed in coastal locations that lack the highly visible structures that are most closely associated with human alterations. There are situations where seawalls are buried under dunes during times of sediment abundance and only function once again during erosional phases of long-term cycles (Kraus 1988). Dunes may be removed to accomplish a specific time-dependent objective and restored to appear similar to the way they appeared prior to alteration. Examples include accommodating landings of offshore pipelines (Ritchie and Gimingham 1989) and removing unexploded ordnance (Demos 1991).

Alteration of the character of the previous natural sediment by introducing exotic materials in beachfill may eventually prevent interpretation of provenance and ancient clastics, obscuring interpretation of natural coastal evolution (Galvin 1991). The resemblance of sand color to local sediments has been used as a criterion for acceptance of material for beach and dune nourishment (Ulrich 1993), and there are calls for compatibility of beachfill materials for aeolian transport, not just for beach stability (Löffler and Coosen 1995). Problems with contamination by exotic sediments prevent use of beach nourishment in locations where the sediment characteristics are important to the integrity of the landform, as at Chesil Beach, UK (Bray and Hooke 1995). In some cases, native sand may be stockpiled and later mixed with nourished sand to create a color that more closely approximates the pre-nourishment sediment (Ferrante *et al.* 1992). In contrast, sediments with visual qualities that are dramatically different from local materials may be preferred on aesthetic grounds (Bodge and Olsen 1992), and exotic sediments have initial potential value as tracer material for interpretation of rates and pathways of sediment transport by waves and currents (Wiegel 1992b; Dornhelm 1995).

Losses associated with truncating the temporal dimension

The conversion of landforms by humans often can go unnoticed. Humans have a capacity to remain satisfied with an environment, even as it changes or degrades around them (Dustin and McAvoy 1982). So much of the vegetation of the earth has already been modified or destroyed by biological invasions that we have become accustomed to accepting secondary or artificial vegetation and landscapes as though they are present norms (Heywood 1989). As time passes, more of the natural landscape becomes susceptible to conversion; human memory of natural landform characteristics fades; and documentation of these characteristics becomes more difficult to find. The landscape may remain of considerable value to humans, but the interactions of processes and the resulting landforms that characterized the natural system are modified, and the intrinsic value of the physical landscape is gradually lost (Nordstrom 1990). Every reduction of nature has the potential for augmenting and reinforcing other reductions, and landscape conversion results in a downward spiral in which our ideas about it are influenced as much by the state of the world as vice versa; objections to the prevailing losses are viewed as unrealistic to the extent that the prevailing reductions are self-validating (Weston 1996).

Human-altered dunes are stabilized in any form people perceive to be appropriate for them, whereas natural dunes become stable only at the fully mature stage of succession. Much of the variety of natural features is associated with cycles of growth and decay, commonly seen in coastal dune fields (Wiedemann 1984; Wanders 1989). Part of the value of natural areas is their causal history, their linkage with the past (Katz 1992). What some managers view as instability is really a valuable glimpse of geomorphic evolution. When the time dimension is truncated, individual identity is lost, and the linkage between process and form is obscured (Nordstrom 1990).

The political dimension

Undeveloped dunes reveal great local variability alongshore, related to differences in distance from inlets, sediment sources, beach widths and kinds of vegetation, and they are characterized by rapid mobility and a variety of shapes and sizes within short distances (meters to kilometers). Human-altered landforms exist at a variety of spatial scales, but their boundaries are inevitably defined by human values. Thus, spatial differences in landform characteristics along the shoreline may conform to jurisdictional boundaries reflecting sociopolitical differences at the local scale (Mauriello and Halsey 1987; Guilcher and Hallégouët 1991) or even international scale (De Raeve 1989) rather than to geographical variation in natural processes or sediment budgets (e.g., Davies 1972; Psuty 1992).

The effects of differences in political factors on the characteristics and evolution of dunes at large scales (international and intercontinental) is identified by Nordstrom and Arens (1998) in a comparison of dune management in The Netherlands and the USA (focussing on New Jersey). Foredunes in the USA, are perceived by planners and managers to have great value for shore protection, but the required level of protection is lower than in The Netherlands. In the USA, the the transmittal of political power to the communities (identified in chapter 7), and the need to levy funds on a site-specific basis while accommodating the needs of residents, results in a great variety of activities conducted in the dune zone, manifested in great variety of sizes, shapes and methods of construction and alteration of dunes. In the absence of wide protective beaches, cycles of foredune destruction and rebuilding occur frequently and often on an *ad hoc* basis. There is reason to expect that residents and municipal managers in communities in the USA will continue to act on their personal perceptions of the resource value of dunes. Foredunes in developed areas are maintained as artifacts and are frequently manipulated, so humans are considered intrinsic agents in their formation and evolution (Nordstrom and Arens 1998).

Larger foredunes occur in The Netherlands due to the greater frequency of onshore winds (that have greater potential to deliver sediment to the dune) and the greater emphasis on dunes as sea defense. Emphasis on sea defense ensures that: (1) beaches will be maintained through nourishment; (2) higher standards of protection will be retained; (3) responsibility for maintenance of foredunes will be vested at the national level; and (4) regular funding for protection projects will be provided. As a result, the foredunes in The Netherlands are higher, wider, better vegetated and not subject to change at the hands of residents and municipal managers. The large-scale dune-building projects that have occurred in the recent past may not be conducted frequently in the future, because the required level of protection has been achieved in most locations, and future protective efforts will likely come in the form of beach nourishment (van Bohemen and Meesters 1992). Designation of foredunes as backup protection allows managers great flexibility in their approach to management. The wider beaches and dunes provide greater potential for ecological value and allow for greater acceptance of modification by natural processes. Human actions affecting foredune evolution in The Netherlands may now be extrinsic at the scale of the landform itself, but the conditions that affect growth, maintenance and ability to survive wave attack are a direct result of human input, and humans may be considered intrinsic at the landscape scale (Nordstrom and Arens 1998).

Summary conclusions

Developed landforms, left unmanaged, respond to energy inputs according to the same laws as natural landforms, but differences occur between the two landform types in terms of the mechanisms of change, freedom of movement, locations of sources and sinks for sediment, internal structure, outward appearance, spatial relationships and the temporal scales associated with cycles of change that are identified in chapter 6. Much of the variation in beach volume and width along developed coasts is the result of human action, not the function of natural factors or processes. Preservation of coastal landforms should focus on the processes that produce landforms rather than the landforms themselves.

The criteria that define a dune on a human-altered coast are somewhat arbitrary; they vary from place to place and over time, attesting to the integral role that humans have in altering both coastal landforms and coastal landform functions. The qualities that differentiate natural from human-modified landforms may be expressed as a continuum, but it is difficult to specify location along this continuum because of the complex interaction of human and natural processes. Indeed, dunes can be viewed as forming a continuum on several axes of differences, including the relative dominance of process as either natural or human or the functional role of these landforms as habitat or amenity.

Return to a truly natural system is likely to be an elusive goal (as noted in chapter 7), but there are many ways that human actions can be made more compatible with natural processes and with the landforms and ecosystems created by them. This chapter identifies characteristics of human-altered beaches and dunes that have already been created. It is also important to focus on coastal locations where natural landforms no longer exist but could exist if they were allowed to form. Suggestions for human actions to create viable coastal landforms in these locations are presented in chapter 8.

6

Temporal scales of landscape change

Introduction

Many of the scale effects of differences between natural and human-altered landforms are treated in earlier chapters, where the emphasis was placed on the initial modification to the landforms or to their characteristics at one point in time. This chapter discusses the evolution of landforms and highlights changes associated with major meteorological and climatological inputs, contrasting the ability of natural processes to cause coastal changes and the ability of human agency to reconstruct the cultural landscape. Temporal cycles of change related to major storms, seasonal changes in wind and wave intensities and long-term changes in sea level are contrasted with cycles related to human uses and management. Alternative models (or scenarios) of change for developed coasts are then examined to indicate the different ways that the role of humans in landscape change is perceived and evaluated.

Effects of storms on evolution of developed coasts

Perspective

Types of studies

There are many studies of the geomorphological and engineering implications of specific storms (Hayes 1967; Dolan and Godfrey 1973; Morton 1976; Penland *et al.* 1980; Dean *et al.* 1984; Nakashima 1989; Finkl and Pilkey 1991; Kraus 1993; Finkl 1994; Stone and Finkl 1995), and at least one major study of shoreline changes is published following each hurricane in the USA (Morton 1976). Most of these studies are reconnaissance-level investigations that are conducted within a few months of storm passage, and descriptions of post-storm recovery are often limited to a paragraph or two near the end of a litany of damages (Nordstrom and Jackson 1995). Post-storm studies often monitor the beaches for a period of no more than 1 or 2 years (Morton *et al.* 1994), and the elapsed time following the storm usually is too short to obtain meaningful conclusions about the recovery process (Nordstrom and Jackson 1995; Valiela *et al.* 1998).

Dramatic effects of storms

The emphasis on storm damage is not surprising because of the great economic costs of the losses and the spectacular nature of alterations to the physical and cultural landscape. For example, 36 percent of the 1056 buildings in Gulf Shores, Alabama were destroyed by Hurricane Fredric (Rogers 1991b); Hurricane Hugo resulted in losses exceeding US$7 billion on the southeast mainland coast of the USA (Finkl and Pilkey 1991); Hurricane Andrew resulted in property losses greater than US$10 billion in the southern USA; and the Halloween Storm of 1991 caused property damage greater than US$1.5 billion on the Atlantic coast of the USA (Dolan and Davis 1994).

The effects of Hurricane Hugo on the coast of South Carolina reveal the ability of storms to alter the physical and cultural landscape. Data in Stauble (1991) indicate that overwash sand from this storm was transported 15 to 250 m inland. Buildings in one area were transported an average of 120 m inland. Flotable debris was transported across coastal barriers to the mainland shore 1000 m landward. Most of the primary dune was eroded to the elevation of the beach. Several incipient inlets formed in the narrow (<150 m) barrier island segments. Most shore protection structures were not designed to withstand a storm of this magnitude, and a large proportion of these structures were heavily damaged or destroyed, even up to 140 km away from the eye of the storm. Overwash occurred between gaps in protection structures, causing flanking scour and damage to structures behind them (Stauble 1991).

Storm debris is a conspicuous element in the landscape following storms (Armstrong and Flick 1989). The chaotic landscape revealed in post-storm photographs of areas impacted by Hurricane Hugo (see Finkl and Pilkey 1991) reveals the disequilibrium between human facilities and natural processes that occur during major storms. The immediate effect of storms can be elimination of most evidence of pre-storm morphology (Dolan 1987; Fowler et al 1993).

Alternative view of long-term effects

A concentration on the post-storm landscape, rather than the post-reconstruction landscape provides a distorted view of the dominance of natural processes *vis-à-vis* human agency in the long-term evolution of coastal landscapes, making storms seem more important than they are. Destruction of natural landforms, coastal vegetation and fauna by storms is usually viewed as part of a cycle of events that includes restoration by natural processes (Godfrey and Godfrey 1973; Gardner *et al.* 1991). Destruction of buildings, in contrast, is viewed as a disaster and evidence of the incompatibility of human alterations, despite the evidence that the cultural landscape is restored, often at a more rapid rate than natural processes can restore natural features (Nordstrom 1994b; Nordstrom and Jackson 1995). Without a perspective on the way the coastal landscape is converted to a human artifact following storms, conservationists and managers may overestimate the ability of coastal storms to restore critical natural features or overlook the opportunity or the need to restore natural values during post-storm reconstruction programs (Nordstrom and Jackson 1995).

Case study of landscape change following storms

Nordstrom and Jackson (1995) extended the time scale of analysis of storm effects to three decades to determine whether storm processes or human actions are dominant over this time frame. They examined multiple storm events and impacts of restoration efforts on susceptibility of the human-altered landscape to changes during subsequent storms. They examined two sites (Harvey Cedars on Long Beach Island and Whale Beach on Ludlam Island) that have been among the most vulnerable locations in New Jersey to storm damage in recent decades. The major storms during that interval occurred 6–7 March 1962, 28–29 March 1984, and 11–12 December 1992.

Effects of March 1962 storm

The greatest geomorphological changes and damage to buildings in New Jersey during the 1962 storm occurred on Long Beach Island. A veneer of fresh sand was deposited everywhere on the surface of this barrier, with major overwash fans occurring where street ends were located. Overwash penetrated to the bay; five breaches occurred in the barrier (four at Harvey Cedars); 5361 residences were damaged by flooding; and 998 had structural damage. A total of 2272 residences were damaged by flooding on Ludlam Island, with 668 suffering structural damage. Nearly all of the dunes along the entire island were destroyed, and all public utility systems failed (USACOE 1962, 1963). All of the buildings on the seaward side of the main shore-parallel road at Whale Beach were destroyed; overwash penetrated up to 185 m inland from the shorefront road; and underlying peat layers were exposed on the beach (Nordstrom and Jackson 1995).

Post-storm reconstruction activities indicate how rapidly restoration efforts are conducted. The President of the USA declared coastal New Jersey a disaster area only 2 days after the storm reached its height. Emergency activities to clear debris and repair public facilities were in progress after only a few days. Closing of the breaches at Harvey Cedars was accomplished in 2 days; beaches and dunes were restored in about 4 months (USACOE 1963). The Corps of Engineers placed 547 000 m³ of sediment on Long Beach Island, much of it in the vicinity of Harvey Cedars (USACOE 1993), completing the project within 6 weeks of the storm. New groins were built by the Corps at an average spacing of every 300 m along the entire shorefront of Harvey Cedars and adjacent communities. Eighty-three of the 110 groins in existence on Long Beach Island by 1972 were built or rebuilt following the 1962 storm (Everts and Czerniak 1977). Just 8 years after the storm, many houses had been built or rebuilt; many of the houses were farther seaward than in 1962; and the beach was 20 m wider in 1970 than it was after the post-storm nourishment in 1962 (Nordstrom and Jackson 1995). Post-storm activities by the Corps of Engineers on Ludlam Island included emplacement of 6 924 000 m³ of fill. No attempt was made to rebuild houses back in locations seaward of the shorefront road at Whale Beach, but a dune was built using fill materials from a source outside the area.

Effects of March 1984 storm

Whale Beach had some of the worst erosion and washover of any community along the New Jersey coast during this storm. The artificial dune system there was almost completely destroyed; the shorefront road was inundated with sand; numerous residences were inundated and had structural damages; and flooding occurred up to 1.8 m above the ground surface in low-lying areas (USACOE 1985). The post-storm cultural landscape at Whale Beach revealed on aerial photographs resembled the landscape after the 1962 storm (Nordstrom and Jackson 1995).

The President declared the New Jersey coastal counties a disaster area the following month, although storm damage was slight. Sand washed onto roads and storm debris was quickly cleared, and the majority of the affected communities in New Jersey were ready for the tourist season, only 2 months after the storm. A beach fill and dune construction project was later implemented, involving 453 000 m³ of fill (USACOE 1985), and the new artificial dune at Whale Beach (Figure 6.1) was larger than the former artificial dune had been prior to the storm (Nordstrom and Jackson 1995).

Effects of December 1992 storm

This storm caused severe erosion of the dune along the entire town of Harvey Cedars, removing about 91 800 m³ of sand. Some dunes were completely eroded, exposing the foundations of oceanfront homes; other houses were in the surf zone. The entire dune was eliminated at Whale Beach, result-

Figure 6.1 Artificial dune at Whale Beach, New Jersey, 1987. The dune was built by dumping sediment following a storm in 1984. Natural aeolian accretion at vegetation on the seaward side and at the sand fence on top of the dune contrasts with the deflated hummocks to the left that have a surface lag layer of shell.

ing in a post-storm landscape similar to the one occurring after the March 1962 and 1984 storms (Nordstrom and Jackson 1995).

The President declared the New Jersey shoreline a disaster area 6 days after the storm. Sand was transported by truck to Harvey Cedars from an upland source (USACOE 1993), and sediment was bulldozed from the beach into the dunes to create a new foredune in only a few weeks. Post-storm activities at Whale Beach included bulldozing sand from the road back to the former location of the dune, and the state subsequently replaced the dune using fill material from outside the area (Nordstrom and Jackson 1995).

Net effect of storms

These storms caused considerable property damage, but little lasting geomorphological effect. There is presently no conspicuous evidence of landforms created by the three storms at Harvey Cedars, and the present seawardmost construction line is nearly at the same location it was prior to the 1962 storm. The net effect of the March 1962 storm at Whale Beach was elimination of numerous buildings seaward of the shorefront road and creation of new substrate over the marsh landward of the shorefront road, but the storm did not re-establish the dominance of natural processes. The overwash platform created at Whale Beach by the 1962 storm remains a conspicuous feature in the coastal landscape, given its elevation, but it does not have a natural surface cover or function. This feature is not as dynamic as it would be under natural conditions, and the higher substrate has been used as the site of several new houses. The artificial dune, in its several iterations, limited further modification of the barrier by both overwash and aeolian transport. It has not migrated inland, as it would under natural conditions, and there have been no new cycles of landform evolution or vegetation growth. The overwash platform that is not covered in houses or parking areas is colonized primarily by *Phragmites australis*. This species is not commonly found at this density at this location on naturally migrating barrier islands, and it appears to be a result of human occupancy (Nordstrom and Jackson 1995).

The large number of shore protection projects implemented after the 1962 storm at Harvey Cedars and at other sites in New Jersey (revealed in Figure 1.6) reduced the likelihood of pronounced geomorphological effects of future storms. Many of the new buildings at Whale Beach and Harvey Cedars are now more elaborate than those built prior to 1962, and they are elevated on pilings to reduce the potential for damage during storms. The elevated buildings provide less interference with natural processes than buildings on the ground. This aspect makes them more compatible with restoration to a natural setting, but active human alterations to the ground surface (especially those designed to minimize sedimentation in the vicinity of the buildings) override any ecological or geomorphic benefit of constructing the buildings above the surface of the ground (Nordstrom and Jackson 1995).

Landscape changes following storms

The time scale of reconstruction efforts

The normal sequence of events following storm alteration of natural systems is creation of a post-storm landscape followed by evolution of landforms according to natural processes. Morton *et al.* (1994) identify 4 time-dependent stages in recovery of natural beaches after erosion by a severe storm. The stages include: (1) berm reconstruction and beach accretion (lasting a few months to a year); (2) backbeach aggradation by flooding and storm berm creation during small storms and aeolian transport; (3) dune formation; and (4) dune expansion and vegetation recolonization. Forebeach recovery was similar in undeveloped and developed areas in their study area because filled lots on the backbeach did not interfere with berm construction, but natural recovery of the backbeach and dunes was impeded by houses (Morton *et al.* 1994).

Many human alterations are accomplished at a far more rapid pace than

Table 6.1 *Temporal scales of selected storm changes and post-storm alterations in New Jersey and other locations as noted*

Natural alterations	Restoration time
Beach recovery	several weeks to 1 year (Davis *et al.* 1972; Sexton 1995); >1 to 2 years (Rodríguez *et al.* 1994; Fletcher *et al.* 1995)
Filling of erosional scarps	weeks to months (Sexton and Hayes 1991)
Formation of small dunelets	several months (Sexton and Hayes 1991)
Creation of dune ridges	1 year (Sexton and Hayes 1991)
Recovery of dunes resistant to overwash or dunes of pre-storm size	5–10 years (Hosier and Cleary 1977; Ritchie and Penland 1988); 3 years (Sexton 1995; Hesp and Hilton 1996)
Re-establishment of incipient vegetation	<1 year (Sexton and Hayes 1991)
Re-establishment of pre-existing cover	3 to 4 years (Judd *et al.* 1991; Sexton 1995)
Inlet closure	<1 year to >4 years (Sexton and Hayes 1991; Sexton 1995; Valiela *et al.* 1998)

Human-induced alterations	Restoration time
Removal of sand from roads	Within 3 weeks (Nordstrom and Jackson 1995; Meyer-Arendt 1991)
Emplacement of sand bags	During storms
Emplacement of rip-rap	During storms and subsequent weeks (Griggs and Johnson 1983)
Litter cleanup	<2 months (USACOE 1985)
Beach scraping to create berms and dunes resistant to overwash	Days to weeks (Katuna 1991)
Replanting vegetation	Months
Re-establishment of pre-existing cover	5 to 10 years (van der Putten and Kloosterman 1991)
Installing sand fences	Months (Katuna 1991)
Construction of groins, bulkheads	<2 years
Inlet closure	Days to months (USACOE 1963; Terchunian and Merkert 1995).

would occur under natural processes (Table 6.1). Cultural debris and sand washed onto roads is removed almost immediately; inlets are closed artificially; sand lost from dunes is replaced by sediment scraped from the beach or sand fences and vegetation plantings are used to trap wind blown sand; sand removed from the upper beach is replaced by sand scraped from the lower beach or nearby accreted areas; and nourishment projects restore the width and volume of the beach (Bush 1991; Katuna 1991; Nelson 1991; Stauble *et al.* 1991).

These human modifications lower the degree of vulnerability to changes during subsequent storms, especially those of lower magnitude. Although a new berm and dune may build up in undeveloped areas by natural processes after a year, these natural features may be smaller than their pre-storm size (Morton *et al.* 1994), and they may be ineffective in stopping periodic flooding that reactivates and increases the landward extent of washovers formed during the earlier storm (Sexton and Hayes 1991). The slower recovery time of natural landscapes to loss of dunes during large storms can perpetuate overwash conditions during smaller subsequent storms or contribute to continued net loss of dunes during subsequent storms (Hosier and Cleary 1977; Morton *et al.* 1994).

Long-term implications

In human-altered systems, the post-storm landscape is quickly restored to characteristics suitable to human perception of the value of the resource. Events in New Jersey indicate that cycles of storm damage and reconstruction of both human structures and human-designed or human-enhanced geomorphic landscapes can be as little as 1 or 2 years for catastrophic storms. The human-altered post-storm landscape dominates over the natural post-storm landscape whether storm effects are prevented by human alterations or the storm obliterates the human-altered landscape and it is subsequently repaired.

Reconstruction of storm-damaged landscapes may be considered impractical in some areas, but it is likely that there are few such locations. Severe damage to beachfront homes by storms in 1878 and 1884 at Atlantic City did not prevent post-storm reconstruction (USACOE 1990). Even in the nineteenth century, the scale of alterations resulting from damaging storms was small enough that human action could restore buildings and infrastructure before natural processes could re-establish natural landforms that could dominate the landscape or control coastal evolution (Nordstrom 1994a). Highly vulnerable and isolated barrier islands may not have been rebuilt following catastrophic storms in the Gulf of Mexico (Davis 1993), but, even there, devastating storms have not been a deterrent where population pressure existed (Byrnes *et al.* 1993).

Speed of reconstruction efforts in areas developed for recreation are a function of the economic importance and size of the market area for tourists (Meyer-Arendt 1991). Restoration of the value for tourism can be accomplished within a year after major storm damage with massive inputs of capital and labor, but activities as routine as removal of rubble can take several years where investment level is low (Meyer-Arendt 1991). Restoration of completely devastated cultural landscapes where economic investment is low may take years or even decades, but even formerly abandoned communities may undergo redevelopment when social and economic forces become more favorable (Meyer-Arendt 1985, 1992).

Prognosis

Each major storm engenders a new suite of shore protection projects (Watanabe and Horikawa 1983; Nordstrom and Jackson 1995). The new seaward-most construction line is often at the same location it was prior to the storm, and damaged communities are often rebuilt at larger proportions (Dolan 1987; Waldrop 1988; Fischer 1989; Meyer-Arendt 1990; Bortz 1991; Beatley *et al.* 1992; FitzGerald 1994; Nordstrom and Jackson 1995). Severe storms may decrease levels of development in locations where there is little money for reinvestment (Meyer-Arendt 1991), but they have increased levels of development in many locations, including Gulf Shores, Alabama following Hurricane Frederic (Meyer-Arendt 1990), Grand Isle, Louisiana following Hurricane Betsy (Meyer-Arendt 1985), south Padre Island after Hurricane Allen (Meyer-Arendt 1991), and Galveston after Hurricane Carla (Meyer-Arendt 1993a).

Many studies have suggestions for design of structures that will reduce damages during the following storm (Morton 1976; Rogers 1991a; Saffir 1991; Nichols *et al.* 1993), contributing to the likelihood that structures will persist after major storms. There is an increase in the size of the new buildings and amount of support infrastructure (utility lines, widths of right-of-ways), making them less easy to be moved, and setbacks, if applied, often have inadequate widths (Rogers 1993). Given past precedent, and considering the development pressure and level of investment in many barrier islands in the USA, it is likely that structures damaged by future storms will be replaced (USACOE 1989) and there will be an increase in the level of development and protection (Nordstrom 1994a).

The reasons why redevelopment occurs are numerous. Buyers of real estate in beachfront communities may not be aware of the nature of the hazards; they may discount the impacts; or they may take for granted that there is a vast array of government programs to compensate them (Hillyer *et al.* 1997). Rapidly developing areas may have a large number of visitors who have never experienced coastal hazards, and many buildings and protection structures are approved and built by planners, engineers and contractors without first-hand experience with storms (Griggs 1994). Coastal policies may facilitate rebuilding of destroyed structures in essentially the same form and location as the original structures by eliminating the need for development permits, and rebuilding does not undergo the same scrutiny as new projects (Griggs *et al.* 1991b). Lessons of previous storms are often forgotten or ignored (Podufaly 1964; Coch 1994). Even if former owners decide that rebuilding is too risky, someone else is willing to develop the site (Dolan 1987). Meaningful post-disaster redevelopment policies may be lacking, and it may take extraordinary effort in some locations to change historical trends related to beachfront development following major storms (Schmahl and Conklin 1991).

Effects of climate change

The conditions for change

Many of the problems in coastal management that are presently being faced will be complicated by changes in climate that are expected as a result of changes in atmospheric composition due to increase in greenhouse gases. Changes in atmospheric composition will affect storm paths, storm severity, sea level rise, conditions for growth of coastal vegetation and human actions taken in response to these changes. It is difficult to predict future effects because of so many uncertainties related to: (1) predicting changes in greenhouse gases; (2) predicting the influence of these gases on climate, given the huge range of influences at a variety of spatial and temporal scales; (3) identifying the complex relationship between greenhouse gases and climatic parameters resulting from the existence of threshold conditions, synergies and the complex influence of oceans and their circulation patterns; and (4) lack of sufficient knowledge of ocean–atmosphere coupling and the associated difficulty of developing useful computer models (Jones 1993). Warrick and Barrow (1991) suggest that the climate will be 0.7 to 1.9 °C warmer by 2030, with mean winter temperatures about 1.5 to 2.1 degrees warmer, and a much higher probability of occurrence of extremely warm years with increased precipitation. Titus and Narayanan (1996) suggest that there is a 50 percent chance that the average global temperature will rise 2 °C by 2100. This warming will contribute to sea level rise through thermal expansion of ocean water and melting of glaciers and ice sheets. Recent estimates of expected amounts of sea level rise by 2100 range from 0.45 to 0.66 m (Gornitz 1995; Titus and Narayanan 1996).

The above figures represent global estimates. Relative sea level rise (thus vulnerability) could be greater at some sites than these mean values (Gornitz 1995). The uncertainty in how climate change will affect regional weather patterns and the magnitude and frequency of extreme events is especially troublesome, given the great amount of geomorphic change associated with low-frequency, high-magnitude events (Lee 1993). Many local planning authorities already consider coastal hazards difficult to predict and feel in danger of making decisions that will result in indefensible legal challenges (Lee 1993).

Recent studies have begun to move away from broad-brush global scale doomsday predictions to predictions at the local scale, where factors such as climate, topography, subsidence, geology and land use affect landform changes (Jones 1993; Nicholls and Leatherman 1996; Bray *et al.* 1997). Sea level rise, *per se*, does not cause geomorphic change; it is the extreme wave activity, surge and swash associated with it that does, and several links have to be made between sea level rise and processes and with these processes and landform change before effects of sea level rise on landforms can be made (Orford *et al.* 1995).

Implications for coastal evolution

Climate change and associated changes in sea level have important implications for the way landforms will evolve in the future. There are many studies of the effects of past changes in the coastal zone, particularly in Europe (e.g., Klijn 1990; Tooley 1990; Carter 1992), and there is an increasing number of studies of future changes (e.g., van Huis 1989; van der Meulen 1990). Some of these studies predict dramatic changes in the future. A simulation of the effect of climatic change in the Coto Doñana National Park in Spain indicates an extension of the dry season, a decrease in summer soil moisture, and an increase in potential evaporation in the winter, leading to an increase in sand drifts; climate in Slowinski National Park in Poland may change from a subcontinental to a more Atlantic-type climate, with an increase in winter temperatures, a decrease in occurrence of frost, and an increase in rainfall, extending the vegetation cover over barren parts of the dunes and decreasing wind erosion (van Huis 1989; van der Meulen 1990). The effects of climatic change and sea level rise may reduce or eliminate coastal habitats (Carter 1992) or result in abandonment of coastal settlements and a return to more natural coastal characteristics (Corre 1989). Many of the future rates of environmental change will fall within limits of past periods of natural changes, but will exceed the recent rates to which humans have become accustomed. Global warming and potential sea level rise will likely heighten the need for more effective coastal management arrangements, rather than generate new problems (Lee 1993). It is likely that even some of the more dramatic of the effects may be obscured in developed areas by human efforts to protect property and maintain a stable or predictable resource base (Titus 1990, Nordstrom 1994b).

Potential human actions

The potential scale of human influences on landforms easily exceeds the scale of likely change resulting from climatic alteration. The climatic, sea level rise and biospheric consequences of global warming may provide the ingredients for geomorphological change, but the scale, extent and pace of resultant changes will be largely determined by human activity (Jones 1993). The types of responses and scales will differ greatly due to global differences in national economies (Bird 1993), and there are many alternatives within existing management frameworks (Klarin and Hershman 1990). Prediction of future scenarios of change for developed coasts will have little meaning unless these human inputs are placed in context. Responses to sea level rise can be classified into a series of human adjustments that can then be used to identify potential effects on beach and dune resources. One of the most useful of these classification systems is the one used by Titus (1990). His scenarios (Figure 6.2) are useful because he combines physical processes and human action, including economic, legal and engineering approaches, and he focuses on change on developed barrier islands, that are dynamic and play important roles in evolution of estuarine resources.

The scenarios by Titus include: (1) no protection; (2) constructing a levee around the barrier; (3) island raising; and (4) engineered retreat. These options were examined in regard to Long Beach Island, New Jersey (Figure 1.10) because it was near the middle of the spectrum in regard to development density. Bird (1993), Nicholls and Leatherman (1996) and Bray *et al.* (1997) cast their scenarios somewhat differently (as retreat, accommodation, adaptation and protection) but many of the components of their scenarios are compatible with those of Titus and are considered here in that context.

No protection

The no protection (on an island scale) alternative can lead to accommodation to the sea level rise by individual residents and municipal actions through changes in land use practice compatible with a more dynamic coast. These small-scale actions could be insufficient in the long-term, leading to eventual abandonment. Accommodation could include elevating buildings on pilings (or making existing pilings higher or deeper) and converting usable surfaces to those that can tolerate more frequent inundation by salt water. Initial problems associated with this alternative include loss of recreation beach through accelerated erosion, loss of bayside land through erosion and inundation of low-lying ground and stranding of greater numbers of buildings and infrastructure on the beach (Figure 6.3) (Titus 1990). Some of the problems associated with this alternative will be delayed where coastal con-

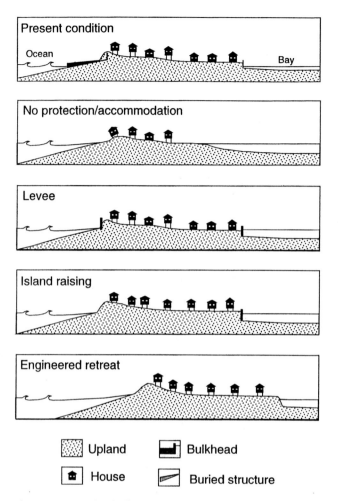

Figure 6.2 Scenarios for human response to sea level rise on a shoreline with an intensity of development similar to that depicted in Figure 1.10. Modified from Titus (1990).

struction setback lines are in place, but the effectiveness of land use controls will be diminished due to the greater erosion rates (Titus 1990). There will also be competition for land and living space as coastal residents move inland, with the potential threat of eliminating natural resources in backbays (Titus 1990; Bird 1993).

The role of coastal foredunes under this scenario is not clear. Presumably, they will continue to be used for a time to provide protection to buildings that are not elevated and to roads (where it is not cost effective to elevate them). The severe restriction to beach width will limit the value of the beach as a source of sand for dunes and as a space available to accommodate dunes. There are

Figure 6.3 Houses on beach in Delaware, USA.

presently a few locations where the dune is landward of shorefront houses (e.g., Fire Island, New York), and this scenario may be somewhat more common in the future.

On a large scale, the strategy of no protection would result in a dynamic barrier island that could migrate inland rapidly and would be susceptible to breaching by inlets. These inlets would result in greater fluctuations in beaches and dunes on ocean shorelines adjacent to them and greater fluctuations in water characteristics in the backbays. Eventual abandonment of the barrier islands is a possibility under this scenario. This is not a likely option in the near future, even on dynamic and lightly developed barrier islands, because of the need to maintain their integrity as barriers to flooding and inlet creation to provide protection to houses bordering the estuary and to maintain inventories of commercial shellfish (Bokuniewicz 1990).

Levee construction

This scenario includes constructing seawalls and flood protection structures around the barrier, creating a ring-levee enclave and using pumping systems to remove rain water. The scenario is an extension of the most common way of protecting extensive developed areas from flooding and rising sea levels in past centuries (Titus 1990). There seems to be a preference for engineering solutions to the problems of a rising sea level that may translate into building more seawalls and better seawalls (Bird 1993), and this approach has the advantage of not requiring major institutional changes regarding land use (Titus 1990). Coasts protected by artificial structures are expected to have increases in water depths, breaking wave energy, storm runup, beach loss and probability of structural damage, requiring construction of improved structures (Bray *et al.* 1997). Lots, houses and roads would not have to be raised, and the beach could be maintained by artificial nourishment, although at greater cost (Titus 1990). The present methods employed for protecting the Dutch coast indicate that environmental and recreational values need not be sacrificed on oceanside sites, but recreational use of bayside sites on barrier islands may change considerably due to the loss of views and restrictions to access caused by the new barrier (Titus 1990). This option appears most desirable for wide islands that are urbanized, such as Atlantic City, New Jersey (Figure 1.9) because the scale of development would make the options of abandonment and engineered retreat more costly and because the option of raising the island would require an inordinate amount of sand (Titus 1990). The degree of urbanization that would justify this type of protection would restrict the use of the island for natural values.

Island raising

Raising the barrier island in place is accomplished by placing sand on the beach and concomitantly raising buildings and support infrastructure. Institutional problems associated with the no protection and engineered retreat scenarios are avoided in that no one is prohibited from building or rebuilding, and the government does not have to buy property (Titus 1990). Raising the island in place would cause bays to become deeper and wider (as in the levee scenario) and would increase the likelihood of bayshore erosion.

A modest version of island raising was attempted at Galveston, Texas after the 1900 hurricane, when 8 360 000 m^3 of dredged sand was used to raise the elevation of a portion of the city (McComb 1986 in Meyer-Arendt 1992). The cost of dredge and fill projects, complicated by environmental problems in borrow areas may prohibit this option from being implemented on a large scale in many locations (Titus 1990).

Engineered retreat

Engineered retreat mimics natural retreat by artificially filling the bay sides of barriers while the oceanside erodes. The fill material would not be beach sand, as it would under natural conditions, because the resulting accelerated erosion of the beach would be considered unacceptable (Titus 1990). The fill option has already been exercised, albeit for a different purpose, in places where artificial fill has been placed on the bayside of barriers to accommodate construction or transportation (Bokuniewicz 1990; Titus 1990). The engineered retreat option involves land acquisition by public authorities, use of setback lines and prohibition of reconstruction after damage, requiring new legislation; costs of protection are avoided, but economic and social costs of land loss and compensation are potentially high (Bray *et al.* 1997). Engineered retreat overcomes the detrimental effect of steepening of the offshore gradient, and is compatible with continued economic use of the islands through lease arrangements rather than outright ownership (Bokuniewicz 1990). The biggest problem with this option may lie with present owners of the land who may have to be compensated. Managed retreat evokes a strong political and public opposition at the local level (Gares 1989; Bray *et al.* 1997), and there is often no means of compensating those affected in any case, even though the costs might be much lower than those incurred by providing protection (Bray *et al.* 1997). One advantage of engineered retreat is that it has the potential of assisting long-term survival of wetlands by migration of barriers inland and maintenance of sediment sources (Bray *et al.* 1997), but the need to provide interim protection to houses on the barriers (by minimizing wave overwash and flooding or preventing new inlets from forming) may restrict natural processes from creating marsh environments on the bay side. At worst, the new substrate on the landward side of coastal barriers would be used as sites for new buildings rather than natural habitat. The requirements to ease former restrictions on bayside filling and the associated opposition by environmental advocates, combined with legal and ownership issues, make implementation of this alternative difficult, although it can result in the least long-term change in the form and function of coastal barriers that are presently developed (Titus 1990).

Prognosis

It is easy to envision a single barrier island that includes many of the elements of the four scenarios identified above. Individual houses will be destroyed and removed; others will be raised as they are threatened; individual buildings and roads may be constructed or reconstructed on substrate built to higher elevations than the surrounding unimproved terrain; seawalls, bulkheads and dikes will be built and rebuilt in segments until they eventually may form a continuous barrier. Many of these alterations are likely to be incremental because of the difficulty of funding large-scale projects. There will be economic and environmental losses and gains in coastal communities affected by sea-level rise, but overall, losses are expected to exceed gains (Bird 1993). The abandonment of areas that are rendered unsuitable for development is not likely to prevent development of locations where the new water/land contact creates coastal landscapes considered desirable for recreational use. Better land use controls (setbacks, post disaster plans) may be in place for those sites, but there is no guarantee that these controls will be effective (see chapter 7). The abandonment of unusable human facilities in some areas and the construction of new human facilities in other areas is likely to displace the location of human activities and structures, but not diminish the human presence at the coast or re-establish naturally functioning landscapes. The abandonment strategy may create a series of lines of coastal defense structures that are allowed to degrade as new ones are built landward of them (Vinh *et al.* 1996).

Cycles of landform change

Beach cycles

Human alterations can affect beach processes at time scales as short as seconds. This occurs when incident waves are affected by waves reflected off shore protection structures and when beach pumping changes the nature of swash uprush and backwash. Effects at semi-diurnal and daily periodicities are found where onshore structures eliminate upper portions of the beach or truncate wave effects at high stages of the tide.

Natural cycles that occur at longer periodicities may be driven by: (1) changes in weather patterns that result in differences in wave approach and longshore transport direction (days) and occurrence of small storms (days to weeks); (2) differences in the seasonal climatology of storms (semi-annual); and (3) periodicities of intensive storms with a long recurrence interval (years). Human periodicities may be determined by: (1) recreational needs that are often seasonal (semi-annual); (2) periodicities of natural processes (especially coastal storms that rearrange cultural landscapes); or (3) periodicities reflecting political and economic cycles.

Shore-normal protection structures locally alter locations of accretion and erosion when longshore transport directions are reversed due to changes in weather patterns. These changes can be observed over periods of days. Protection structures can greatly increase the amount of seasonal change and can reverse the direction of seasonal erosion–accretion cycles that would occur under natural conditions. The latter can occur due to trapping on updrift (considering net transfers) sides of structures during high-energy conditions in winter, followed by erosion on downdrift sides in the summer; these changes alter the process of seasonal change from one induced by cross-shore transport to one induced by longshore transport (Leidersdorf *et al.* 1993, 1994).

Beach nourishment is cyclic in that it must be considered a repetitive process to retain its desired effect on shoreline maintenance over the long term under the influence of sea level rise (Stive *et al.* 1991). Nourishment projects introduce cycles of beach evolution that have a time scale determined by human decision-making processes. Their timing depends on perceptions of the need for wider beaches, the political climate that affects the availability of funds and the delays associated with beaurocratic response at higher levels of government. The timing and locations of nourishment operations in most locations are highly variable because of the need to involve so many participants and levy funds on an individual basis. Some beaches are nourished at cycles related to maintenance dredging projects. Timing of other projects may be related to major storm events, when the political climate favors post-disaster aid and when economic resources are readily forthcoming, but the timing can be erratic (from the standpoint of natural events) at other times.

Changes in inlet cycles were introduced in chapter 5 in the context of eliminating cycles or diminishing the magnitude of their phases of erosion and accretion. Construction of jetties and armoring of the shorelines adjacent to these jettied inlets may change the cycles from periodicities of major storms (that breach barriers or increase hydraulic efficiency of new channels) to those of seasons, caused by trapping of sediment moved by longshore transport. The periodicity of some cycles may be deliberately altered by humans by relocating channels, as described in chapter 2. These cycles are difficult to predict because they are unique operations.

Dune cycles

The periodicities of natural changes in the morphology of dunes are not less than the periodicities of wind events capable of moving sediment (usually associated with near surface wind speeds greater than about 5 m s^{-1}), whereas human actions can alter morphology over periods of hours or days using earth-moving equipment. As on beaches, the most common cycles associated with human action are related to protection and recreation, but dunes also are affected by stabilization and destabilization cycles related to changes in surface cover.

Shore protection cycles

Cycles in dunes that are related to shore protection efforts may have periodicities driven by storms, seasons or years or by economic or political constraints. The economic and political constraints to modifying dunes are easier to overcome than the constraints to modifying beaches. Dunes are smaller and are in a less energetic environment, permitting major changes using locally available resources.

The initial construction phase of dunes that are eliminated during major storms is determined by the recurrence interval of storms of that magnitude, but the time it takes for reconstruction of the dune is dramatically less than the time it would take under natural conditions. Dune reconstruction may last days to weeks (using earth-moving equipment), a season (using sand fences) or several years (using artificial vegetation plantings). The periodicities of cycles related to repair of dunes by earth-moving equipment following wave attack by small storms is less than the periodicity of major storms and may be greater or lesser than cycles related to seasonal differences, depending on the perception of the vulnerability of the remaining portion of the dune (Nordstrom and Arens 1998).

Repairs to breaches or access gaps that are not perceived to be immediately vulnerable are often annual. Yearly repair of dunes for protection often occurs in the spring in The Netherlands and in the early fall in New Jersey. The destruction phase may be accomplished by natural processes (usually winter storms) or by recreation activities (trampling or deliberate creation of access gaps); the construction phase may be accomplished solely by human action (bulldozers) or a combination of human action and natural processes (vegetation plantings or sand fences) (Nordstrom and Arens 1998).

The periodicity of the sacrificial dune cycle (where aeolian accretion is encouraged seaward of the designated protective foredune) is less than the periodicity of the storm for which the landward foredune is designed. The sacrificial dune cycle may follow a seasonal periodicity or it may occur more than once per year, especially in communities where beach widths are narrow (Nordstrom and Arens 1998).

Recreation-related cycles

Dunes created in the autumn to provide a barrier against storms may be eliminated in the summer to accommodate bathers, resulting in an annual cycle determined by seasonal use patterns. The destructive (to dune) phase is normally accomplished at the end of spring by bulldozing and can last days to weeks. The constructive phase can be accomplished using bulldozers or fences (Mauriello and Halsey 1987), the latter taking several weeks to months (Nordstrom and Arens 1998).

Clearing of aeolian deposits resulting from small wind events from municipal recreation surfaces (boardwalks, parking areas) behind the beach and foredune is often one of the initial activities of lifeguards hired for summer work, resulting in an annual cycle. Clearing of nuisance dunes from individual properties generally follows a cycle related to the recreational life style of the owner. These dunes are usually formed by the strong winds that occur in the winter and survive until late spring, when residents and managers of commercial properties return to actively use the buildings in the tourist season. Year-round residents may keep pace with aeolian accretion, resulting in cycles of deposition and removal at the temporal scale of strong-wind events.

Disposal dunes, created by dumping of sediments removed in beach cleaning operations, usually follow recreation-related cycles (day, weekend) in the summer tourist season. They may have periodicities related to storms during the winter, when only hazardous or aesthetically displeasing flotsam delivered by these storms is considered worth eliminating (Nordstrom and Arens 1998).

Stabilization–destabilization cycles

These cycles may be introduced to enhance dunes to provide a better means of protecting human facilities from wave erosion or wind blown sand, or they can be instituted for the purpose of improving habitat or economic value of dunes. Several species of exotic vegetation have followed cycles of introduction and elimination with stages determined by human decisions and actions. The stages include: (1) accidental (or intentional) introduction; (2) rapid colonization; (3) acknowledgement of the advantages of the species for stabilization; (4) institution of planting programs; (5) recognition of undesirable side effects; (6) attempt to eradicate the exotic; and (7) attempts to reintroduce native species that institute a new cycle of changes. The exotic Bitou bush, for example, became well established in the dunes in Australia in the early 1900s and was identified as a good species for sand-drift control in the late 1940s; it was successfully utilized in this role until its detrimental effects were discovered (development of monospecific stands, visually boring landscape, threat to vegetation variety and biological heritage); recommendations for its use in reclamation were withdrawn in 1971; and steps were taken to eradicate it (Chapman 1989).

Aforestation–deforestation cycles show a similar long-term, single cycle change, where trees are first planted to stabilize the dune and then removed to create a more mobile landscape characterized by greater diversity of vegetation. The elapsed time between introduction of the plantings and their removal can be hundreds of years. Interim cycles of removal and regrowth of forest cover may occur, where the planted trees are exploited as an economic resource, but this temporary change in ground cover need not change the morphology, function or genesis of the dune.

The stabilization–destabilization cycles can occur without using exotics or commercially important species. For example, during the 1970s, management efforts in dunes in the Oxwich National Nature Preserve in Wales were primarily targeted toward stabilization of mobile dunes using sand traps, fences and planting of marram grass (*Ammophila*). Most of the mobile frontal dune areas had been stabilized by 1981; by the early 1990s, the frontal dunes exhibited symptoms of diversity decline through overstabilization. Current management now consists of repeated mowing, grazing, cutting of shrub, turf stripping and excavation (Mullarda *et al.* 1996).

Relationship between landform size and longevity

Individual landforms frequently reveal a relationship between size and longevity. In natural coastal environments, smaller dunes close to or on the beach are more frequently destroyed by wave action than larger ones, but they can also be rebuilt again much faster because they are near the source of sediment available for aeolian transport and there is no vegetation to interfere with sediment delivery. Larger forms with longer lifetimes exist at greater distances from the beach, where the energy level of destructive forces is lower. Investigation of dunes in New Jersey (Nordstrom *et al.* 1999) indicates that the relationship observed for natural conditions is changed under the influence of people (Figure 6.4), because small features can survive between houses or on lots as long as dunes that are larger but farther seaward, and foredunes can survive on the beach in places where they would neither form nor survive under natural conditions. The dunes survive because humans use sand fences, dune vegetation and machinery to accelerate the process of dune building (Nordstrom *et al.* 1999).

The comparison of dunes in New Jersey before and after the 1992 storm shows that all large and most medium size dunes survived. The smallest dunes that did not survive were located on the beach (and were eroded by waves) and the sand sheets that formed landward of the beach (and were eliminated by human efforts). Small forms can exist for longer periods where they are accepted by humans as part of the landscape. Many of these dunes are remnants that are decoupled from the beach and protected by bulkheads (Figure 5.1). Protective foredunes are designed to be large enough to provide protection against storms and thus survive periods of years. Dunes of medium size, that would form on the narrow New Jersey beaches in the absence of human efforts would last a few years, but foredunes of this size can last decades where they are artificially enhanced (Nordstrom *et al.* 1999).

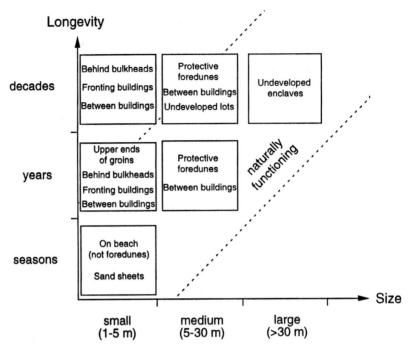

Figure 6.4 Relationship between size and longevity of dunes in developed areas in New Jersey. Modified from Nordstrom *et al.* (1999).

The space–timescale relationship between size and longevity is well known in geomorphology (Cowell and Thom 1994; Brunsden 1996). For dunes, this relationship describes a linear or log-linear trend under natural conditions that becomes a triangular distribution in a human-altered system. The dune types exhibiting the greatest departure from the size–longevity relationship (upper left of Figure 6.4) are the small dunes that are behind shore-parallel protection structures, in front of buildings and between buildings. These dunes can have a longer lifetime and have a denser vegetation than normally found on dunes of this size. The closer these dunes are to human facilities, the more humans influence their shape, internal sediment structure and vegetation. Dunes behind bulkheads can grow naturally at some distance from houses, but they are influenced by accretion and erosion related to lot cleaning when located close to houses. Dunes between houses are the most unnatural dunes because their location close to buildings and away from the beach makes them more susceptible to human alteration by lot cleaning operations. The most natural dunes are the types in the center of the corridor marked as naturally functioning in Figure 6.4. They provide the best types of dunes to use as a basis of restoration of a more natural coastal image and naturally functioning habitat as described in chapter 8.

Conceptual scenarios of evolution of developed coasts

Scenarios of evolution for beaches and coastal foredunes on a transgressing natural shoreline include a landward shift of these features, with cycles of destruction and rebuilding but restoration of their equilibrium form and cross-sectional area as they are displaced inland. The foredune may maintain its form and dimensions in the long term by continuous migration (Psuty 1989) or re-achieve its form and dimensions through breaching of weaker portions of the dune by storm waves, followed by buildup landward of the former crest line (Godfrey *et al.* 1979). Scenarios of change for eroding developed coasts that do not include beach nourishment as a management option are driven by limitations in sediment availability (Nordstrom 1994a). The ability of these sediment-starved human-altered systems to adjust their forms and dimensions to major process changes is impaired relative to natural systems; low-magnitude, high-frequency self-maintaining systems are changed to high-magnitude, low-frequency systems where self-maintenance may not occur (Carter 1980). Low-magnitude, high-frequency events on developed shorelines eventually convert wide beaches and wide and topographically varied dunes to truncated forms and convert multiple dune crest lines to single ridges that decrease in width and eventually are eliminated as erosion proceeds (Pilkey 1981; Nordstrom *et al.* 1986).

Closed-system scenarios

Large seawalls without a fronting beach (Figure 4.3) are considered the final stage of the progression of human alterations to maintain a fixed position on a migrating shoreline (Pilkey 1981). In this stage, beaches and dunes are replaced by shore-parallel protection structures; shoreline position is static; sediment availability is minimal; and the waves are incapable of creating an equilibrium feature. The scenario of change for unjettied inlets (Figure 4.5C, and discussed in Nordstrom (1988a)) has similarities to Pilkey's seawall alternative, in that it is based on a reliance on structural protection and is driven by limitations in sediment availability.

The two scenarios identified in the previous paragraph have assumptions that are similar to the no-action scenario used by the US Army Corps of Engineers to determine the benefit of future protection projects. The assumptions are: (1) the present shore protection structures will be maintained or improved; (2) sediment budgets will be diminished in locations where the structures interfere with sediment transport; (3) shoreline mobility will be reduced in areas protected by structures; and (4) inlets that are now dredged will continue to be dredged, with existing channels maintained in place. Scenarios like these describe a closed system of coastal evolution. They are grounded in the physical processes operative on developed coastal barriers, but they are simplistic in evaluating future human responses (Nordstrom 1994a).

Cyclic (open-system) scenarios

Rejuvenation of features characteristic of a natural landscape can occur with massive inputs of beach nourishment or changes in regulations that prohibit or limit construction or reconstruction of protection structures. At present, rejuvenation by beach nourishment is the more likely of these adjustments, given human reluctance to abandon development sites, but abandonment occurs in isolated locations where regulations are in place (Psuty 1989). These human adjustments re-establish the potential for an open-system (cyclic) scenario of change for developed coasts (Nordstrom *et al.* 1986; Titus 1990; Nordstrom 1994a).

The re-establishment of the potential for cyclic change by nourishment is the result of conscious human decisions (and the availability of funding). Whether natural geomorphic features will form on or adjacent to new beaches is dependent on human values for the resource. Although restored beaches are likely to evolve under natural cycles of wave action in the short term (at least between nourishment operations), transfers between the beach and locations farther inland may be prevented by human actions. A lack of linkage between the beach and the remainder of the coast can occur under both the beach nourishment scenario and the seawall scenario (Nordstrom 1994a).

The scenarios presented by Titus (1990) may be compared to the two responses postulated by Psuty (1986) for the New Jersey barriers (without regard to human action) that include drowning in place and migration through overwash, wind transport and inlet sedimentation. The no protection and engineered retreat scenarios in Figure 6.2 would mimic the migration model of Psuty (1986), although the engineered retreat scenario would occur by human efforts. The other two scenarios of Titus address the potential for drowning in place, although drowning would be prevented by human efforts. Titus uses a greater number of assumptions and more explicit assumptions about human actions than are usually employed in scenarios of evolution for developed coasts, considering factors such as preference and precedent, rights of property owners, legal implications, economic motivation, environmental interests and constraints of government programs. The most telling finding of the economic evaluation is that the rental value for the moderately developed barrier where Titus conducted his study (Long Beach Island) exceeds the protection cost for the most expensive option (raising the barrier), leaving the relevant question "how," not "whether," to protect the island (Titus 1990). The technology is now available for these protection options (Bokuniewicz 1990). Given the economics of protecting Long Beach Island, many developed barriers could justify protection for almost any conceivable rate of sea level rise or unit cost of sand for nourishment (Titus 1990). The resulting response of the barriers to accelerated sea level rise may mimic the natural response by migrating or remaining in place as water level rises, but the role of human agency in the process could be greater than at present (Nordstrom 1994a).

Assessing the degree of human input

Human action may be considered as an external factor that perturbs or changes the natural system or as part of the system (Phillips 1991). Human action can be considered exogenous if the temporal or spatial scale under consideration is an order of magnitude greater than the temporal or spatial scale of human influence, or an endogenous factor if the scales under study and the scales of human agency are similar in magnitude (Phillips 1991). Human alterations on many developed coasts are endogenous in that they are persistent, recurrent and are more frequent than changes made by natural processes or equal or exceed the spatial scale of natural changes.

Scenarios of change for developed coasts may be constructed following several different approaches that may determine, a priori, the extent to which human alterations will be considered an aberration or an integral part of the coastal system. These may be described as the overlay method, the no-action method and the active human-input method (Nordstrom 1994a). The overlay method compares and contrasts a developed area with an undeveloped area that is assumed to have the same process controls. The undeveloped area may be the same location prior to development, or it may be a different location (Gares 1990). Use of the same location prior to development may result in data bases that differ in methods of collection, level of detail or quality of measurement. Use of an undeveloped shoreline segment that is distant from the developed shoreline is problematic because of differences in setting, such as wave climate or tidal regime. An undeveloped shoreline segment that is adjacent to a developed segment may have a similar wave climate and tidal regime, but may be affected by the developed segment through sand starvation as a result of protection structures or sediment surplus due to beach nourishment. Care also must be taken to ensure that a segment that lacks buildings or infrastructure actually represents a natural condition. Use of sand fences, vegetation plantings and introduction of exotic species by humans has converted thousands of kilometers of dune landscape in the USA (Cooper 1958; Godfrey and Godfrey, 1973; Wiedemann 1984), and these coastlines may have no human structures to provide clues to human influence (Nordstrom 1994a).

The no-action method assumes that the kind of shoreline change that occurred in the recent past will continue unabated by local actions (and often implicitly defines a closed system on an eroding armored shoreline). This method is useful to evaluate potential changes in the absence of future adjustments so that these potential changes can be used as the basis for selecting the optimum solution for protection (Nordstrom 1988a), but it is unrealistic for long-term prediction because of the likelihood of future human action. The active-human-input approach bases predictions on probabilities of human action, calibrated with knowledge of physical processes in a multiple scenario format. This approach implicitly defines an open system (Nordstrom 1994a).

Many geomorphologists may use the overlay approach because they feel that it is necessary to understand undeveloped systems before superimposing the complexities of human modification (Sherman and Bauer 1993). It is likely that the active-human-input approach, that considers human action endogenic rather than exogenic (Phillips 1991), is more realistic for intensively developed areas and will have increased usefulness in the future in areas now developing (Nordstrom 1994a).

Human influence is sufficiently important to change both the space and time frameworks within which geomorphological processes should be examined (Trofimov 1987). For example, long-term erosion data normally have been considered more reliable for calculating future erosion rates and shoreline positions than short-term rates, but exceptions to this rule must be made where recent human alterations have a demonstrable effect on the beach (Byrnes et al. 1993). The past is no longer key to the present in many coastal areas, and it will have less applicability in assessing changes in the future.

Prediction of natural processes and coastal evolution in many developed areas according to existing geomorphological principles now is restricted to locations seaward of the berm crest on nourished shorelines and seaward of the seaward-most line of pioneer vegetation on most of the non-nourished shorelines. The history of landform changes in developed systems is determined by human action, and the linkage between natural process and form is obscured (Nordstrom 1990). Although human alterations obscure the effects of natural processes, these human alterations provide the best clues to the future evolution of coastal systems where human influence is an endogenous factor (Nordstrom 1994a).

Summary conclusions

Appreciation of the scale and primacy of human alterations to the coast may be obscured by studies of hazards and landscape change that concentrate on the post-storm landscape rather than the post-reconstruction landscape. Concern about the effect of climate change on coastal regions is widespread and well publicized, but human action has greater potential for altering the coast than natural processes. Scenarios of climate change do not have real predictive value if human activity is not included in the modelling. Engineering solutions to shoreline erosion and natural hazards are preferred for several reasons, but chief among those are the speed with which they can be implemented (in part, because they require no institutional change in land use policy) and the rapid results they produce in contrast to the long-term results of retreat and relocation. As discussed in chapter 7, emergency actions pose some of the most serious threats to the long-term preservation of beaches and dunes.

Human scales of time and space have been superimposed on geomorphic processes along the coast, making the past an inappropriate predictor of the future in this setting. The relationship that exists in nature between size and longevity of coastal landforms under human influence also is altered. The primacy of human alteration constitutes a rationale for focussing studies of change on human activity in the coastal zone, and the consequences of emergency and short-term management practices call for greater design and planning in coastal management practice to retain environmental values.

7

Management programs

Introduction

There is a vast literature on issues in coastal management, and overviews and critiques are available (Born and Miller 1987; National Research Council 1990; Griggs *et al.* 1991b; Godschalk 1992; Jackson and O'Donnell 1993; Beatley *et al.* 1994; Knecht *et al.* 1997; Sorensen 1997; Kay and Alder 1999). This chapter concentrates on program components that have applicability to maintenance of beaches and dunes. Other management issues that may be more pressing (e.g., pollution) are avoided unless they affect landforms. Legislative and administrative frameworks are restricted to a few recent programs and policies of national governments and selected state/provincial and local governments. Opportunities and constraints are discussed, revealing the difficulties of maintaining beach resources in a way that preserves elements of the natural system. Future actions that can be taken to maintain or enhance these resources while accommodating the need to protect private developments are presented in chapter 8.

National programs and incentives

Coastal management policy is usually a tiered structure of national, state/provincial and local programs. In the USA, for example, national policy establishes broad mandates for state-controlled coastal programs and sets restrictions on investment of national funds in the coastal zone. States use permitting procedures and site reviews of large-scale developments, and local jurisdictions employ land use controls and health ordinances to control small-scale development.

The objectives of many government programs are similar and are summarized in Table 7.1. Programs may differ in terms of level of government that

Table 7.1 *General objectives expressed in national programs that have impact on evolution of beaches and dunes*

Acquire sites in order to ensure their long-term conservation and meet projected needs.
Maintain the image of the coast and build up a heritage to be passed on to future generations.
Plan and manage sites to make them accessible to a public respectful of the natural environment.
Create protection easements to prevent or control type and density of construction close to the beach.
Shift the concept of coastal property to public uses that require no fixed installations.
Institute regulations for declaring accreting areas as public property.
Provide strict limits for locating concessions and installations on beach.
Prevent mining of beach materials and mining of materials on tributaries.
Provide for hazard mitigation and protection against natural disasters.
Protect existing beach and dune systems from human-induced erosion.
Regulate use of fill or disposal of incompatible materials.
Give priority to water dependent uses.
Avoid exploitation that threatens coastal resources.
Provide for environmental quality, recreation and property protection.
Carry out cost-effective strategies.
Develop a program to pay for the management strategy and obtain commitments for financing.
Provide a long-term (e.g., 20 to 50 year) strategy.
Use strategies that mimic natural processes when they are equal to alternatives in cost effectiveness.
Adopt proactive planning programs rather than case-by-case decisions on permit applications.
Give consideration to cumulative impacts.

Sources: Brindell 1990; San Diego Association of Governments (1995); Fischer *et al.* (1995) Meur-Ferec (1995).

plays the primary role, comprehensiveness, amount of funding and degree that policies are legally binding or enforceable. Some countries do not include legal provisions for general protection of the coast, but provide guidelines, incentives or directives for individual states, provinces or local jurisdiction to implement.

Programs within the USA are discussed in greatest detail in this chapter. Short discussions of programs in other countries that differ substantively, (e.g., United Kingdom, The Netherlands) are included for perspective.

Programs in the USA

National acts and programs in the USA include the Coastal Zone Management Program, the Coastal Barrier Resources Act, the National Flood Insurance Program, federal disaster assistance, activities of the US Army Corps of Engineers, national parks and nature reserves and threatened and endangered species programs. There are also technical advisory programs within national agencies.

The Federal Coastal Zone Management Act

This act assists the states in developing and implementing comprehensive plans that did not previously exist. The Coastal Zone Management Improvement Act of 1980 added provisions to help conserve beaches by encouraging and assisting states in implementing programs that provide for the protection of natural resources, including beaches and dunes. This national law is seen as advantageous in that it puts coastal management on the national agenda and insures that broad policies considered beneficial to the nation are incorporated into state programs (Jackson and O'Donnell 1993). The result of the Coastal Zone Management Program is that states in the program have a comprehensive set of resource and development policies to enhance the likelihood that consistent and predictable decisions on allocation and use of coastal resources will be made in the best interest of society. However, state programs vary widely in the absence of national program standards (Jackson and O'Donnell 1993). There is no single agency at the federal level in the USA that has control over coastal management; and there is no single or unified national coastal zone plan or strategy to guide or co-ordinate national actions or programs (Beatley *et al.* 1994).

The Coastal Barrier Resources Act

Federal economic incentives that open coastal areas to development pressure are provided by: (1) funds for highway construction or infrastructure development; (2) loan guarantees and tax codes that allow deductions for uninsured damages; (3) interest and property tax deductions for second homes; and (4) allowance for depreciation of seasonal rental properties (Beatley *et al.* 1994). Acknowledgement that use of these funds resulted in unwise development of barrier islands led to passage of the US Coastal Barrier Resources Act of 1982. This act was designed to help minimize loss of human life, expenditures of federal revenues and damages to natural resources by restricting federal financial assistance for most private development, including buildings, roads and other support infrastructure or static erosion protection structures on barriers within the Coastal Barrier Resource System. The effect of the act is to place financial risk associated with development on those who live on or invest in the coastal barriers (Watzin and McGilvrey 1989). Barriers included in the system still will be subject to beach loss and environmental degradation as a result of private construction that is not subsidized by the federal government (Watzin and McGilvrey 1989; Essig 1993).

The National Flood Insurance Program

This act, passed in 1968 and then amended by the Flood Disaster Protection Act of 1973 was designed to encourage prudent land use planning and to minimize property damage in flood-prone areas by encouraging state and local governments to restrict development of land subject to flood damage and ensure that construction materials and techniques were used to minimize potential flood damage. Subsequent changes were made to this program to: (1) consider wave damage in addition to flooding; (2) establish criteria for considering the effects of storm-induced erosion on sand dunes; (3) include protection structures in calculating degree of vulnerability; and (4) (with passage of the Upton–Jones Amendment) provide a mechanism for funding the relocation or demolition of structures threatened by erosion (Davison 1993).

New structures in coastal high-hazard areas are evaluated on their potential to withstand the impact of wave action and scour (e.g., all buildings must be elevated on adequately anchored pilings, placed above the height of the waves cresting above the 100-year flood elevation). The requirements of the program can have a beneficial effect on maintaining the viability of beach resources, in that the performance standards for constructing buildings could eliminate the need to implement static protection structures that could replace the beach, although federal program requirements do not ensure that all erosion control structures are passive and conducive to maintenance of the natural system.

The present design standard for protective dunes in the USA is a dune volume provided by the Federal Emergency Management Agency (FEMA) for protecting against the 100-year flood (Hallermeier 1987). This level of protection is only likely to be adopted by municipalities if required as a condition of economic support from FEMA. Communities that do not seek funds from FEMA to rebuild damaged dunes are not required to build dunes of the recommended size. Even where dunes are built to FEMA standards, they may bear little resemblance to natural dunes (chapter 5).

US Army Corps of Engineers activities

The Corps conducts activities related to: (1) beach erosion control; (2) hurricane protection; (3) navigation improvement; (4) regulation of structures or operations in navigable waters; and (5) discharge of dredged or fill material in waters and wetlands. The Corps designs and implements beach nourishment and dune building projects as part of erosion control, hurricane protection and dredge and fill projects (including those associated with navigation) but only after funding is authorized by Congress. The Corps has become increasingly involved in environmental restoration. These projects have mostly been conducted in marsh environments, but the potential exists for restoration in more energetic environments. The degree to which Corps activities enhance naturally functioning coastal landforms is determined by the degree to which environmental engineering (van Bohemen 1996) is reflected in the project designs.

National parks and nature reserves

The purpose of national parks in the USA is to conserve the scenery and natural and historic objects and to provide for their enjoyment in ways that will leave them unimpaired for future generations. The US Federal government owned virtually no land on coastal barriers at the beginning of the twentieth century but has purchased land using a series of statutes to create 10 national seashores plus national recreation areas in more developed locations. The national seashores include about 1000 km of ocean-facing shoreline plus inland areas and receive about 30 million visitors annually (Platt 1985). Some of the national seashores are managed as relatively natural areas with minimal visitor impact (e.g., Cumberland Island, Georgia), but several of the national seashores consist of scattered units in federal ownership adjacent to state, local or private holdings, resulting in undesirable impacts from adjacent areas and debate over decisions involving land use and shore protection. The Park Service generally favors letting nature take its course within its boundaries but has employed beach and dune stabilization and hard structures (often designed as temporary) to protect facilities (Platt 1985). The requirement to provide access and recreational opportunities means that inevitably there will be some human alterations to landforms, but parks usually have a low-intensity-use zone where landforms and biota can be conserved.

Executive orders

New policies can be implemented as a presidential executive order, such as the Floodplain Management (E.O.11988) and Protection of Wetlands (E.O.11990) orders of 24 May 1979 or the executive orders on use of off-road vehicles on public lands (E.O. 11644 and E.O. 11989). Executive Order 11988 directs federal agencies to avoid the adverse impacts associated with the occupancy and modification of floodplains and to avoid support of floodplain development. Executive Order 11990 is designed to avoid adverse impacts associated with the destruction or modification of wetlands and to avoid support of new construction in wetlands. The potential impact of orders such as these is greater than generally realized (Scott 1982), and it may be relatively easy to adopt new beach conservation policies in this form.

National programs in other countries

The Netherlands

Some nations can muster the economic resources to carry out virtually any policy deemed suitable for management of the coastal zone. The coastal defense policy for The Netherlands reveals what can be done to protect, augment or redesign beaches and dunes given a strong national interest in these features (van Bohemen 1996). No comprehensive coastal zone management policy exists for The Netherlands, but the national policy on physical planning and integrated water management, together with the national nature policy plan and the policy on dynamic natural preservation form powerful tools for sustainable development in the coastal zone (Hillen and Roelse 1995).

Design criteria for coastal protection in The Netherlands are the responsibility of Rijkswaterstaat (Ministry of Transport, Public Works and Water Management). Coastal protection in The Netherlands was practiced on an *ad hoc* basis until 1990, with a focus on the raising of foredunes to Delta Strength, as presented in the Delta Act of 1958. This refers to a specified minimum volume of sand, termed the border profile (*grensprofiel*). In 1990, the national policy was formulated to preserve the coastline at its 1990 position, while allowing the natural dynamics and character of the coast to be preserved through large-scale nourishment projects. The result is a dynamic dune that will change shape but remain in a predictable location through time, and the spatial scale of management has shifted from a line to a zone, with the foredune considered an important element in a sea defense zone (Nordstrom and Arens 1998).

United Kingdom

Coastal management in the United Kingdom is interesting for its long-standing reliance on charitable and voluntary efforts aimed at protection, rather than attempts to manage competing uses (Williams 1987; Jackson and O'Donnell 1993). There is no national law governing coastal management, but there are national policy guidelines and several national conservation programs (Jackson and O'Donnell 1993; Nordberg 1995). The planning system is two-tiered with statutes and guidelines provided by Parliament and local governments drafting local plans providing more specific directives for implementation (Jackson and O'Donnell 1993). Local authorities must prepare plans for their area that take account of national policy guidelines; their guidelines adopted in 1992 declare new development should not be accommodated in undeveloped areas if it can be accommodated inland or in developed areas and require that areas identified as Heritage Coasts must be incorporated in the plans (Nordberg 1995).

The management objectives for Heritage Coasts are to conserve the quality of scenery and foster leisure activities based on informal recreation that rely on scenery and not on man-made activities (Williams 1987). Heritage Coasts are undeveloped, although they may be farmed. They are designated because they possess special characteristics deemed worthy of preservation by the Countryside Commission, a government agency that provides funding and management advice relating to scenic and recreational lands. The Commission collaborates with local authorities that retain responsibility for the management of the Heritage Coasts. The Heritage Coasts appear in statutory structure and local plans and thus are able to gain some protection in the planning process, although control of development rests entirely within the hands of the local authorities (Jackson and O'Donnell 1993). The success of Heritage Coasts demonstrates that a management philosophy can be successful when it brings recognizable benefits to both visitors and local residents at a low cost but high benefit (Williams 1987). The key to management of these coasts is the establishment of a close relationship with farmers, landowners, residents and visitors with emphasis on voluntary agreements and persuasion (Williams 1987; Morgan *et al.* 1993).

About 44 percent of the coast of England and Wales has protective designations related to landscape quality, and substantial lengths of the coast are protected by statutory designations relating to nature conservation (Nordberg 1995). About 70 percent of the coastline of Wales (and about 83 percent of the surviving sand dune resource) is protected under some kind of designation including statutory protection for important biological and geological features and designation to preserve outstanding coastal landscapes (Smith *et al.* 1995). The Heritage Coast idea has spread to India, France and South Africa (Williams 1987). The British initiative to protect relatively undeveloped portions of the coast appears to work well, but the lack of a national focus on the entire shoreline has taken attention away from more developed areas of the coast (Jackson and O'Donnell 1993).

Other countries

Many countries have components of their coastal regulations that help preserve a naturally functioning beach and dune environment by direct preservation, by conserving sediment or by reducing interference with natural processes. Many national acts regulate use and protection of the coast by restricting construction within a prescribed distance of the water. In European countries, the regulated strip is often 100 m or greater, with additional regulatory zones extending 0.5 to 3 km inland (Nordberg 1995). Denmark is considered to have perhaps the most advanced coastal legislation in Europe, with a Nature Conservation Act (1992) that protects a coastal strip 100 to 300 m from the position where continuous vegetation begins landward of the shoreline, and a Physical Planning Act (1993) that: (1) declares that undeveloped coasts shall remain essentially national natural landscape resources; (2) declares that the coast shall be kept free from construction that does not require a coastal location; and (3) establishes a coastal zone (generally 3 km in width) where all land use planning is regulated (Nordberg 1995). The French Coastal Decree of 1986 requires protection of natural habitats of ecological or scenic interest that include active sand dunes within a 500 m strip of the shoreline (Favennec 1996).

Acts designed to protect coastal resources may have components designed for specific purposes that allow for landforms to be created but do not affect them directly. The Spanish Shores Act (Ley de Costas) prohibits the use of sand dredged from navigation projects for purposes other than beach nourishment (Gómez-Pina and Ramírez 1994) and explicitly protects coastal dunes from destruction by mining, development and changes in land use (Sanjaume and Pardo 1992). The German Nature Conservation Act of 1987 contains a list of protected habitats, leaving it up to individual states to implement protection through their own laws (Nordberg 1995).

State/provincial programs

States within federal systems of government are given great latitude in developing their own programs, resulting in considerable differences in program components within states. The great variety is not necessarily a problem because a national management plan imposed in all states in large coastal countries would probably not be feasible, given the widely varying physical nature of the shoreline (Jackson and O'Donnell 1993).

States in the USA provide comprehensive resource and development policies that can be adopted as administrative rules and used to regulate types of development and ensure consistent and predictable decision-making in the coastal zone. The policies may be implemented as local regulations or imposed on development proposals or activities regardless of local regulations. Location policies guide development toward the most appropriate, least environmentally sensitive sites; use policies assure that a proposed use is appropriate for a site; and resource policies establish performance standards to protect coastal resources (Mauriello 1991). States can also provide non-statutory advisory documents that contain policies and working plans to guide the development and management of local governments. These can be prepared by state planners or by contractors for statewide application or for local communities (e.g., Western Australia Department of Planning and Urban Development 1993). Summaries of recent state initiatives and their impacts on local activities are presented in Brindell (1990), Beatley *et al.* (1994) and Good (1994).

States use waterfront development laws, wetlands and tidelands management acts, water quality planning and maintenance programs, critical area laws for site plan reviews for new developments, erosion control setback lines, hazard and shore protection programs, parks and recreation programs, laws that guarantee access to or along the beach, programs and policies for specific areas of concern and education and outreach (information) programs. A major benefit of state regulatory programs is that there has been increased scrutiny of permit applications for shorefront construction under the authority of many of these laws.

Waterfront and water quality programs

State waterfront development laws now regulate construction or alterations to structures in intertidal lands at or below the mean high water line, and development is usually subject to permits. The principal potential benefit to beach resources under these laws is the restriction of recreational structures that could interfere with processes on the beach.

The State of New Jersey regulates dredging and beach nourishment projects under the Waterfront Development Act, using its Rules on Coastal Zone Management as the basis for review. The state must also certify that proposed projects are consistent with state water quality standards under Section 401 of the Clean Water Act. The issuance of approvals is based on a review of the proposed project by several agencies within the Department of Environmental Protection including agencies dealing with wildlife, water resources and coastal resources. These agencies evaluate the impacts to (among other things) shellfish, fishing areas and endangered species habitat (Mauriello 1991). These approvals are designed in large part to minimize adverse impacts, but additional values, such as habitat restoration or enhancement can be added as conditions for approval of the permit. For example, the State of New Jersey requires that all filled beach areas be stabilized with sand fencing and vegetation to retain the greatest amount of sand on the beach as a protective dune (Mauriello 1991). This requirement can restore dunes to locations where they have been eliminated by past actions of developers and residents.

Critical area laws

Critical-area/site-review laws regulating activities below mean high water are generally strict and may apply to single family residences as well as shore protection and recreation structures. These laws provide a mechanism for regulating commercial structures and large dwelling units above mean high water, but control of minor development, is often left to communities (Blair and Rosenberg 1987).

Control over activities in the dunes in the municipalities in New Jersey derives from the 1993 legislative amendments to the Coastal Area Facility

Review Act (CAFRA). This regulation prohibits direct disturbance to dunes that would reduce their dimensions. Sand can be added to dunes by bulldozing, and vegetation may be planted. Construction of walkways across the dune is permitted, and construction is permitted seaward of the dune if the structures are used for shore protection, for seasonal or recreational uses (and are movable) or for permanent water-dependent recreational use (and are elevated on pilings). There are no state regulations controlling activities inland of the landward toe of the dune and no requirement for control of activities adjacent to the dune that affect transport of sand to and from it (except near high-rise structures). Recent cuts in human resources in the state have severely restricted the ability to enforce regulations in the field, and control is largely accomplished through initial evaluation of permits.

Setback laws

Erosion control setback lines are designed to prevent people from locating structures too close to eroding shorelines in order to: (1) minimize loss of life and property; (2) reduce public costs from poorly located development; (3) protect public access and use of the beach; (4) protect natural features; (5) provide a natural buffer area for beach mobility; and (6) preserve visual openness and aesthetic values (Owens 1985; Houlahan 1989). Setback lines are delineated on the basis of: (1) a multiple of the annual erosion rate; (2) a distance from a prominent natural coastal feature, such as a foredune or vegetation line; or (3) an arbitrary value. The North Carolina minimum oceanfront setback, for example, is the farthest landward of: (1) a distance equal to 30 times the long-term annual erosion rate measured from the vegetation line; (2) the first dune with an elevation at the 100 year storm plus 1.8 m; (3) the landward toe of the first dune with substantial protective value; or (4) 18 m landward of the vegetation line (Owens 1985).

Hazard and shore protection programs

State programs and policies oriented toward shore protection play a major role in influencing the likelihood that strategies compatible with beaches and dunes are implemented. States usually have the authority to conduct their own shore protection projects, reimburse municipal projects and share the cost of national projects. These initiatives will only produce naturally functioning coastal landscapes if the solution includes nourishing beaches and constructing dunes and ensuring that local activities are compatible with naturally functioning beaches and dunes once these landforms are constructed.

States usually have the authority to pass laws regulating development in the interest of reducing the degree of coastal hazard, including development permits for rebuilding substantially damaged structures (Beatley *et al.* 1994). The permitting process may result in significant improvements to shore protection and beach maintenance projects that would not have otherwise been possible without the interests brought to bear by the involved parties, but there may be considerable costs involved in the added requirements (Bodge and Olsen 1992).

Parks

State park systems often reflect the dual purposes of national parks, namely resource conservation and public enjoyment. As in the case of national parks, there is difficulty operationalizing these policy directives because it is difficult to determine the relative importance of each purpose (Morgan 1996). Some parks are managed as natural environments with isolated areas of intensive use, whereas other parks may be managed to accommodate large numbers of visitors and favor a variety of cultural features and amenities. The significance of the differences in management are discussed at greater length in chapter 8.

State programs for acquisition to protect or maintain natural features or restore or promote development for recreation outside of the state park framework, such as The Green Acres Program in New Jersey or the Coastal Facilities Improvement Program in Massachusetts, are potential ways coastal resources can be retained or enhanced. These kinds of programs provide opportunities to prevent future shorefront development that would eventually result in beach loss, and the resulting undeveloped enclaves can provide local demonstration areas to serve as reminders of the way the undeveloped coast functions.

Access laws

Laws governing rights of access to or along beaches provide a means of regulating coastal structures on the beach. The beach is considered to be a public highway in Oregon, for example. Public rights to beach access in Texas have been used to preclude rebuilding of coastal structures that impaired lateral access when the public/private ownership line changed through erosion (Schmahl and Conklin 1991). An access law can preserve large segments of beach when this landform is specifically designated as the route of access, but an access law without this designation does not guarantee preservation of natural landscapes because access can be provided by structures, including promenades, paths on seawalls and groins and stairways built on coastal cliffs.

Areas of concern

States can formulate plans for managing areas of special concern and develop guidelines for municipalities in development of their own plans. Areas of concern are most likely to be based on the need to preserve historic or archaeological sites or target species, perhaps in support of national or international initiatives. Preservation of cultural sites does not mean that coastal landforms will be preserved or restored unless the natural environment is considered part of the cultural heritage that is preserved. Programs to manage target species hold more promise in protecting natural environments. These programs are discussed here in terms of actions at higher levels of government.

State management programs may include policies that indirectly protect landforms by protecting scenic areas. New York State has a program to identify, evaluate and recommend areas for designation as scenic areas of statewide significance and another policy that requires actions just outside these areas to protect, restore or enhance the overall scenic quality of the coast. The effect of human activities on geologic forms is included in this policy (Nugent and Hart 1989). Programs such as these have similarities to the Heritage Coast concept, operative at the level of state government. The likelihood that they will be implemented will probably vary according to the way the image of the coast is valued by the state recreation industry.

Planning goals and mandates

State management acts may require each local government to prepare shoreline master plans based on state policies or goals (Born and Miller 1987; Ortman 1987; Burby and Dalton 1993; Good 1994). The Oregon Land Conservation and Development Commission's Beaches and Dunes Goal is designed to conserve and, where appropriate, restore the resources of coastal beach and dune areas and reduce hazards to life and property. Local jurisdictions are required to consider these factors in their plans and ordinances by: (1) prohibiting residential developments on active foredunes and foredunes subject to wave undercutting and overtopping; (2) regulating actions to minimize erosion; and (3) regulating dune breaching.

Laws specific to beach and dune management

Florida's Comprehensive Beach Management Law is considered the first real effort in the USA to try to manage the beaches of an entire state (Tait 1991). It is intended to prevent erosion by: (1) eliminating causes of human-induced erosion, especially sand loss at inlets; (2) establishing beach restoration and nourishment as a primary tool to repair eroded beaches; (3)

reinforcing state regulations that insure that construction and reconstruction do not damage remaining beaches and dunes; (4) recognizing the need to purchase large tracts of remaining pristine beaches before they are developed; and (5) directing the state Division of Beaches and Shores to undertake comprehensive coastal studies on erosion rates, offshore and onshore environmental conditions, type and density of coastal development and economic impact of beaches in different areas (Tait 1991). The initiative includes a comprehensive beach restoration management plan to recommend projects and set annual priorities to nourish eroding beaches (Tait 1991). The majority of the beach erosion in Florida is thought to be related to ocean inlets (McLouth *et al.* 1994), and the state requires a management plan for each altered ocean inlet to evaluate how the inlet contributes to beach erosion and identify remedial measures for sand bypass (Tait 1991; McLouth *et al.* 1994).

Some states have imposed restrictions on building of erosion control structures, and some, such as North Carolina, South Carolina and Maine, have banned new shore-hardening structures. The South Carolina Beachfront Management Act of 1988 (amended 1990) was designed to protect, preserve, restore and enhance the beach dune system, and it has policies designed to: (1) severely restrict the use of hard erosion control devices; (2) replace structures with soft technologies; and (3) promote carefully planned beach nourishment where economically feasible (Van Dolah *et al.* 1993). State permitting requirements can also contain requirements that structures be designed to allow for adjustability in the future, leading to modular construction (Olsen 1996).

Information programs

States may have the resources to establish major public information programs and to produce guidelines for management of resources that can be implemented by communities or local residents. State programs, in turn, are often funded by national programs or initiatives. These information programs have great significance in determining public attitudes about environmental issues and legislation and adopting environmentally friendly land use actions. Some of the major state programs are identified here. Future efforts at all levels of government are a focus of attention in chapter 9.

Some of the best examples of state information programs are from Australia. The Western Australia Department of Agriculture produced a Coastal Rehabilitation Manual that outlines the techniques available for dune construction and rehabilitation. The Beach Protection Authority of Queensland has produced a newsletter *Beach Conservation*, that is useful for making the public aware of ways that beaches change and how they can be managed as well as leaflets on specific topics such as the formation and function of coastal dunes and description of dune plants. The Soil Conservation Service of New South Wales has a manual of coastal dune management and has initiated a program involving voluntary community participation called Dunecare that focusses primarily on dune rehabilitation using vegetation management, sand-drift control, community awareness and education. Ingram and Chapman (1993) identify the characteristics of what makes a successful Dunecare group.

Regional commissions and plans

Coastal ecosystems and sediment exchanges often do not follow jurisdictional boundaries, suggesting the need for multi-governmental bodies (Beatley *et al.* 1994). Regional plans can be developed for a number of local governments that share similar demands, opportunities and constraints; these plans may have wide-ranging scope like state plans and may take the form of comprehensive strategies or relate to specific environmental issues, such as tourism or management of sand-drift areas (Chapman 1989). Regional groups can help overcome the lack of resources in communities by pooling them, for example, preparing a "best practice" guide to define roles of key players, give examples of ways to help resolve competing pressures in the coast and clarify how different elements of management interact (Holgate-Pollard 1996).

Counties in the USA enforce local regulations for unincorporated boroughs. This function allows them to take the individual voices of the communities and turn them into a strong voice for the preservation of the regional shoreline (Goss and Gooderham 1996). County governments are often more likely to respond to the need to preserve open space than municipalities, and they have the authority to establish health regulations that regulate septic systems, thus providing a means of controlling development of beaches and dunes.

Other examples of regional organizations and plans are found in Western Australia Department of Planning and Urban Development (1994a and b), Clarke (1991) and San Diego Association of Governments (1995). The San Diego, California regional plan assesses the extent of the current and future shoreline erosion problems and identifies potential solutions, including beach nourishment, shore protection structures and policies and regulations on use and development of the shoreline, including building setbacks for bluff tops. This strategy is put together by a committee made up primarily of elected officials from the coastal jurisdictions in the region, relying on advice and opinions of technical advisory members representing agencies involved in managing the shoreline and individuals and interest groups who review and comment on the preliminary strategy (San Diego Association of Governments 1995).

Informal voluntary regional coastal defense groups have recently been formed in England and Wales to encourage co-ordination and exchange of information between neighboring authorities (Lee 1993; Bray *et al.* 1997). One example is the Standing Conference on Problems Associated with the Coastline (SCOPAC) that comprises 29 local authorities and agencies covering a 200 km segment of the south coast of England (Bray *et al.* 1997). These groups have great potential for wise management of beach and dune resources, and they can take the lead in evaluating local effects of sea level rise and options for implementation at local level. To be most effective, they must be based on littoral drift cells and seriously consider other aspects of shoreline management beyond coastal defense (Lee 1993; Purnell 1996).

Perhaps the most powerful regional authorities, in terms of activities in the dunes, are the Dutch Water Boards. They are the main organizations involved in the maintenance of the sea defense function of dunes in The Netherlands. The water boards are responsible for reshaping the dune and vegetating it and all subsequent repairs. They provide permits for use of the dune and for construction of recreation facilities in accordance with provincial plans. The outer appearance of the dune in a given region is essentially determined by the perceptions and preferences of managers in the water boards. The Dutch situation is unusual because the municipalities and residents play a minimal role in affecting changes to the dune. Municipalities in The Netherlands may act as consultants, but they do not have specific tasks related to dune construction or maintenance (Koster and Hillen 1995).

Local programs

Great freedom is given to municipalities and residents in some countries in managing foredunes. In the USA, two dogmas of the political system obstruct national initiatives to protect environmental resources; these are privatism, whereby owners are entitled to use their land largely as they wish, and localism, whereby planning and management of coastal resources is considered within local government purview (Platt 1994). Controls over development landward of mean high water are executed through municipal zoning regulations, subdivision controls and building permits.

Periodic maintenance of beaches and dunes is often performed at the municipal level. The resources available to municipalities in the USA often include beach cleaning equipment, bulldozers and other types of earth-moving equipment. Promenades, boardwalks and piers are built and maintained by the municipalities. Some municipalities have comprehensive programs for maintaining dunes and have strong ordinances regulating activities in them (Godfrey 1987). They may have a municipal budget for sand fences, and they may have programs for planting vegetation utilizing volunteer labor (Mauriello and Halsey 1987; Mauriello 1989). Most dune management issues are local in nature, and community groups are best placed to control access and implement and monitor rehabilitation and protection works (Western Australia Department of Planning and Urban Development 1993). Volunteer activities using local residents, school children and youth organizations are best mustered at the local level. Activities include beach cleaning (Breton and Esteban 1995), "adopt a dune" programs (Carlson and Godfrey 1989) and construction of boardwalks through dunes (Mullard 1995).

Regulations vary considerably among municipalities in details about permissible activities. Some municipalities have regulations that significantly reduce the number of people and value of structures at the shoreline, facilitate removal of threatened structures and limit the amount of impervious surfaces (Bortz 1991). Some municipalities in California require a complete geologic report, written according to specific guidelines and subject to peer review, and have minimum setbacks ranging from 15 to 46 m, but at least one community has only a 6.1 m setback and no requirement for a geologic report (Plant and Griggs 1991).

Local officials are familiar with local interests and are directly accountable to the landowners most affected by planning decisions. Local governments are potentially the most closely linked to coastal management at the scale of individual landforms. Cause-and-effect factors can be understood within the geographical, ecological and institutional scope of concern (White *et al.* 1997). Some local communities can exhibit a surprisingly high level of public interest and consensus for a holistic systems approach to managing sediment and minimizing shore protection structures that would interfere with beach use (Bray *et al.* 1997). There are some excellent examples of documents that develop management and development policies with recommendations to ensure a comprehensive approach to dealing with coastal problems at the local level (e.g., City of Stirling 1984).

Municipal level actions can greatly expedite implementation of initiatives taken at higher levels of government, especially if they are informed and organized and formulate beach management plans (Smith 1991). It is questionable whether local authorities can always defend natural values by resisting pressure for commercial exploitation unless they are given clear rules or minimum standards that must be followed (Nordberg 1995), although it has also been claimed that preparation of non-statutory management plans by local authorities could play a valuable part in effective management at the local level (Holgate-Pollard 1996). Municipalities can go beyond minimum state requirements for reducing coastal hazards and protecting resources (Beatley *et al.* 1994), including implementing more stringent requirements for coastal construction (Yazdani and Ycaza 1995). They can fund their own comprehensive beach management programs and can designate their own preservation areas, such as the 1.5 km stretch of Catalonian coast administered by the council of El Prat (Breton and Esteban 1995). They can also conduct their own restoration projects, such as the project to restore the Devesa of Saler, Valencia (Sanjaume 1988). Large-scale projects are possible in large jurisdictions because of the greater resources available. The town council in Valencia invested more than 2 000 000 pesetas for their restoration (Sanjaume 1988). The Borough of Avalon in New Jersey is less intensively used by tourists than many resorts in the state, but it has managed to fund its own 304 000 m³ beach nourishment project and mechanical recycling program (Mauriello 1991).

Intergovernmental programs

Although not designed as such, intergovernmental programs for preservation of wildlife hold promise for protection of beaches and dunes. Preservation of migratory species depends on establishing habitat networks at international scales. Actions to conserve waterfowl are increasingly focussed on collaborative international conservation efforts (Davidson and Stroud 1996), as are actions to find solutions to atmospheric and ocean pollution. The adoption of nationally integrated planning and management has been promoted by numerous international organizations, such as the Intergovernmental Panel on Climatic Change, the Council of the European Communities, the United Nations Conference of Environment and Development and the Organization for Economic Co-operation and Development (Jones 1996b).

Designation as Special Area of Conservation under the European Union Habitats Directive will ultimately create a Pan-European network of key nature conservation sites on the coast; it will provide an additional tier of protection, increase awareness of the importance of a site among local interest groups and authorities, and increase exposure to their efforts (Jones 1996a; Julien 1996). Related measures include the Wild Birds Directive, LIFE funding program and the Conservation of African/Eurasian Migratory

Waterbirds (Davidson and Stroud 1996; Julien 1996). The success of programs designed to protect key wildlife species in protecting beach and dune resources is related to the degree that these programs consider landforms critical to their habitat needs.

Trusts and non-profit organizations

Environmental trusts and private non-profit organizations provide a way of preventing environmental losses associated with development by purchasing and managing sites with natural beauty, ecological value or historical significance to make them available for public use or to restrict them from public use. Environmental organizations can exert a powerful legal influence when united on a regional scale, and they can delay implementation of development plans or cause them to be altered, even where they cannot stop them or purchase the land (Miossec 1988).

A host of private (mainly charitable) organizations exist in Great Britain, headed by the National Trust (Williams 1987). The French Conservatoire de l'Espace Littoral is patterned on the National Trust of Britain, but it is a public entity and is allowed to buy or receive coastal lands and delegate their management to public councils or private societies. This organization was created in 1975 to buy coastal sites of natural or scientific interest and had included 190 sites by 1985 (Guilcher 1985).

Not all activities of environmental groups are designed to perpetuate a completely natural landscape. Justification of some purchases by environmental trusts requires non-intensive uses that may need boardwalks, footpaths, car-parks, visitor facilities and litter clearance or allow grazing (Harvey 1996). Some purchases may be made to enhance specific types of wildlife (especially birds) that may involve transformation of portions of the landscape to accommodate these species, locally altering the natural function of beaches and dunes. This problem is hardly unique to environmentally friendly organizations. The following section identifies many ways that programs and policies designed to enhance beach and dune resources can be diminished in their effectiveness.

Problems of establishing programs favorable to beaches and dunes

Many problems associated with coastal development in the past cannot occur today because there is greater regulation of detrimental activities and increased knowledge of their effects (Shipman 1993). Regulations now make it difficult to alter shorelines indiscriminately, but alterations that destroy coastal landforms still may occur for many reasons that are identified in Table 7.2 and discussed below.

Problems of implementation

Integrated coastal zone management involves: (1) horizontal integration of separate economic sectors and the units of government that significantly influence planning and management of resources and environments; (2) vertical integration of all levels of government and non-governmental organizations that influence planning and management; (3) adopting a perspective that combines land use and sea use; and (4) adopting a program that contains planning, management, education and applied research components (Sorensen 1997). Integrated coastal management at this kind of scale is a goal that takes many years to accomplish, and coastal management in its modern sense is still lacking in some countries (Cambers 1993b; Savov and Borissova 1993).

National and international programs

There may be a lack of coordination and cooperation between government agencies, especially in countries where ministers are beaurocratic, over-centralized, hierarchical and management responsibilities are scattered across many different government agencies (Fermán-Almada and Gómez-Morin 1993; Shah *et al.* 1997). The quality of the environment deteriorates because of the absence of mechanisms allowing the complex relationships between human activities and the environment to be taken into account in a decision-making process where success depends on integration at local, regional and international level, as well as at national level (Julien 1996). A challenge for managers of protected areas is to maintain a unified approach in meeting requirements of different levels of protection due to different management designations within their jurisdictions; the Gower Area of Outstanding Natural Beauty and Heritage Coast, for example, has eight different kinds of statutory designations with the purpose of conserving or protecting natural and cultural features (Mullard 1995).

Problems can occur in the same agency administering different programs. The US Army Corps of Engineers constructs and maintains navigation and beach nourishment projects, but the costs and benefits of the two types of activities have not been considered jointly to make use of dredge spoil as beachfill. The occasional placement of sand on beaches has been conducted as

Table 7.2. *Problems of establishing planning and policy programs compatible with maintenance of natural beaches and dunes*

Problems of implementation

Many participants cause difficulties of co-ordination and co-operation.
Modern management is still lacking in some countries.
Regulations may apply only to new development, not improvements.
Court-ordered penalties may be too low.
Illegal activities occur despite regulations.
Existing environmental policies may be rescinded or amended.
Support for rebuilding damaged structures favors property owners.
Value of shorefront property argues against preventing development.
Requirement to purchase threatened properties at market value cannot be met.
Approved initiatives may lack funding for implementation.

Problems of conflicting goals

Incompatible or contradictory policies occur in different regulatory agencies.
Residents and developers perceive resources differently from planners.
Uses that eliminate the beach may otherwise be considered compatible with a coastal location.
Uses that do not require a beach or dune may have priority.
Co-operation at local level is often dependent on personalities, not optimal solutions.
Programs may favor public user facilities over natural values.
Individual species rather than landscapes are often targets of conservation.
Sites of geomorphic interest are less significant than sites of ecological interest.

Problems of spatial coverage

Management may emphasize stability, not sustainability or spatial and temporal flexibility.
Control zones may not coincide with physiographic units or coastal dynamics.
Policies may not establish coastal construction setbacks, or setbacks may be too small.
Degradational activities may be displaced to jurisdictions where there are no controls.

Problems in technical expertise

Jurisdictions usually lack the staff to make technical and scientific evaluations.
Undeveloped environments may be managed as natural, but they may not be natural.

Problems in timing

Land use management may take decades to reveal benefits.
The process of nourishing a beach can take up to 15 years.
Waiting for erosion to become an emergency often results in structural solutions.
Prescribed lifetimes of structures do not reflect their longevity.
The life of engineering projects exceeds programs of local sponsors.
Timing of nourishment projects is determined by administrative factors, not beach width.
Politicians respond readily to emergencies but lose interest in long-term projects.
Long-term study of effects of projects is unappreciated by politicians.

Source: Modified from Nordstrom 1998.

a least-cost disposal option rather than a planned action to minimize disruption of the littoral system (National Research Council 1995). Countless millions of m³ of sediment dredged for maintenance of harbors and navigation projects have been lost in many locations as a result of failure to combine harbor and beach erosion control projects, although some jurisdictions have instituted effective programs to ensure that beach-quality sand is not lost (Gómez-Pina and Ramírez 1994; McLouth *et al.* 1994).

State programs

State programs can have different focuses. Differences that are due to spatial variations in process controls and natural landforms need not be detrimental to environmentally compatible programs, but differences in topical coverage or levels of rigor can lead to inconsistent protection for coastal resources viewed from a national perspective.

In Australia, Queensland's program focuses primarily on erosion control and setback lines, whereas New South Wales uses a variety of planning and development controls and focuses on balanced use of the coast and protection of scenic and recreational areas in addition to erosion and safety (Jackson and O'Donnell 1993). The nature of coastal conservation and associated legislation and regulation can be expected to vary even more greatly at the international scale. Even across Europe, where international initiatives are already in place, there is great variation (Nordberg 1995).

Local programs

Planning and management initiatives are also difficult to coordinate at the local level, and effective state policies may be rendered ineffective because of lack of commitment at that level (Burby and Dalton 1993). For example, the inlet management initiative of the Florida beach management law was delayed because the 37 improved inlets were under local government control; some seemed to have no one in charge; and many inlet authorities were slow in coming up with required plans (Tait 1991).

Pre-storm planning for post-storm construction can occur in some municipalities (Bortz 1991), but many local governments pay little attention to programs to build citizen awareness of natural hazards (Burby and Dalton 1993). Many communities still lack regulations to prevent development in coastal high-hazard areas or mitigate effects of hazards (Fischer *et al.* 1995). Local officials tend to rely on tourism for making decisions affecting their coastal zones, and they are reluctant to restrict development in response to coastal hazards, although they are increasing aware of the conflicts they face between public and private concerns (Fischer *et al.* 1995).

Co-ordination and co-operation at the local level is often dependent on personalities involved, resulting in sub-optimal solutions (Ballinger *et al.* 1996). It is difficult to envision local governments and property owners accepting solutions that opt for retreat from coasts; most local residents would advocate an option that approaches the status quo (Titus 1990), and for most communities, prevention of coastal construction or restrictions to reconstruction following storms based on geomorphic principles is an elusive goal (Nordstrom and Jackson 1995).

Problems of conflicting goals

There are many government agencies responsible for management and regulation of activities in the coastal zone due to piecemeal legislation, leading to fragmented structure, difficulties of coordination and cooperation and incompatible policies (Ballinger *et al.* 1996; Knight *et al.* 1997). Many authorities have contradictory policies within their coastal zone management programs because of the need to accommodate diverse resources and interests. There are 16 interest groups involved in the issue of coastal defense in Wales (Ballinger 1996). There are 38 coastal resource policies in the state of Massachusetts alone (Born and Miller 1987). Permissible uses may vary between jurisdictions, and modifications are allowed at any time to reflect changing local circumstances. Regional authorities and commissions are task-oriented and may have goals and responsibilities that conflict with other programs that are designed to enhance a different aspect of the public good.

Environmentally friendly goals

International conventions can be concerned with protection of individual species and neglect coastal conservation and landscape protection (Nordberg 1995). Migratory bird and coastal habitat conservation are still often approached separately (Davidson and Stroud 1996), and target species may not represent the full range of potential species, even in international programs such as the European Habitats Directive (Davidson and Stroud 1996; van der Maarel and van der Maarel-Versluys 1996). World Wildlife Fund and Society for the Protection of Nature can identify locations suitable for conservation, based on the significance of beaches and dunes as breeding grounds, but their value does not mean that the value of these areas will take precedence over development for tourism with incompatible facilities and support infrastructure (Demirayak 1995). Sites of geomorphic value are treated even less comprehensively than sites of ecological interest (Mather and Ritchie 1977).

Sites of landscape interest may be mentioned in some coastal regulations (Miossec 1993) but not all. Eight of the 20 municipalities evaluated by Fischer *et al.* (1995) did not emphasize particular features of the coastal zone, although they possessed coastal landforms of considerable natural or scenic interest. Acts for the protection of natural beauties of the landscape, even where they include important geologic sites, may be directed only toward preservation of the external appearance of the landscape (Mura 1989). Part of the problem is that there are alternative and subjective interpretations of what constitutes remarkable or characteristic landscape or what is natural coastal heritage (Miossec 1993). Objective evaluation of physical characteristics and recognition of the structural importance of the physical features are more successful elements in a strategy of real nature conservation (Mura 1989).

Competing non-environmental goals

A major problem of management programs is the lack of emphasis on preservation of the natural aspects of beaches and dunes (Healy and Zinn 1985; González-Yajimovich and Escofet 1991; Guilcher and Hallégouët 1991; Sanjaume and Pardo 1992). Ocean beaches are destroyed or truncated to accommodate alternative human uses such as recreation, transportation and residences. Management programs may stress public access or public utility structures such as walkways and promenades (Anthony 1997), and programs for acquisition of resources are often designed to improve public facilities (Born and Miller 1987) rather than maintaining or enhancing beaches and dunes for their natural values or resource values. Elements of a development or use plan that inevitably result in considerable degradation of landforms and biota may pass intensive scrutiny because they appear beneficial to the public. Provision of car-parks that change levels and patterns of human use have great potential for this kind of damage (Capelli 1991). These alternative uses are compatible with a coastal location according to many policies.

Local land use controls in shorefront communities usually are more responsive to the need to preserve beaches and dunes as economic resources rather than natural resources. Local governments are often pro-development, and increase of regulatory powers at the local level is seen as disadvantageous to reduction of coastal hazards and protection of coastal landscape features (Miossec 1988; Gares 1989; Good 1994). Permission to construct buildings is an outcome of the policy of encouraging residential and commercial development. Loss of beach area occurs as a result of construction of new buildings or protection structures that replace the beach or that result in subsequent beach loss when the shoreline erodes to the location of the structure. Development of coastal resources for greater human use is a major emphasis in most nationally supported coastal management programs, but water-dependent uses that do not require a beach or naturally functioning dune also may have priority, leading to loss of beach to accommodate those uses.

Relationship to shore protection programs

Programs that provide for protection from coastal hazards or for disaster relief may conflict with programs for protection of natural resources by encouraging development in hazardous coastal environments (Beatley *et al.* 1994). Requirements of the US National Flood Insurance Program do not ensure that all structural measures designed to control erosion are passive and conducive to maintenance of the natural system, but states and local jurisdictions could augment the requirements of the program to achieve this goal. The US Army Corps of Engineers has created beaches and dunes where none existed previously and has increased the dimensions of some, but the design of these features has been based on considerations for shore protection rather than maintenance of a natural sediment exchange system or natural values.

Coastal protection strategies are often incompatible with preservation of landforms, where static structures are used as a substitute for natural forms of protection or where structures prevent migration of landforms. Coastal protection strategies are adaptable to provide protection against sea level rises of 1.0 m or more, but these strategies have a high likelihood of interfering with the targets of nature conservation; conservation–benign strategies such as managed retreat will require explicit statements about natural features considered important as heritage for future generations (Kunz 1993b).

Cost constraints

Problems in funding programs

Locations that are presently undeveloped often have a different status in terms of degree of erosion hazard and therefore priority for funding measures to mitigate erosion. A somewhat confusing aspect of coastal erosion is that the rate of erosion may be less important than the type and value of the development landward of the beach (Schmahl and Conklin 1991). For example, Cape San Blas on the Gulf coast of Florida has a rate of shoreline erosion estimated at >9.0 m a^{-1} but is not considered critically eroding by the state because it is undeveloped and primarily in public ownership, whereas Flagler Beach on Florida's northeast coast, has a low recession rate, but the area is considered a critical erosion area because there is a public roadway within about 15 m of the dune vegetation line (Schmahl and Conklin 1991).

One of the greatest impediments to preservation of natural coastal features is the requirement to purchase portions of the coast at market (development land) value to preserve them from exploitation (Nordberg 1995) or to compensate owners forced to abandon land under scenarios of landward retreat (Bray *et al.* 1997). These costs can be considerable. For example, the estimated purchase price of a 5.5 km tract in Walton County, Florida is US$100 million (Schmahl and Conklin 1991).

Failure to fully fund national and state programs and projects can lead to suboptimal solutions involving beach resources. The Florida beach management law may be the best in the country but one of its major components – the inlet management initiative – is being held back by lack of funding. The legislature recommended committing $35 million annually when the law was passed in 1987, but they have never funded to that level, and the 1991 state legislature appropriated only $2 million of the $6.7 million requested for inlets (Tait 1991). Sand bypass operations are often included in federal projects designed to control inlets with jetties, but they are not always implemented because of the costs involved. The sediment deficit may be overcome by using beach nourishment as a mitigating measure, such as at Cape May, New Jersey (Figures 3.4 and 4.1), but the delay between construction and nourishment (that can take decades) can result in considerable expenditures for interim shore protection structures and loss of natural geomorphic features.

Questions about the high cost of beach nourishment, the appropriateness of using public funds and the long-term effectiveness of this strategy have resulted in an array of technical, fiscal and permitting obstacles to successful completion of projects (Stauble and Nelson 1985), along with calls for increased benefit–cost ratios and lower federal cost sharing (Finkl 1996a). One of the greatest constraints to implementing large-scale federal beach nourishment projects in the USA is the current policy that excludes recreation as a primary benefit of nourishment projects, reducing the amount of national involvement and reducing the likelihood that the highly successful Miami Beach project will be repeated in other locations (Domurat 1987; Houston 1995). The congressional authorizations that enable the US Army Corps of Engineers to undertake coastal projects result in independent and narrowly prescribed projects, and the planning and calculation of benefit–cost ratios are conducted independently of each other; in some cases, this constraint has resulted in ocean dumping of beach-quality sand dredged from inlets (National Research Council 1995).

Problems in preventing development

Arguments to prevent coastal development or convert developed areas to natural environments because of economic costs (e.g., Pilkey 1981) hold little weight because of the enormous value of shorefront property (Titus 1990) and because property owners are not fully accountable for their actions (Boyles 1993). The threat of litigation also appears to cause local elected officials to approve development applications that may be denied at the level of the planning staff (Griggs 1994).

Damage assessments and economic support for rebuilding damaged structures are liberally interpreted in favor of property owners (Beatley *et al.* 1992), and the costs of government actions are borne by the tax-paying public (Boyles 1993). Planners and policy makers in the USA continue to recommend changes in federal policy and state programs to end public subsidies to private development in hazard areas and break the build–destroy–rebuild cycle (Godschalk *et al.* 1989), but the problem of increasing vulnerability continues (Nordstrom and Jackson 1995). Even where governments have the economic resources and mechanisms to buy threatened and damaged property from owners, the owners often are not willing to sell (Davison 1993; Rogers 1993; Schmahl and Conklin 1991).

Limitations to technical expertise

Cost constraints can limit the amount of technical expertise available for planning and management, particularly at the local level (Haward and Bergin 1991; Jackson and O'Donnell 1993). Geologic information required to make permit decisions is often incomplete, and jurisdictions lack the staff to make the necessary technical and scientific evaluations (Griggs 1994) or prepare detailed beach management plans (Goss and Gooderham 1996). Degradation of geomorphological features on human-altered coastlines is often attributed to ignorance of the processes of coastal evolution (e.g., Miossec 1988, 1993), implying that broader application of expertise would help prevent this degradation. In some cases, it may be necessary for higher levels of government to require expert studies prior to new coastal development (Fischer *et al.* 1995). This level of support can result in high-quality plans for some communities (e.g., Western Australia Department of Planning and Urban Development 1993).

Economic opportunities

Greater attention needs to be focussed on environmental opportunities and penalties that can increase the likelihood that beaches and dunes will be enhanced. Programs to pay farmers for an environmental product rather than a commercial one hold promise (Mullard 1995). Increasing the penalties for unauthorized pedestrian and vehicular use through dunes has been suggested, with surveillance and policing functions given to dunecare groups (Koltasz Smith and Partners 1994). Another alternative is to use beach user fees for environmental improvement.

Revenues from taxes associated with tourism can greatly exceed the cost of beach nourishment in important tourist areas; the US federal government receives about 6 times more tax revenue annually from foreign tourists at Miami Beach than it spends to restore beaches in the entire country (Houston 1996a). Large-scale beach nourishment operations, coupled with improvements to recreational infrastructure can transform low-value real estate into intensively used resorts (Dornhelm 1995) that can then justify future nourishment operations. Changes to the federal involvement in beach nourishment projects in the Water Resources Development Act of 1986 increased the commitment to periodic nourishment from 10 years to 50 years and authorized use of fill materials from non-domestic sources (Wiegel 1992b), greatly increasing the long-term viability of projects that are approved.

The economic cost of beach nourishment is trivial relative to the importance of beaches to the tourist economy and the amount of money available, providing that a mechanism can be found to channel the funds into nourishment (Bell and Leeworthy 1985; Finkl 1996a; Houston 1995, 1996a). A greater share of the financial burden of funding large-scale projects is expected to be directed at local communities in the future, but there are more ways to raise the needed funds than are commonly used or perceived (Stone 1991). Transient lodging taxes, parking fees and user fees may be used to provide funding for beach nourishment projects, providing there is a clear management strategy and mechanism in place (Goss and Gooderham 1996). Privately funded nourishment projects using creative financing offer a viable alternative to public financing for communities unable or unwilling to provide the access and facilities that are required to receive state or national aid, or for communities that are unwilling to wait the time required for approval of government projects (Olsen 1982).

Difficulties of enforcement

Illegal activities
The effect of new environmentally friendly laws may be minimized because public officials are unable or unwilling to apply them for political or economic reasons (Fischer 1985; Granja 1996) or because the controlled activity is done illegally. Illegal construction, grading and extraction of resources occurs despite the existence of regulations against them (Mather and Ritchie 1977; Guilcher 1985; Meyer-Arendt 1991; Dettmer and Cave 1993; Ledesma-Vasquez and Huerta-Santana 1993; Marra 1993).

Illegal activities can be the result of frustration with vague policies, prohibitive costs of acquiring required technical expertise, lack of local enforcement or inadequacies of ordinances (Marra 1993). For example, the Mexican Law on National Property deems that all beaches are in the public domain and places the coastal zone from mean high water to 20 m inland under national ownership, but no attempt is made to enforce this edict (Fermán-Almada and Gómez-Morin 1993). The 1993 legislative amendments to the Coastal Area Facility Review Act (CAFRA) in New Jersey prohibits direct disturbance to dunes that would reduce their dimensions, but recent cuts in human resources in the state have severely restricted the ability to enforce regulations in the field, and control is largely accomplished through initial evaluation of permits. Violations to the California Coastal Act include forged permits, overbuilding of permitted structures, grading and construction of roads and seawalls (Dettmer and Cave 1993).

The enforcement of provisions of coastal management acts may place greater emphasis on collection of court-ordered monetary penalties than on violation prevention or restoration, and the developers may merely view the penalties, that are low relative to the value of the development, as the cost of doing business (Dettmer and Cave 1993). Official community response to illegal activities may involve little more than a warning letter (Marra 1993).

Remoteness from population centers and government offices also may prevent the application of existing regulations. There are numerous squatter settlements in Western Australia, for example, and many of these are built on primary dunes (Western Australia Department of Planning and Urban Development 1994a)

Finding loopholes in regulations

Regulations to development may apply only to new development and may not include provisions governing major enlargement of existing structures (Griggs 1994). Some local policies will allow construction on vacant lots that otherwise would be unbuildable, providing owners construct a shore protection structure (Good 1994). Proactive construction projects by development interests can modify coasts prior to obtaining government approval and force the government to redefine spatial limits of regulatory zones (Miossec 1988).

Regulations that prevent adverse use by coastal residents and visitors may not be applied to similar adverse uses by government employees. Use of off road vehicles for official patrol and maintenance may result in severe impacts on vegetation in reserves that are otherwise well managed to reduce impacts of tourists (Carlson and Godfrey 1989). Conversations with municipal managers in New Jersey indicate that it is difficult to get beach maintenance personnel to drive along prescribed seaward routes along the beach rather than routes at the base of the dune that minimize driving time but cause damage to dune vegetation. Government employees may not even be aware of (or conveniently forget) environmentally compatible policies within their own organizations (Griggs *et al.* 1991b).

Problems of spatial coverage

Boundaries of protected areas

The natural features of the coast and the social, political and economic frameworks within which coastal managers operate are dynamic, and no regulatory agency that attempts to achieve environmental goals will be successful if it adopts a static approach (Harvey 1996). Coastal zone management or designated area management may emphasize stability rather than sustainability; sustainable use of the coast requires both spatial and temporal flexibility (Pethick 1996). Lack of appreciation of the dynamism of the coast may be especially apparent in local plans (Granja and Carvalho 1995), but it can extend to dune reserves, where managers attempt to maintain a specific resource inventory rather than a mobile system (Nordstrom and Lotstein 1989).

Identifying and defining control zones

Administrative boundaries controlled by local authorities do not coincide with physiographic units, such as drift cells (Lee 1993; Purnell 1996). Another problem is that local planning authorities often refer only to the land, not the intertidal zone or the sea (Taussik 1995). Planning powers often stop at the low-water mark, and a high proportion of threats to habitats occur below this line (Mullard 1995). A study of municipalities in Cantabria, Spain revealed that only 8 of 20 coastal municipalities define their coastal zones beyond the 100 m required in the Shores Act of 1988 (Fischer *et al.* 1995). In many cases, it is the subaerial beach and dune environments that are unprotected. Prohibition of urbanization close to the beach does not protect dunes landward of the control zone (Haeseler 1989).

Many valuable coastal habitats will still remain outside the core of statutorily protected systems (Jones 1996a). The restrictions to development under the US Coastal Barrier Resources Act, for example, apply only to the barriers specifically included in the system. More can be done with programs for conserving, restoring or developing the natural characteristics of coastal environments by striving for biotic and abiotic variation and diversity in landforms and species outside protected areas (Tekke 1995).

There is always a danger that the use that is controlled may be displaced to locations where controls are less restrictive. Races and off-road testing of all-terrain vehicles have increased in Mexico as a result of increased restrictions in the USA (Espejel 1993), and development of the French Riviera coast increased as a result of greater restrictions on the neighboring Ligurian Riviera in Italy (Anthony 1997).

Setbacks

Setback distances in the USA can be as much as 305 m or as little as 4.6 m (Nordstrom 1992). A setback of 100 m is prescribed in the French coastal law of 1986 (Miossec 1993); the Spanish Shores Act of 1988 specifies a 100 m easement in undeveloped areas and 20 m in urbanized areas, along with a 500 m zone of influence to ensure access, stricter zoning of building density and removal of illegal buildings (McDowell *et al.* 1993; Fischer *et al.* 1995). Other distances include 400 m where the rate of erosion is high due to sediment starvation resulting from use of jetties (Awosika and Ibe 1993) and 60 times the annual erosion rate to correspond to the mean life of a building (Paskoff 1992).

There is still no policy to establish buffer zones or setback lines in many locations, or there is vague wording in existing policies that has this effect (Houlahan 1989; Fermán-Almada and Gómez-Morin 1993; Lee 1993; Granja and Carvalho 1995). In some cases, the typical setback distances are not enough to accommodate the expected accelerated erosion due to sea level rise (Titus and Narayanan 1996). Erosion control setback distances prescribed in state programs can be too short to allow for formation of a dune that is adequate to provide natural resource values, and erosion can eliminate the dune entirely if there is no provision for removal of buildings and structures following shoreline retreat. In the Siletz, Oregon littoral cell, where new construction setbacks met the minimum requirements in the county and city hazard inventory, 40 percent of the sites later required shore protection structures to mitigate erosion hazards (Good 1994).

Statewide guidelines exist for determining the geological instability of blufftop development in California, but there is no state policy establishing safe setbacks from the edge of a seacliff for any type of development (Griggs 1994). Some local jurisdictions use a fixed setback, but this can be as little as 3 m (Griggs 1994). Owners often locate at the minimum required distance, even when more landward locations are available on the lot (Stutts *et al.* 1989).

Problems in timing

Changes in social/political forces

The evolution of public policy is not a linear process and public decision makers may reject or rescind environmentally favorable management initiatives as a result of direct social pressure or changes in government (Nordstrom 1988b; Platt 1994; Fischer *et al.* 1995). The change may be to weaker as well as stronger controls. The Oregon amendment to their dune regulations allowing grading to provide a view of the sea for shorefront residents was viewed as a strengthening of policy because it included provisions for maintaining an effective barrier against flooding, provision of a management plan, increased technical assistance, improved reporting procedures, model enforcement ordinances and volunteer groups to assist in monitoring and maintenance (Cortright 1987; Nordstrom 1988b; Marra 1993). In developed areas, the issue of grading is one of management, not conservation, and the amendment was viewed as an acceptable response to a development situation. The change in policy and practice created a dune that is out of equilibrium with natural processes (Nordstrom 1988b). The resulting landform may have greater value as a protection structure and it may enhance recreation, but it is an artifact, and its appearance and function have been altered to enhance human perceptions of a limited number of values for the resource.

Emergency actions

Emergency actions are among the most serious threats to long-term preservation of natural beach and dune resources. Perhaps the greatest threat is associated with use of static protection structures. Statewide Planning Goal 17 of the Oregon Land Conservation and Development Commission requires that land use management practices and non-structural solutions to erosion problems be preferred to structural solutions, and, if structural solutions are shown to be necessary, they must be designed to minimize adverse impacts on water currents, erosion and accretion patterns (Marra 1993). In practice, the preferred method of mitigation in developed areas is usually rip-rap revetments (Marra 1993; Good 1994). This action may occur because the emergency situation has been reached by the time structural protection is requested or because alternative solutions have not been demonstrated to provide effective hazard mitigation (Marra 1993). There are no consistent criteria identifying the time when emergency permits for shore protection structures are warranted, and permits have been issued when no actual threat was warranted (Good 1994).

Emergency actions may use environmentally compatible alternatives, but they may be under-designed. Operations performed in New Jersey following the March 1962 storm (described in chapter 6) provided a beach and dune designed to protect against a storm with a recurrence interval of 1 in 10 years. This level of protection was considerably less than the level of the 1962 storm and was chosen to provide interim protection prior to the summer hurricane season. Plans called for a 15 m wide beach berm at 3 m above mean low water with a 6.1 m wide dune at 3.7 m above mean low water. Despite the logic of designing these emergency landforms and the clear rationale justifying their low level of protection, subsequent dune rebuilding programs conducted by municipalities have rarely increased the level of protection beyond that provided by these prototypes.

Perception of time horizons

A significant problem with environmentally compatible land use management measures is the length of time for benefits to be apparent; unlike stabilization structures, land use management is perceived by residents as a negative form of public activity and benefits may take as long as 2 or 3 decades to come to fruition (Chapman 1989). Another problem is that structures that are designed for a period of use (e.g., 30 years) actually survive and affect coastal processes and land use decisions long after their prescribed lifetimes (Griggs 1994).

Local officials can ignore erosion problems until there is an uprising from the locally affected community, in which case it may be too late to provide needed funds or take advantage of cost-effective solutions. Inaction on the part of government on matters of post-disaster redevelopment is especially problematic in that it encourages unwise or inappropriate uses (Schmahl and Conklin 1991). Politicians respond readily to emergency actions that focus on repair but quickly lose their interest in support for long-term projects or the continuity of data bases or technical staff (Smith and Jackson 1990). More effective management can be achieved through standing committees that encourage a proactive approach rather than reactivity (Goss and Gooderham 1996).

One of the greatest problems in retaining a viable suite of coastal landforms and associated biota in developed areas and adjacent sand-starved natural enclaves is the inability to conduct beach nourishment projects before natural features are lost to erosion (Figure 3.6). The timing of a beach nourishment project should be appropriate to conditions at the site rather than as prescribed by the administrative time of government projects (Kana 1993). The federal process for renourishing a beach in the USA usually takes 10 to 15 years from the reconnaissance study through the first nourishment; the time is 5 to 6 years for projects that are already authorized and require only preconstruction engineering, design and acquisition of real estate or rights of way (National Research Council 1995). The implication is that action must be taken before resources are threatened so that a solution is available when they are threatened.

The life cycle for a US Army Corps of Engineers project is typically 50 years. This time frame is rarely paralleled by a similar long-term program by the public or local project sponsors, and nourishment operations cannot be considered bona fide programs unless long-term planning commitments are in place (National Research Council 1995). This long-term commitment should be extended to monitoring as well. The long and expensive work involved in studying nourishment projects after many years is now unappreciated by politicians, particularly newly elected ones who do not understand the importance of beaches (Smith and Jackson 1990).

Summary conclusions

Coastal management policy is usually tiered, with national, state or provincial and local governing bodies sharing responsibility and authority. In some cases, this tiered quality is appropriate to the management of a natural resource that crosses political boundaries and has global environmental significance but represents a local source of income and recreational benefit. It is often difficult to integrate management at different administrative levels and properly address environmental issues because of conflicting goals, coordination problems and differences between the temporal and spatial expression of natural and human processes.

National and international laws and agreements can set standards and increase standards of coastal environmental protection, but their usefulness is limited in areas where coastal management is still in the inception stages and where local interests are allowed to dominate. Management at local levels is rendered difficult because the administrative boundaries of the shore controlled by local jurisdictions usually do not extend as far as the boundaries of natural geomorphic systems and because local authorities are unable or unwilling to enforce environmentally compatible restrictions to land use. Higher levels of government can provide perspectives, resources and regulations that are often lacking in local governments. The social, political and economic systems within which coastal managers operate are constantly changing, and policy must be sensitive to that change and evolve accordingly. Recognition of economic considerations as the current basis for decisions about coastal erosion and the applicability of scenic and natural resource protection laws to coastal areas is essential for effective management of coastal areas.

8

Maintaining and enhancing natural features in developed landscapes

Introduction

Previous chapters underscored the potential for eliminating landforms and ecosystems. It may be inevitable that human involvement in developed areas will continue to affect processes and landforms, but it is not inevitable that this involvement will be harmful to the beach (Flick 1993). Ways must be found to develop the shoreline to have both natural value and non-consumptive human use value. This chapter discusses ways that natural processes, landforms and ecosystems can be conserved or enhanced in relatively unaltered areas or restored in locations where natural environments have been severely altered or eliminated.

Values vs. degree of naturalness

The potential for modification of natural beaches and dunes to artifacts and the potential for restoration of coastal landscapes to more naturally functioning systems depends on human values for coastal resources and the perceived role of natural components in providing uses related to these values. The most commonly occurring human uses and their associated alterations may be placed in a continuum (Figure 8.1) to highlight those that are most natural. The headings on the vertical axis of Figure 8.1 are arranged according to the likelihood that the human alteration selected will result in natural or naturally functioning landforms. The human alterations are presented in the figure as mutually exclusive categories for simplicity of portrayal. They would not generally be considered mutually exclusive by coastal planners, although only one value may be emphasized in management programs, especially those programs conducted by local jurisdictions in small parcels of land.

The values represent nominal data and cannot be ranked quantitatively, but they have been placed in the figure according to the degree to which they

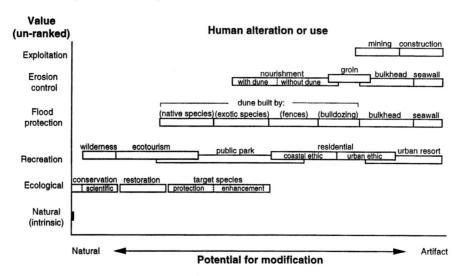

Figure 8.1 Potential for modification of coastal landscapes from natural to artifact based on perception of values for coastal resources.

contribute to, or enhance, "naturalness" to simplify interpretation. If managers of a segment of the coast wish only to address one value, the most natural of these will be to the left of each row on the diagram. If alternative values for beaches and dunes are considered, the most natural of these will be on the lowest row considered appropriate and as far left as possible.

Intrinsic values

Intrinsic (inherent or essential) value refers to the value that a component in nature has in itself. It is not a human use value, although it is a useful concept for humans in developing a management ethic about natural landforms and ecosystems (Nordstrom 1990). Consideration of intrinsic value provides a safeguard against manipulation of environmental meanings and can be used in the deconstruction of prior conceptions of the environment. The concept can play a more prominent role in environmental debate and carry weight in pragmatic decisions in coastal management if it is defined and refined so it retains its original meaning but can be better approximated by arguments in human terms (Nordstrom 1993).

Ecological values

Conservation and nature study

Use for scientific study of nature is distinguished from conservation value in Figure 8.1 in that landscapes that are preserved for study of nature are used directly by humans, although this use may have minor impact. Wilderness recreation and ecotourism (discussed below) also can have minor impact on landforms.

It is necessary at least to make some attempt to preserve relatively unaltered landscapes to function as templates against which the natural values of restored and artificial landscapes can be measured. The advantages of preserving relatively large tracts of undeveloped coastal land as reserves or as protected segments within national parks cannot be overemphasized. The first step in a comprehensive protection strategy is to focus on increasing the size and number of protected wilderness areas (Grumbine 1994). The lack of emphasis on natural landscapes in this book should in no way detract from the pressing need to ensure preservation of these environments and to obtain inventories of landform types and species (especially natives), dimensions of landscape components, spatial relationships (juxtapositions, exchanges) between these components, degree of mobility of the landscape and differences in all of these characteristics through time. The temporal variability is especially important because restored and artificially maintained landscapes will lack this evidence of landscape history.

Natural environments that exist as enclaves in developed areas may not be ideal templates for what occurred in the past along other sections of the shoreline or what should be done in those areas in restoration efforts. The portion of shoreline that appears to best fulfill the conservation function in New Jersey, for example, is Little Beach, an undeveloped barrier in the Brigantine National Wildlife Refuge. That barrier island is too short and too close to inlets to represent the kind of environment that would be typical in the central portion of the other barrier islands in New Jersey. Landscapes in other undeveloped enclaves in the state are human altered in the sense that dunes have been modified because: (1) sand fences have been used; (2) exotic vegetation has not been removed; (3) recreational use has trampled vegetation on the beach and dune (through illegal access in the latter case); and (4) sediment budgets have been altered by shore protection programs conducted in updrift segments. These locations have great value as templates for restoration in developed areas, but they may have limited value as templates for reconstruction of natural landscapes of the past.

Primary consideration for conservation is usually based on biological value, especially for target species, but geomorphological processes merit conservation much the same way as biological systems (Nordstrom and Lotstein 1989; Bray and Hooke 1995; Jones 1996a). The complexity and dynamic characteristic of coastal landforms are not easily defined in time or space, and change may be seen as the inverse of sustainable use and perceived as degradation or a hazard (Bray and Hooke 1995; Pethick 1996). This mobility is worthy of conservation, requiring more flexible approaches towards conserving landforms as dynamic components of natural coastal process systems (Bray and Hooke 1995). For example, some sand-drift areas are now considered a threatened resource, and management plans may recognize dynamic components of the dune landscape as worthy of protection, such as the Leba barriers (Piotrowska 1989) and the Oregon dunes (Pinto *et al.* 1972). This strategy of reverting to a dynamic system is likely to be perceived as acceptable only where human use facilities are not threatened.

Criteria for specifying value for terrain characteristics include: (1) rarity of geomorphological structures and of vegetation; (2) diversity of plant species within plant communities; (3) maturity of the ecosystem; (4) presence of complete assemblages of ecosystems that have developed and still exist; and (5) presence or potential for large-scale gradients between wet and dry and young and mature coastal ecosystems (Louisse and van der Meulen 1991). Habitat protection should be provided for viable populations of all native species (Grumbine 1994). Fragmented enclaves, although rare, could be of less importance than complete sequences of coastal ecosystems such as zonations from beach to inland dunes (incorporating all parts of the cross-shore transect), and valuable abiotic systems without actual, but with potential, values for biota should be included (Louisse and van der Meulen 1991). Other major considerations for conserving coastal sites for natural geomorphic and habitat value include minimal interference with natural processes occurring within these areas and minimal effect on natural inputs to these areas from adjacent areas.

The concept of dynamism in landforms includes short-term fluctuations in response to waves (and winds) and the evolution of landforms themselves (Bray and Hooke 1995). Landforms that are sustained by (or have access to) abundant sediment have the capacity for self-regulation and can be considered infinite in their ability to adjust; those that are isolated from sediment sources have a finite capacity for self-regulation (Bray and Hooke 1995). The greatest threats to protected areas may result from off-site activities that affect longshore sediment supply (Shipman 1993), indicating that the boundaries may have to extend well beyond the landforms themselves or the zone into which they are expected to migrate within the planning horizon.

These spatial requirements indicate that conservation areas be of substantial size and be large, undivided units where vegetation and landforms can undergo cycles of growth and decay (De Raeve 1989; Wanders 1989; Westhoff 1989; van der Meulen and Salman 1996) while restricting access and placing any buildings and support infrastructure required for public use outside the mobile areas (Nordstrom 1990). The requirement for minimal interference with natural processes means that attempts by managers to preserve environments by stabilizing them should be prevented, and attempts should be made to preserve the process of environmental evolution that requires a dynamic system.

Restoration of landscapes

Restoration involves human actions to create desired landscapes by reconstructing the physical characteristics that formerly occurred there or changing the vegetation by re-introducing native species or eliminating exotic species. Examples abound of restoration of degraded dune landscapes through a combination of acquisition by environmentally friendly agencies or conservation foundations, followed by restoration efforts, often using volunteer labor. Intervention management is now being advocated to sustain natural characteristics in coastal locations where natural processes can no longer create dynamic dune environments, for example, to replace those lost to scrub encroachment (Edmondson and Gateley 1996).

Ethical issues

Some scholars would question the view that humanity has the ability to repair or reconstruct damaged natural systems because natural reality is misperceived thereby as a human construct, and they may express discontent that an ahistorical, technologically created nature is substituted for reality (Katz 1992; Cowell 1993). In turn, some restorationists adopt a position of unity between humans and the environment and reject the premise that all restorations are fakes (Cowell 1993). Restored nature should not be used as a basic policy goal to take emphasis away from loss of natural landscapes, but it is a policy that makes the best of a bad situation by restoring previously disturbed locations (Katz 1992).

Many scientists feel that we have crossed a threshold beyond which human activities play the major role in evolution of coastal landforms (Morton 1979; Nordstrom 1994a), and nature conservation in some countries is more often perceived to be the protection of a human-made scenery (Klomp 1989). It can be argued that an integrated nature management strategy for beaches and dunes can no longer consist of solely striving for a natural situation or evaluating human activities as negative factors in a pristine environment, but it should try to safeguard or promote an optimal diversity of landforms, species and ecosystems, compatible with human values, even while using controlled disturbance (van der Maarel 1979; Westhoff 1985, 1989; Nordstrom and Lotstein 1989; Roberts 1989).

Restoring dynamic processes

Scientists working in coastal dunes have recently presented compelling arguments for maintaining coastal landforms in a dynamic, migrating state, based on the significance of these features for ecological productivity, diversity of habitat and preservation of rare species (Doody 1989; van der Meulen and van der Maarel 1989; Jones *et al.* 1995). Suspension of sand fencing programs in foredunes that were formerly stabilized has occurred in some areas, while, in others, foredune blowouts initiated by wave action or wind action have been left to develop unhindered (DeKimpe *et al.* 1991; Gares and Nordstrom 1991; Jones 1996a). More radical measures for restoring dynamic habitat characteristics, such as soil profile stripping and destabilization of recently quiescent dune blowouts, are being considered for some areas (Geelen *et al.* 1995; Jones *et al.* 1995). These options are expensive and may be politically sensitive where past problems of large-scale dune instability occurred (Jones *et al.* 1995).

In many locations, the former perception of sand-drift areas as waste lands or a source of danger has changed (Sherman and Nordstrom 1994), and bare sand in dunes and migrating sand areas are now recognized for their ecological and recreational values (Wanders 1989; Westhoff 1989). Ecological concern for bare sand and mobile dune systems reflects the shift in scientific interest to population dynamics and the effects of disturbance (De Raeve 1989). Scientist now often call for abiotic diversity in dune reserves to ensure ecological value and richness of the dune area, even if promotion of this diversity prevents retention of a fully vegetated dune system (Tinley 1985; Westhoff 1989; Jones 1996a). Stabilization measures, such as aforestation, now are viewed as problematic (Doody 1989), and some areas of woodland are already being removed from dunes (Sturgess 1992).

Despite the new interest in mobile components of dunes, many bare-sand areas still are stabilized categorically, although the fear that sand drifts will get out of control does not appear justified, given the stabilization techniques that are now available (Sherman and Nordstrom 1994). The major problem confronting managers of dune reserves is that the optimal landscapes for species diversity are dynamic ones that may occur for a relatively short time during a long evolutionary landscape continuum from mobile bare-sand landscapes to completely forested areas; many small-scale processes, including creation and movement of bare-sand areas cannot be separated from the large-scale evolution (De Raeve 1989). As a result, Sherman and Nordstrom (1994) call for management of dune reserves in ways that are compatible with new attitudes of scientists toward bare sand in coastal dunes.

The need for compromise in developed areas

Restoration of landscapes in developed areas is more problematic than restoration in areas set aside as reserves because of greater competition with human uses and the need to find compromise solutions. The need to compromise may require that restorationists reject the dualism that indicates that wild nature is good and human-manipulated nature is sullied (Light and Higgs 1996) and set aside the goal of complete recovery from population declines in target species to focus on what can realistically be achieved given an expected continued physical and functional loss of habitat due to development and recreational activities (Melvin *et al.* 1991). Nearly 30 percent of the US Atlantic coast piping plover nests are on municipal and county-owned beaches; multiple use philosophies must be developed for these beaches so that plovers and other beach-nesting birds are protected from human disturbance while also providing adequate opportunities for recreation (Melvin *et al.* 1991).

Target species protection and enhancement

Protection of target species may be distinguished from enhancement of target species because protection, if conducted properly, involves less direct alteration of the natural system. Enhancement may involve artificially creating a natural habitat type or even an exotic environment (e.g., establishing fresh-water ponds in a brackish environment) and is therefore considered a less natural alternative than species protection in Figure 8.1.

Target species protection offers a way to overcome public inertia or antagonism against efforts to return parts of a severely altered system to a naturally functioning one, although the naturally functioning portion of the landscape may be severely restricted in size or lack all of the components required for the target species. Bird species on the northeast coast of the USA that are important in efforts to return to a more naturally functioning beach system include least terns (*Sterna antillarum*) and piping plovers (*Charadrius melodus*) (Burger 1989; Melvin *et al.* 1991). An example of the opportunities and constraints associated with species protection, in a beach management context, may readily be seen in the changes that have taken place at Ocean City, New Jersey, USA (Figure 5.6) as a result of use of a nourished beach as nesting area by piping plovers.

Case study, Ocean City, USA

Piping plovers are listed as endangered and threatened by the US Fish and Wildlife Service because of their rarity and declining populations (Melvin *et al.* 1991). They nest on beaches above the high-tide line, on gently sloping foredunes, on washovers or in blowouts, and nests are placed on open sand or in patches of sparse to moderately dense beach grass and other dune vegetation (Melvin *et al.* 1991). Adults and chicks feed on invertebrates found in intertidal sand, in the wrack line and on upper beaches and dunes (Melvin *et al.* 1991).

Spotting of piping plover nests by the New Jersey Division of Fish, Game and Wildlife led to establishment of a protected enclave to prevent trampling by recreational use and loss through beach cleaning operations. The site (Figure 5.6) was protected by warning signs and a symbolic fence that allowed the birds (and natural processes) to pass unimpeded but restricted human access. The result of this management practice was growth of a low (0.65 m high) incipient dune and colonization of the backbeach by eight native species of plants, creating a hummocky topography that contrasts greatly with the adjacent flat, mechanically groomed beach around it. The nesting site portrayed in Figure 5.6 does not function in a fully natural way because the dimensions of the protected enclave only correspond to observed nests, and the enclave does not extend far across or along the shore. The natural features are not contiguous to the existing foredune, due to human use of the backbeach landward of the incipient dune as a transport corridor for beach vehicles. The site does not represent the kind of environment that could occur if the target environment were fully protected, but it reveals the potential for reestablishment of a naturally functioning system on a nourished beach that otherwise had limited natural value (Nordstrom *et al.* 1999).

Integrating species protection and landform restoration

Examples abound of the need to target habitat protection and enhancement rather than species protection (Waks 1996). Protecting nesting habitat as well as specific nest sites is important because of fledgling mobility after the nesting stage (Watson *et al.* 1997). The potential for achieving these goals and returning to a naturally functioning system in the case study presented above can be realized by extending the zone of protection across the beach to include the upper wrack line and the zone between the nests of target species and the former foredune. Recreational use can still be made of the seaward portion of the beach, and raking, if it must occur, can be seaward of the upper wrack line. Sand-trapping fences tend to create steep dune faces that are incompatible with plover nesting, and planting of beach grass can quickly result in a foredune that is too thickly vegetated to be used by nesting plovers (Melvin *et al.* 1991). Restriction of revegetation efforts and use of symbolic fencing instead of sand-trapping fencing can accomplish species protection and development of natural landforms. An advantage of allowing dunes to be built up naturally on nourished beaches is that there will be growth of natural vegetation and fungi throughout the dune that will contribute to its resistance to erosion. Beach grass does not thrive in artificially stabilized areas where burial by sand cannot confer its selective advantage on this species (van der Putten *et al.* 1993). The establishment of successional species depends on the richness of the soil micorrhizal fungi and other soil nutrients (Koske and

Gemma 1997) and on the physical soil structure resulting from a healthy pioneer growth of beach grass in wind blown sand (Maun 1993).

Recreation values

Significance of levels of use

Recreation can be accommodated with minimal impact on natural beach and dune environments (e.g., wilderness use) or it can result in their complete elimination (e.g., in urban resorts). Visitation to recreational areas (tourism) initially relies on the quality of the environment, but the development that accompanies tourism may threaten the future of the quality of the recreational experience and the tourism industry (Hawkins 1996). Beaches have been turned into an object for consumption under mass tourism and recreational use (Breton and Esteban 1995). The challenge to tourism is to find approaches to coastal management that guarantee business profitability and long-term resort viability by promoting environmentally responsible tourism and environmentally compatible development (Hawkins 1996). Sustainable development is now a concept that is pervasive in effective management plans (Holgate-Pollard 1996). The concept of sustainable landscapes is a more appropriate image for sustainability than sustainable development because it removes the focus on human projects while describing a place for all species, and it includes human use without excluding non-human beings and their needs (Grumbine 1994).

Ecotourism and nature-based tourism

Ways can be found to develop or use the shoreline in a manner that maintains the natural sediment budget and preserves the mobility of landforms and their tendency to grow and be altered. One possibility is to use ecotourism to incorporate environmental conservation and tourism development in a single strategy (Cousins 1991; Pearsall 1993).

An important consideration in nature-based tourism is that nature is not the product being produced, and the level of tourism must remain small scale, requiring a high cost for the experience (Whitlock and Becker 1991). Visitors to the coast may be willing to pay a premium for the privilege of exploring relatively undisturbed coastal landscapes (Pearsall 1993). This kind of tourism can deliver economic benefits with minimum infrastructure development, and local interests could be recipients of income and serve as stewards for sites (Whitlock and Becker 1991; Pearsall 1993). Support infrastructure in these non-intensively used areas can be limited to small huts (Wong 1993).

Carlson and Godfrey (1989) indicate that, if managers were to reduce visitation in coastal landscapes with great natural value to their natural carrying capacity, only a few visitors per day would be allowed. Impact control measures are required to allow for greater numbers of visitors while minimizing adverse effects. Suggested control measures include self-guided interpretive trails with accompanying pamphlets and instructional signs (Carlson and Godfrey 1989; Breton and Esteban 1995; Mullard 1995). The trails should be plainly marked by brightly colored posts and walkways that are elevated above the surface (0.5 m high) and have sufficient spacing between planks (30 mm) to allow light to penetrate through them for the survival of plants. Fences are likely to be needed, but these can be symbolic fences (Carlson and Godfrey 1989). Walkways and observation platforms that may be expected to be an eyesore may subsequently be regarded as attractive focal points because of the scenic views they provide (Carlson and Godfrey 1989). Negative messages (e.g., keep off the dunes) can be replaced by more positive and instructional ones to increase their value to visitors and make them more likely to participate.

The concept of ecotourism is normally applied to relatively undeveloped areas, but it has potential for restored landscapes that have value to tourist and resident populations who have lost appreciation for the natural environment through mass use (Breton and Esteban 1995). Established walks and heritage trails emanating from developed areas with provision of lookouts and interpretive signs have been suggested to help tourists gain the most from visits to municipalities designed for beach recreation (Western Australia Department of Planning and Urban Development 1993). Koltasz Smith and Partners (1994) recommend municipal landscape protection areas for key areas of remnant vegetation adjacent to principal tourist routes in resort municipalities (with appropriate density and development provisions to maintain landscape amenity) in addition to the more typical conservation zones designed to function as wildlife corridors.

At present, it is unclear how small a beach or dune can be and still have potential for ecotourism. Evaluation of fragmented ecosystems has demonstrated the feasibility of integrating development and preservation of biologic richness and ecologic functionality by establishing biologic stations in small patches left over after development where preservation goals can be combined with environmental education and recreation (Espejel 1993). These components of a kind of urban ecotourism could be incorporated into modest dune construction and beach protection programs to ensure the survivability of the protected landform and the likelihood that programs will be instituted in nearby sites. Sand dune vegetation can be used to demonstrate specific dune-related phenomena or more general biological principals, and the clear zonation and obvious and rapid response to coastal environmental changes make dunes excellent subjects for ecological observation and teaching (Chapman 1989). Environmental education and well-controlled ecotourism have to be intensively supported; diversification of economic activities and recreational activities can help reduce the fragility of coastal systems that depend on a single source of income and integrate a wider spectrum of the local population into planning and management (Espejel 1993).

Public parks

Public parks include national, state/provincial, county and municipal. These parks vary greatly in balance between human and natural components. Many public parks were formed on the basis of preserving unique or rare natural resource features, but they may be distinguished from conservation areas in that they are biological preserves for people; supporters of public enjoyment generally do not view natural resources as more important than the benefits they provide to people (Morgan 1996). The level of development, in terms of amount and appropriateness of facilities, is one of the most important policy issues facing parks (Morgan 1996).

Size and level of management

Size is often critical to the ability of a park to protect natural environments. Larger parks can divide areas according to specific uses, including natural areas and recreation areas, managing each as a discrete unit. Larger parks are usually established at national and state levels because of the need to levy large sums of money to purchase the land and pay personnel to manage it, but the degree to which a park favors development of a natural system is not necessarily related to the jurisdictional level. Some municipal parks may be managed to include urban recreation activities, such as paved surfaces for playing ball, whereas other municipal parks may be managed for environmental values (with no recreation facilities provided) but are accessible to the public. Many municipal parks now have restrictions on direct use of dune environments, even those locations where non-coastal uses are provided landward of the dune. State and national parks may have greater restrictions of vehicular use of the beach than municipal parks, and they may minimize beach cleaning operations that would otherwise eliminate incipient dunes and remove the vegetation colonizing them, preventing creation of new dunes or extension of existing dunes seaward. Some parks may remove exotics, whereas exotics are tolerated, valued or ignored in others.

Attracting and accommodating users

The likelihood of conversion from natural environment to artifact increases with actions to increase access and make modifications to enhance non-environmental use of the beach. Promotion of state parks in statewide tourism campaigns and the resulting attraction of tourists who lack traditional coastal values creates a large user group who may demand entertainment and service over nature appreciation. These users may already be adapted to the congestion or environmental degradation of urban environments. As a result, they have no loss of satisfaction when visiting parks (Dustin and McAvoy 1982; Morgan 1996). Opinions and perceptions of beach users surveyed on-site generally indicate that facilities and levels of commercialization are appropriate (Morgan *et al.* 1993). Use of visitation or activity data as a guide for beach management in public recreation areas inevitably leads to a management response to provide "more of the same"; this is a problem because modifications made to the beach environment by management may be more important in determining the type of visitor than natural physical characteristics of the setting (Chapman 1989). The provision of additional comforts in public recreation areas may not be necessary. Chapman's (1989) survey of beach users indicated that, once the basic facilities of water, toilets and litter bins are provided, the majority of users demonstrate concern for the quality of the natural environment. The highest level of satisfaction in his study was expressed by users at beaches where provision of cultural amenities was lowest, whereas the lowest level of satisfaction was expressed by users of beaches where provision of amenities was highest (Chapman 1989).

Residential use

Landscape variability

The beaches and dunes managed by municipalities in the USA can represent a spectrum of landforms and habitats, ranging from large natural environments to virtually featureless sand platforms. Privately managed lots landward of municipally managed beaches may have some semblance of a dune, although not always built by aeolian processes or colonized by indigenous coastal vegetation (chapter 5). Some lots may be graded flat and kept free of vegetation, whereas other lots may be well vegetated but bear no similarity to a coastal landscape (Figure 8.2).

Residents directly alter the characteristics of the shore within the limits of their properties according to personal preference, and indirectly alter the characteristics of municipally maintained segments by means of their collective participation in municipal level decisions. Many owners of coastal proper-

Figure 8.2 Characteristics of shorefront properties at Ocean City, New Jersey, modified to conform to suburban perceptions of landscape taste.

ties are not residents of jurisdictions in which those properties are located (Stronge 1995). In locations where individual property rights are held in high regard and the owners are seasonal users of the property and not imbued with a coastal ethic (e.g., much of the USA), the result can be a pronounced divergence of landscape types from natural ones and pronounced differences between individual lots. In many municipalities, residential use is mixed with commercial use to the point where the distinction is blurred and management practices used in intensively developed coastal resorts (eliminating dunes, raking the beach) are applied in adjacent moderately developed residential areas.

Restoration potential

The potential for restoration of natural and human values differs by site and through time and ranges from re-establishment of large-scale landscape components by beach nourishment and dune building to small-scale actions, such as removing exotic species from dunes migrating onto properties. According to present regulations in many municipalities in the USA, dunes designed for flood protection may be constructed using fences, vegetation plantings or bulldozing. Vegetation planted to build dunes may be natural species or exotics, and, once formed, dune mobility may be tolerated or restricted. Candidate locations for restoration of dunes with natural value include existing beaches, beaches that can be built through nourishment, existing dunes, locations where new dunes can be built, undeveloped lots, undeveloped portions of developed lots and lots that are fully developed but may be purchased and restored. Actions that can be taken to enhance the natural value of these dunes include suspending non-compatible management programs (such as pedestrian use that tramples vegetation, raking of the beach to eliminate litter, excessive use of barriers to stop migrating sand and planting exotic species).

Suspension of incompatible actions

Pedestrian and vehicular uses that trample vegetation are often restricted in foredunes but are uncontrolled on the beach. Restrictions to pedestrian use of the beach are probably unnecessary in sparsely developed municipalities and difficult to implement in intensively developed residential municipalities except in specific nesting areas. The need for vehicles to use the beach is less easily demonstrated than the need for pedestrian access and is potentially more damaging, even at low levels of use. Municipalities often have regulations preventing private recreational use of vehicles or restricting this use to the winter season, although restrictions are often based on safety and enhancement of recreational opportunities rather than enhancement of ecological or geomorphological features. Even where private use is restricted, the regulations often allow municipal employees to use the beach as a thoroughfare. Problems arise when drivers take routes along the backbeach near the dune in the interest of saving driving time, instead of driving directly toward the foreshore before turning in the longshore direction. The result is destruction of incipient vegetation accompanied by prevention of seaward dune extension and isolation of geomorphic features and habitat on the beach from related features in the dune (Figure 5.6). The solution involves careful instructions to employees and monitoring to ensure instructions are followed.

There is a dramatic difference between raked and unraked beaches in terms of topography and habitat value. Figure 5.6 provides an indication of the potential that can be achieved with suspension of this activity. Breton and Estaban (1995) indicate how the vegetation cover of the beach at El Prat, in Catalonia increased dramatically in 7 years following a change in the method of cleaning, leading to the beginning of natural reconstruction of the dune system. An important step in adopting management strategies that accommodate natural vegetation and topography is recognition of the conceptual difference between natural beach litter and cultural litter and adopting cleaning strategies for the latter while accommodating the former. Studies have called attention to the need to restrict beach cleaning to allow vegetation litter to decompose (Hotten 1988) or use manual cleaning of the beach rather than mechanical cleaning as a way to allow native plant communities to establish themselves. The greater percentage of cultural litter in urban resorts and the limited potential for establishment of viable natural environments may argue for mechanical cleaning in those areas, but mechanical cleaning in residential areas should be resisted to the extent possible. The flat, raked beaches that many visitors and resident of coastal municipalities see are not acceptable representations of nature, and it is questionable whether adoption of true environmentally compatible alternatives is possible with this perception in their minds. Raking beaches to remove litter ranks relatively high in the amenity preferences of beach users (City of Stirling 1984; Chapman 1989), indicating that suspension of raking may be difficult to achieve in some areas without a major change in the culture of the municipality. The willingness of residents to adopt environmentally friendly strategies is related to their perception of the value of natural components of the landscape and their role as stewards of the landscape. These concepts, and their application to local management practice, are elaborated later in this chapter.

Urban resorts

Urban resorts often emphasize infrastructure or recreational uses that are enhanced by a shorefront location but do not require one, including shopping for elegant clothes, dining in gourmet restaurants and gambling. There is little question about the importance of a viable beach in a tourist location, and some major urban tourist locations did not have the sandy beach environments for which they became famous after nourishment (Wiegel 1992b; Woodell and Hollar 1991; Flick 1993). However, an urban beach can be totally artificial. There is usually little attempt to provide for natural habitat or for a physical or visual barrier between urban functions and the beach (Figure 8.3), and dune environments are eliminated in favor of additional beach space, shorefront roads, promenades or boardwalks. The urbanization process continues seaward as a result of extension of human use structures onto the beach, including cabanas, bars, restaurants and piers. Natural process are limited by the direct manipulation of landforms and the destruction of vegetation to accommodate use of structures as well as the indirect effects of the structures acting as barriers to winds, waves and sediment transport.

Coastal development and over-use of public beaches can result in environmental losses beyond the developed area. Sediment starvation as a result of

Figure 8.3 Platje Barceloneta in Barcelona, Spain, showing wide, flat beach created by artificial nourishment and graded and cleaned to enhance recreation. The trees are alien to the beach environment, but they have a visual appeal that compensates for the absence of natural variability and beauty.

construction of navigation and protection structures are the most conspicuous effects, but there are other, as yet poorly documented, effects that likely occur. For example, loss of coastal strand communities in developed areas may lead to reduction in seed sources, reducing the resiliency of plant communities in undeveloped areas following loss by storms (Cunniff 1985). Perhaps the biggest problem is the non-natural image of the coast that millions of tourists take home with them and becomes the basis for decisions on coastal environmental issues.

Noise and light remake the land, and the sounds and lighting of the human-altered landscape depart from those of the natural landscape (Weston 1996). The omnipresent urban infrastructure may prevent the return of a natural feel to reconstructed coastal landscapes in urban areas, but beaches and dunes can provide visual and sound buffers that enhance recreational values while retaining an image of the coast that conveys the impression that natural features belong there.

The least natural urban resort is one that has been constructed or has evolved to the point where there is no longer a fronting beach. As long as beaches are the principal *raison d'être* of coastal resorts, there is likely to be at least an attempt to maintain this feature through artificial nourishment.

Whether the beach is allowed to function naturally or whether dunes are allowed to form and be reshaped by natural processes depends on the perception of managers, residents and users of the landscape of the value of the natural features in the coastal setting.

Flood protection

Flood protection involves maintaining a continuous barrier at prescribed height above the level of inundation by natural processes, and any means of protection designed to achieve these characteristics will depart from the height and topographic diversity of natural landforms. The governing height for protective dunes is often a compromise between the need for protection, the need to accommodate buildings and their use and the need to retain views of the sea. These factors affect dune development even where dunes are considered integral to shore protection plans.

Planting of stabilizing dune vegetation is often accomplished at the municipal level, and the success of maintaining this cover varies considerably. Some dunes in the USA are devoid of vegetation because they are considered sacrificial features that are readily manipulated using earth-moving equipment, and investment in vegetation projects appears unjustified. Monocultures often occur in vegetation programs because the dune is planted using a single species that is perceived to have the greatest utility for stabilization (usually *Ammophila* in mid-latitude regions).

Native species will create dunes that are more natural in appearance and function, but these species are likely to be less diverse on artificially maintained foredunes than on natural foredunes because the protective dunes are usually narrow; they have little topographic variability; aeolian transport is deliberately restricted; and they are prevented from migrating landward. These conditions reduce the potential for variability in micro-environments. Exotic species used to stabilize dunes often create a dune with unnatural looking vegetation that may form mono-specific stands that dominate the natural vegetation and alter the nature and function of habitats (Andersen 1995b).

Dunes created using sand fences result from aeolian processes, like dunes formed around vegetation, but they differ from natural dunes because of the external shape. Bulldozed dunes may mimic natural dunes in their location on the beach profile and in certain functions, but they have a non-aeolian origin; they have a different shape; they have no internal stratification; and they may contain sediments that are too coarse to be moved by aeolian processes. The degree to which vegetation permeates dunes built by sand fences and bulldozing will also differ from dunes built using vegetation plantings alone (Nordstrom and Arens 1998).

Dunes created with the help of sand fences or vegetation plantings are composed of well-sorted, fine- to medium-sized sand that is transported and deposited as a result of aeolian processes, and these dunes may have bedforms and strata that are characteristic of these processes. Vegetation is the best alternative for creation of dunes where beaches are sufficiently wide for incipient dunes to survive the storm season. Sand fences may be required to augment use of vegetation where beaches are narrower. Bulldozing should be considered the last resort.

There is little that can be done to enhance the natural value of dunes that have been created for the primary purpose of protection until they are perceived to have reached the level of protection required. Much of the problem with implementing management programs to enhance natural characteristics involves perceiving the dune, once formed, as more than a structure to mitigate flooding and erosion. At that point, the ecological and recreational values can be enhanced. Suggestions for accomplishing this goal are provided in the section on strategies for maintaining and enhancing natural features presented later in this chapter.

Bulkheads on high-energy coasts are built for backup protection, and they are fronted by beaches that provide sand for aeolian transport and sufficient space for dunes to form. Most bulkheads are low enough in elevation that sediment can pass over them by aeolian transport and they can be buried beneath dune deposits (Figure 5.2). Bulkheads thus may be considered compatible with natural processes, providing that this accretion is not removed from the seaward side of the structures by earth-moving equipment before it forms an effective transport surface.

There is limited potential for formation of natural landforms where seawalls are the principal form of protection. Seawalls are built as a last resort to make up for loss of the beach. They restrict the landward migration of the beach and dune and prevent the upland from functioning as part of the dynamic coastal system. Narrow beach widths (that are either the initial reason seawalls were built or an end result of seawall placement) limit the potential for aeolian transport to occur and the space available for dunes to form.

Erosion control

Beach nourishment

Beach nourishment has great potential for reinitiating natural evolution of coastal landscapes, but this potential is usually not realized.

Nourishment in many developed areas is perceived as a means of providing protection to shorefront buildings and providing a recreational platform for summer visitors rather than a means of restoring ecological values or providing aesthetic values. Most nourished beaches are graded into "slabs of sand."

Most previous applied science projects for evaluating nourishment have been devoted to assessing the level of protection required, the suitability of fill materials, the rate of loss, the detrimental effects on the environment (rather than potential gains) and the cost effectiveness of the projects; projects generally are not conducted to restore the beach for indigenous biota or biota that use the beach for foraging or nesting, although this benefit is cited occasionally (Houston 1991; Schmahl and Conklin 1991; Nelson 1993; National Research Council 1995). It is often difficult to quantify benefits of beach nourishment operations beyond the defense function (Townend and Fleming 1991). The procedure for calculating costs and benefits of nourishment operations in the USA is overly restrictive and allows for only storm damage reduction and limited recreational benefits; the full complement of recreational benefits and benefits outside the project area are not included (National Research Council 1995).

A critical component of restoration efforts is selecting the appropriate target environment. Kana (1993) suggests using an "ideal present profile" as a basis for evaluating beach nourishment projects. The ideal present profile is typical of the current status of a nearby existing beach (in width, slope and unit volume) where that beach is considered healthy (with a foredune, dry-sand berm and a wide intertidal profile). This profile can be matched against existing profiles in developed areas and used to establish where the shoreline would be in the absence of shore protection structures. It can also be used to provide an estimate of sand deficit by comparing the ideal volume with the existing volume seaward of shore protection structures or erosion scarps.

The locations where dunes are placed following nourishment operations are rarely considered in terms of natural processes, and the ideal profile could be used to suggest where the dune should be relative to mean high water on the foreshore during erosional and accretional stages. Superimposition of the ideal present profile on existing beach conditions or on projected nourished conditions calls attention to the departure of both existing and design conditions from a more natural shoreline that should be the target of restoration efforts. The ideal present profile in the context used by Kana represents conditions at an existing, rather than improved, beach. Design for an improved beach could use an "ideal historical profile" that represents a return to a previous condition lost to erosion or an "ideal enhanced profile" that represents a new condition considered desirable for multiple uses, including natural habitat.

Hard structures

Bulkheads and seawalls were discussed earlier, in the context of flood control methods. Those structures are considered more responsible for the lack of beach than groins (Hall and Pilkey 1991). They are less compatible with natural landforms than groins because they inevitably replace, rather than alter, the characteristics of these landforms. The wider beaches updrift of groins often provide more suitable locations for dunes to form than the original beaches (Figure 4.2), and the presence of groins restricts the beach raking activities in these accretion areas, allowing incipient dunes to increase in size and develop a vegetation cover.

Groins are artifacts, and the beaches and dunes that accumulate as a result of their placement reflect human impact in terms of their location and shape. However, the mechanisms of sediment erosion, transport and deposition mimic natural processes, and the resulting landforms are more natural than bulldozed landforms or nourished landforms that are graded and raked (Figure 5.8). The perception of groins as structures to be avoided stems from their local effect on the sediment budget and as eyesores, but they have considerable value as aqueous habitat and as features that enhance the likelihood that dunes will form and survive.

Attitudes to coastal management in some countries still indicate that engineering structures are better than natural features as a form of protection (Carter *et al.* 1993), but there is increasing concern that engineering solutions to coastal erosion should be "soft" or "geomorphically compatible" alternatives (Rijkswaterstaat 1990; Kiknadze 1993; Kos'yan and Magoon 1993; Basco and Shin 1996; van Bohemen 1996). They must also be based on economical expediency and social necessity (Kiknadze 1993). Many features that are considered adverse when judged from the pragmatic or utilitarian viewpoint are also adverse from the aesthetic or ecological viewpoints (Mather and Ritchie 1977). This convergence of opinion facilitates selection of strategies that accommodate both natural values and tourism values. The following section identifies management strategies for eroding developed shorelines that attempt to balance natural and human use values in ways that produce a more naturally functioning system.

Strategies for maintaining and enhancing natural landforms

It is assumed that ways can be found to develop or use the shoreline in a manner that maintains the natural sediment budget and preserves the mobility of landforms and their tendency to grow and be altered. Dunes need not be leveled to provide views of the sea or be displaced seaward to make room for houses. Buildings can be elevated on pilings, so they do not interfere with sand transported by wind and overwash. Buildings can be made more aerodynamic and built in lower densities to minimize interference with airflow. Uses of the shoreline environment can be restricted to only those uses that require a shorefront location, and access can be restricted to modes that do not require extensive support infrastructure (Nordstrom 1990).

To achieve these goals, actions must be taken to provide human use values and protect investments while creating features that resemble natural environments in many ways. Some of the strategies for achieving environmental goals (e.g., setting aside conservation areas) have proved successful and are well represented in the literature. Other strategies (e.g., instituting ecotourism or establishing management zones) show promise, but case studies of successful implementation on a large scale are still to be worked out. The following sections discuss some of the potential advantages of programs that may have value in locations where there is a light or moderate intensity of development (retreat and relocation, establishing comprehensive management districts) and in locations where a greater intensity of development requires a focus of attention largely on dune environments on the seaward portion of the backbeach. Success of these programs may hinge on public acceptance of more dynamic landforms. Accordingly, suggestions are made at the end of the chapter for establishing an appreciation for naturally functioning landscapes.

Retreat and relocation

Withdrawal from the shoreline in lieu of protection is an option that has been advocated and demonstrated as workable under certain conditions (Manoha and Teisson 1990; Rogers 1993). This alternative has been formalized in the USA in the enactment of the Upton–Jones amendment that altered the US National Flood Insurance Program to authorize payment to relocate or remove undamaged but erosion-threatened buildings. The alternative is described in detail in Buckley and Rhodes (1989) and evaluated in Davison (1993) and Rogers (1993). Relocation of threatened structures can be a cost-effective alternative where: (1) investment in structures is low; (2) relatively inexpensive undeveloped land is available nearby; (3) regulations prevent erosion control structures and explicitly favor or require relocation; (4) low density of development makes it difficult to demonstrate cost effectiveness of beach nourishment; and (5) the size and method of construction (e.g., wood frame on pilings) facilitate relocation (Rogers 1993). The Upton–Jones initiative resulted in removal of 178 buildings in 3 years in North Carolina, where large numbers of buildings can be expected to be moved in the future, largely because the state oceanfront erosion management program relies heavily on removing threatened buildings (Rogers 1993).

Abandonment or managed retreat become more appealing with economic austerity and reluctance to fund protection projects, and removal of buildings has been suggested as a first alternative for developing countries wherever availability of land is not a problem (Matondo 1991). In contrast, the relocation/removal alternative is becoming less viable in areas where it was previously viable, due to increases in the the number and size of new buildings (Rogers 1993) and in intensively developed areas where there is no available shorefront land.

There are still many impediments to implementing programs incorporating options for retreat and relocation at the local level, where prevention of coastal construction or restrictions to reconstruction following damage is not a commonly used option (chapters 6 and 7). Changes in regulations that would prevent reconstruction of damaged structures can result in violent criticism by local interests, and laws that seek to prevent rebuilding at the municipal level have not proved effective in denying property owners rights to rebuild (Gares 1989; Stoddard 1995). In practice, the retreat option is not seriously considered as a viable policy in most developed areas (Lee 1993; Schmahl and Conklin 1991; Hooke and Bray 1995)

Establishing shoreline management and regulatory programs

Comprehensive management programs can be established to use control zones and performance criteria for land uses to favor natural shoreline environments that can also provide some degree of protection to coastal buildings and support infrastructure. Examples have been developed for shorelines of low relief, in the form of Dune Management Districts (Gares *et al.* 1980) and shorelines of high relief in the form of Coastal Cliff Management Districts (Nordstrom and Renwick 1984). The objectives of a comprehensive regulatory program are: (1) make people aware of the dangers of locating in a coastal hazard area; (2) reduce population and property and thus reduce costly expenditures for repair and restoration of damaged coastal structures; (3) preserve and restore natural resource systems for future generations; (4) preserve and expand recreational opportunities; (5) promote aesthetic considerations; (6) favor long-term over short-term benefits; and (7) provide plans for regional protection. The management district concept stems from the recognition that the coastal landforms are essential elements of the system and require protection and management if they are to maintain their characteristics and functions while they and areas inland of them are used.

Comprehensive management models based on the physical processes that shape the shoreline do not provide a means for a complete return to the natural system. They are designed to allow the system to run its course while obtaining a degree of predictability and stability that allows developers to get

Table 8.1 *Principles used in developing a Dune Management District and Coastal Cliff Management District*

The boundaries of management districts and their land use zones are dynamic and migrate in response to rising water levels and a negative sediment budget.

The beach and dunes (or beach and cliffs) represent a zone of sediment exchange; they must be treated as one system.

The management district is dynamic, and boundaries should be reviewed at predetermined times (e.g., every 10 years) or following major storms.

Natural processes should be subjected to minimal interference, and development in control zones should be compatible with the natural processes expected in those zones, including landward shoreline migration.

All construction projects within any part of the management districts should be subject to a review process.

No sediment should be removed from the system.

Dunes and cliffs should be allowed to form and to migrate in a natural fashion, but dune building through artificial means may be encouraged, gaps in dune lines breached by beach access paths should be closed, and soil stabilization and vegetation planting may be used where necessary to reduce human accelerated erosion of cliff tops and slopes.

There should be no interference with natural processes; bulkheads or similar engineering structures are not permitted; but growth of natural vegetation is encouraged.

Existing structures in the management districts that sustain damage as a result of storm action should be removed or rebuilt to environmentally compatible standards. Existing structures in the Imminent Failure Zone that sustain damage should be condemned and removed.

New construction of buildings and support structures in the Dune Management District and Imminent Failure Zone is prohibited (including expansion of existing facilities and construction of public facilities).

New or expanded residential development landward of the presently existing management districts is acceptable provided that implementation complies with the requirements to maintain the integrity of the dune or cliff according to accepted performance criteria (e.g., buildings landward of dunes should be elevated on pilings).

Recreational usage in the Imminent Failure Zone is restricted to access to the beach and non-vehicular activities on the cliff tops. Access to the beach over the top and face of the cliff shall be by designated walkways that are maintained to prevent alteration of the ground surface.

New construction projects should be reviewed on a unit-by-unit basis to insure compliance with performed standards.

Source: From Gares *et al.* 1980 and Nordstrom and Renwick 1984.

a return on their investment and allows shoreline residents and visitors to enjoy the recreational benefits of a beach, dune or coastal bluff. No sand is gained in the process of returning to a naturally functioning system; the sediment used to build up the beach is derived from the dunes or coastal bluffs that provide interim protection. These methods offer a form of protection that is essentially a compromise with nature, as reflected in the operating principles identified in Table 8.1. The strategies may involve few public expenditures and result in considerable public benefit, but the restriction to land use

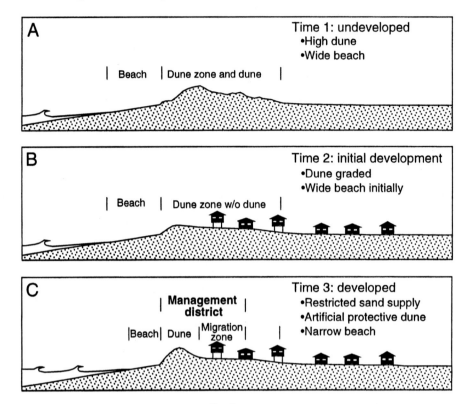

Figure 8.4 Dune management district.

involves other sacrifices to be made by shorefront property owners and commercial establishments.

The Dune Management District concept

A distinction may be made between the natural limits of the location where dunes could occur (or dune zone) and the limits of the regulated Dune Management District. Under natural conditions, the dune zone and the dune are the same (Figure 8.4A). Under conditions of development, the dune may be severely restricted or eliminated (Figure 8.4B). The Dune Management District is a legal construct designed for protection and enhancement of the dunes and is located in the seaward portion of the dune zone (Figure 8.4C). The dimensions of the Dune Management District are wider than the existing dune ridge that exists in many developed areas because it must provide: (1) a space for a dune that is large enough to provide adequate protection against flooding; (2) an adequate reservoir of sand in storage to compensate for beach erosion and provide protection to the buildings to the lee; and (3) a migration zone to allow for inland movement of the dune as it continues to provide protection. The level of protection prescribed for the dune is determined using traditional engineering criteria (e.g., Hallermeier 1987; van de Graaf 1994). Gares *et al.* (1980) present a methodology for identifying the degree of protection afforded by existing dunes and a process for delineating a Dune Management District along the New Jersey shoreline for dunes that could provide enhanced protection. Variables include foreshore and nearshore slopes, dune foreslopes, beach width, deepwater wave height, wave period, storm surge and wave and tide frequency.

Ideally, management districts should migrate along with the shoreline, and their boundaries should be readjusted after a predetermined time interval. This migration aspect is a fundamental difference between the conception of the dune as a static protection structure and as a shifting component of a broader regulatory program that includes application of land use controls in landward areas. The Dune Management District differs from traditional setbacks in that it acknowledges that erosion is a continuing process and it includes a dune as a principal component of regulation. The landward boundary of the Dune Management District is determined by adding the design dune width to the amount of erosion predicted to occur before the next formal readjustment of district boundaries takes place.

The degree to which the dune should function as a barrier to overwash or aeolian transport is a critical concern. If dune districts were created on eroding shorelines and no migration of the dune crestline were allowed, the result could be the elimination of the beach from in front of the dune and eventual dune destruction with elimination of its protective function. A 50-year storm was suggested by Gares *et al.* (1980) for use in designing the protective dune in the delineation procedure for New Jersey because: (1) many dunes had this level of protection; (2) a 50-year dune allows for natural processes to shape the dune; and (3) dune migration would proceed in response to a rising sea level. The design dune still would have the advantage of providing protection against high frequency/low intensity storms. The dimensions of the design dune would also be such that a minimum amount of land would have to be set aside for the Dune Management District. Thus owners of property within the management district would be able to draw some benefit from their investment.

The Dune Management District concept is not an attempt to remove all development from the barrier islands but an attempt to reduce expenditures of public monies on erosion protection structures and on post-disaster relief by taking advantage of natural features. Part of the strategy involves requiring coastal construction standards. The US National Flood Insurance Program already has performance standards for buildings that are designed to minimize the effects of the natural system on houses, and municipalities are often encouraged to develop more prescriptive standards (Nicholls *et al.* 1993). With

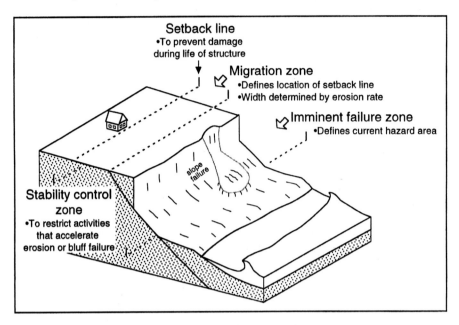

Figure 8.5 Coastal cliff management district.

little modification, these standards could be designed to minimize the effect of structures on dune and the beach, even as these landforms migrate under them through erosion.

Actions taken to lower the density of shorefront structures are also important in minimizing direct impacts on natural landforms, reducing the amount of support infrastructure and enhancing the viability of the retreat and relocation option (Schmahl and Conklin 1991). Many provisions in the approach used by the state of Maine to regulate new construction in dune areas (Lindsay and Tupper 1990) are compatible with the migration aspects of the dune management district. The dune management district concept appeared to be unworkable on the densely developed New Jersey coast when it was introduced (Gares 1989), but it holds promise for locations now undergoing initial development.

The Coastal Cliff Management concept

Management strategies for coastal cliffs that consider local geomorphic processes require the recognition of different zones with varying hazards and potentials for human impact (Springman and Born 1979; Michigan Sea Grant 1982; Nordstrom and Renwick 1984). Potential land use control zones in the Coastal Cliff Management District (Figure 8.5) include: (1) The Imminent Failure Zone, where slope failure can occur at any time regardless of whether debris materials are removed from the base of the bluff by waves; (2) a Migration Zone that recognizes that the area subject to failure will migrate landward in response to removal of materials from the base of the bluff; and (3) a Stability Control Zone, where human activities can have a major effect on slope stability. These three zones will have different dimensions along a coast due to difference in the geologic formations that make up the bluff materials and rates of shoreline retreat (Nordstrom and Renwick 1984).

The Imminent Failure Zone (Figure 8.5) includes the base, face and top of all bluffs that are subject to immediate failure. Facilities of high unit value should not be constructed within this zone because of the high susceptibility to hazard, and activities that would contribute to bluff instability should be precluded. Actual cliff failures are not prevented by protection strategies in this zone because of the natural value of the evolving landforms. For example, many actively eroding landslides on sea cliffs create habitats that are important to colonizing species that thrive on the environmental change resulting from instability, including threatened species such as hoary stock (*Matthiola incaria*) and Scottish primrose (*Primula scotica*) (Lee 1995).

The timing of bluff top failure is difficult to predict because it is related to long-term changes in the strength of bluff materials, gradual changes in slope geometry caused by surficial processes, short-term fluctuations in pore water pressure caused by seasonal or storm precipitation, and variations in wave attack, but the horizontal extent of the cliff top subject to failure can be estimated by deterministic, empirical and arbitrary methods as discussed in Nordstrom and Renwick (1984).

The Imminent Failure Zone reflects the position of the unstable bluff top without regard to long-term shoreline retreat. Bluff top developments should be sited and designed to assure that they are not damaged during their expected economic lifespan. This is the purpose of employing a coastal construction setback line (Figure 8.5). The owner of a building would have use of the structure for the years between time of construction and the time when it falls within the Imminent Failure Zone and is rendered unsafe by slope failure. At that time, the house would have to be abandoned and removed. The distance of the setback line landward of the edge of the Imminent Failure Zone is related to the rate of bluff retreat. Where the buildings are considered to have useful lives of 30 years, the setback line should reflect the amount of bluff retreat expected over a 30-year period.

Bluff top buildings, erosion control structures, and activities such as foot traffic and waste-water disposal must not be allowed to contribute to problems of erosion or geologic instability on the site or on surrounding areas.

These activities may accelerate erosion by creating indentations in the bluff top that spread to adjacent properties by gullying or retrogressive landsliding. Alternatively, the prevention of natural bluff erosion interferes with its value as habitat and may deprive the beach of an important component of the sediment budget (Fletcher *et al.* 1997). Land use controls must be implemented on the cliff face and on the cliff top to control adverse human activities so that natural processes are not prevented but also are not accelerated. This is the rationale for the creation of the Stability Control Zone and the appropriate restrictions to land use designed for incorporation into zoning ordinances (Nordstrom and Renwick 1984).

The sensitivity of the bluff to human-induced failure will be reduced with distance landward, and the landward boundary of the Stability Control Zone is difficult to determine. The determination is made more difficult because not all activities have the same effect on slope failure. Human trampling may have significant effect on stability only within one meter of the landward edge of the bluff, whereas watering of lawns farther landward may affect stability at the bluff face. Jurisdictions may find it convenient to place the landward boundary of the Stability Control Zone at the landward boundary of the Imminent Failure Zone or the Migration Zone to avoid a long and costly delineation study. The latter boundary is suggested because it represents a more conservative estimate of the width of the zone in terms of safety (Nordstrom and Renwick 1984).

Controls designed to prevent accelerated slope failure may include: (1) limitations on the size and density of buildings and support infrastructure, that affect loading; (2) watering of lawns and gardens, that affects moisture levels in bluff materials; and (3) alteration of bluff tops, faces, or bases by excavation, grading, removing natural vegetation, trampling and inscribing graffiti. Controls designed to insure continued sediment delivery to the beaches fronting the cliffs would include limitations on the construction of bulkheads, seawalls and other static defense structures designed to prevent erosion of the toe of the slope and construction of buildings or use of asphalt or gunite (Figure 4.4) or exotic vegetation to stabilize the upper slope (Nordstrom and Renwick 1984).

Local initiatives in areas that are already developed

New municipal level management programs (more modest in design and scope than comprehensive management districts) can be established to restore valuable beach and dune habitat in developed areas to enhance ecological values and strengthen the drawing power of the shoreline for tourism.

There is greater potential for restoring these habitats in intensively developed areas than has been realized, and the restored habitats will have greater human use value than is now perceived in those environments.

Pronounced differences in levels of development, ratios of seasonal to permanent residents and comprehension of the values of naturally functioning components of the coastal system will result in different potential for re-establishment of natural features. Biota and landforms in intensively developed areas where mobile environments cannot be allowed may have to be less natural than in sparsely developed areas, but ways can be found to make the dune look as natural as possible in appearance and function if not in genesis (Laustrup 1993; Favennec 1996). This may be done by creative use of sand fences (Snyder and Pinet 1981) or by bulldozing.

Restoration of dunes to more natural forms by eliminating exotic vegetation and other unnatural elements is often a goal of state and national programs, but there is no reason why these activities cannot be conducted at local levels once residents and municipal authorities realize the values of more natural alternatives. The plan for Horrocks Beach, Western Australia includes removal of introduced vegetation and landfill and reshaping of the dune to a profile similar to local naturally occurring dunes (Western Australia Department of Planning and Urban Development 1993).

Restoration of natural values of coastal dunes in developed areas will be determined largely by human actions, so changes in the perception of the value of coastal resources and implementation of management initiatives to reverse the degradation process are important to restoration planning. Recent initiatives for reducing coastal hazards, protecting target species or encouraging ecotourism provide separate examples of components that can be combined and enlarged into broad-based restoration programs for implementation at the municipal level. These restoration programs can be linked to environmental education initiatives aimed at re-establishing an appreciation of naturally functioning components of the coastal landscape. This will, in turn, lead to pilot programs for restoring environmental values in residential areas (Nordstrom *et al.* 1999).

Creating or enhancing dunes

Figure 8.6 contrasts the characteristics of beaches managed for recreation or protection (using a small, linear dune) with a beach designed to enhance the beach and dune habitat and aesthetic values. The prototypes are found on the shoreline of New Jersey and are described in Nordstrom *et al.* (1999). The suggestions for change in management practice identified below are designed to add value, not trade one value for another. Accordingly,

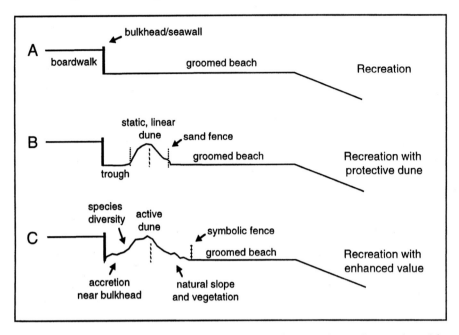

Figure 8.6 Comparison of beaches maintained for recreation and protection with beaches having natural values. Based on Nordstrom *et al.* (1999).

ecological values cannot be maximized at the expense of recreational or protective values. Enhancing values involves compromise solutions that provide more natural functions and appearance but at a smaller scale and on a different position on the profile than would occur in the absence of development. In this case, enhancement refers to an improvement over the human-modified landscape, not the natural one. The management scenario is designed to allow for greater play of natural processes on the upper beach and on the landward side of the dune, so the dune can achieve greater width, more natural contours, greater topographical variability and greater species diversity than the one typically associated with development.

The value of protective dunes as a means of reducing coastal hazards is appreciated by coastal scientists and planners at state and national levels. This value has been unappreciated or ignored in many municipalities, where active recreational use of the beach and maintaining views of the sea take priority. Efforts by the State of New Jersey to restore dunes to provide protection in several communities have led to greater acceptance of foredunes as a means of shore protection. Approximately 20 municipalities received funds passed through to the state from the federal government to repair and restore dunes damaged during a storm that occurred in March 1984 (Mauriello 1989). These dunes were built to Federal Emergency Management Agency guidelines that suggest a dune with a cross-sectional area of >50 m² above the 100-year recurrence interval storm flood level. One of the most challenging aspects of the state dune enhancement program was convincing residents and municipal officials of the need to increase the height and width of dunes (Mauriello 1989), and concerns of local residents and officials about maintaining their views of the sea resulted in actions to keep the new protective dunes relatively small and linear (Nordstrom *et al.* 1999).

The resulting coastal landscape (Figure 8.6B) achieved a state that may be described as somewhere between the optimal condition for a developed community and the worst case scenario (a completely raked beach with no dune). A major obstacle to enhancement of beaches and dunes, as reported by municipal officials, is the attitude of local residents. The state initiative caused residents in the communities that reinstated dunes to accept these dunes as part of the landscape, and it familiarized residents with natural vegetation that was prevented from growing because of beach raking. The next step is to gain acceptance of a more naturally functioning foredune that has larger dimensions, greater topographic diversity and greater amounts of blowing sand. Greater height results in interference with views of the sea from residences and boardwalks. Greater width on the ocean side restricts beach space and may involve restrictions to litter removal; greater width on the landward side would bring the dunes and blowing sand closer to cultural features (Nordstrom *et al.* 1999).

Enhancing ecological and aesthetic values

The conversion of existing narrow, linear protective dunes to dunes with enhanced ecological and aesthetic values (Figure 8.6C) includes removal of sand fences in municipalities that already have dunes. The elimination of fences allows the vegetation to become the primary means of trapping wind blown sand. Growth of the enhanced dune seaward can be caused by accretion within vegetation at the base of the existing dune or at the storm wrack line on the upper beach. A mechanism must be in place to prevent grooming of the beach or trampling of the beach and dune that would destroy new plant growth or eliminate the litter that provides a sand trap or source of nutrients. Species now lost as a result of cleaning and truncation of the foreslope of the dune on the east coast of the USA include sea-rocket (*Cakile edentula*) and beach pea (*Lathyrus maritimus*) that provides food for insects, birds and small mammals (Amos and Amos 1985).

Suspension of beach cleaning operations or manual cleaning have been suggested as environmentally viable alternatives that allow vegetation to decompose and to allow native plant communities and dunes to become established (Hotten 1988; Breton and Esteban 1995). Suspension of cleaning operations is not feasible in most developed areas open to tourism because of the potential for loss of visitors. Manual cleaning is likely to be too expensive. Compromise solutions involve restricting cleaning operations to summer months. Cleaning operations during the summer could be restricted to the portion of the beach below the upper limit of winter storm uprush where most of the beach recreation takes place. The uppermost wrack line could be left intact, with its seeds and nutrients. Symbolic fences that restrict human access but do not interfere with aeolian transport would serve this purpose.

Some sand will pass over the dune crest during high-speed onshore winds, so it is necessary to provide a wind baffle and sand trap on the dune to retain sediment volume and prevent inundation of the bulkhead and properties landward of it. The vegetation in the backdune area can be allowed to develop naturally or be planted with native plants that will provide shelter and food for fauna and have aesthetic value complementary to grasses. On the east coast of the USA, these species include bayberry (*Myrica pennsylvanica*) and beach plum (*Prunus maritima*), the latter being important in providing attractive blossoms in early summer and fruit for birds and mammals in late summer. These plants will require greater effort to improve the local environments in which they are placed than beach grass to ensure survival. Seaside goldenrod (*Solidago sempervirens*) and sand bur (*Cenchrus tribuloides*) can also be expected to grow on the lee side of the dune crest through time in this region.

Sand fences are often used to control visitor traffic as well as build dunes. Exclusion of visitor use through employment of non-trapping fences is crucial on the backbeach where natural transfers of sediment from the beach to the dune and natural colonization of the backbeach from the dune would naturally occur. Isolation of the backbeach using non-trapping fences will allow dunes to form at the uppermost wrack lines and define the location of the foredune on solely natural criteria. The recreational values to be enhanced in locations where dunes provide primary protection must be passive ones (visual and sound buffers, ecological appreciation, aesthetics) to ensure that the structural integrity of the dune is not compromised by pedestrian traffic.

Establishing an appreciation for naturally functioning landscapes

An appreciation of naturally functioning landscapes is required to enhance the recreational experience of tourists and redirect thinking of coastal residents from exercising landscape options more typical of inland suburban development (Figure 8.2) toward favoring coastal landscapes on their properties and seaward of their properties. The concept of re-establishing a culture of the coast at the local level is poorly defined and underutilized, but it has great potential for providing incentive for implementation of zoning regulations by municipalities or independent actions by individual residents to create a local environment that enhances coast-related uses.

Tourist developments and coastal communities may be beginning to show interest in nature preservation but little interest in coastal heritage and resources of the past (Télez-Duarte 1993). Cultivating a sense of the land as a community where humans are biotic citizens offers an alternative to the view of land as a subdividable and consumable commodity and allows us to recognize how much of that community has been destroyed and how the cycle of environmental degradation can be broken (Weston 1996). Aesthetics can play a role in the appreciation of nature by residents, but people must learn how to appreciate and enjoy an object for its own sake by learning about its natural history and development cycle (Thompson 1995). Responsible behavior toward an ecosystem depends on diminishing the gap between people and nature. By sustaining nature in human communities, wildness may explicitly become part of culture, causing people to work to protect biodiversity (Grumbine 1994).

Residents and municipal managers may now think that there is little to be gained in allowing natural processes to form incipient dunes and micro-habitat on backbeach locations that are near the water and are likely to be eliminated in storms with an annual recurrence interval. However, these cycles of growth and destruction of dunes on the beach have great pedagogic value in documenting the natural instability of coastal landforms. This dynamism is now unappreciated in communities where stabilization has been the goal so long that stable landforms are accepted as the norm (Nordstrom *et al.* 1999).

The concept of a natural coastal heritage or image has potential as incentive for implementing zoning regulations by municipalities or independent actions by individual residents to create local environments that enhance the image of the coast and provide templates for landscaping actions taken by neighbors (Nordstrom *et al.* 1999). Dunes on private lots have been virtually ignored as potential ecological and aesthetic resources, largely because of their complexity, their inaccessibility to the public, and the perception that property owners will not be willing to exercise options that are more environmentally compatible. Public participation is critical to the success of restoration programs. It can help reduce apathy toward restored landscapes and decrease the likelihood that specific restorations will produce technically proficient but contextually irrelevant landscapes (Light and Higgs 1996).

Landscape considerations in planning can be designed to minimize changing the image of coasts of scenic or natural significance and alter degraded landscapes to create a new image (Scholl 1991; Koarai *et al.* 1993); in such cases, it is important to discriminate between current images and those that have been lost due to changes in land use (Koarai *et al.* 1993). The most modern form of environmental reduction is not where the land is destroyed outright, but where it is reduced to scenery, a surface facsimile of a pre-existing media-generated nature where its essence is eradicated to ensure that the simulation meets user expectations (Weston 1996). By making landscaping an integral part of building codes (Awosika and Ibe 1993) a municipality may establish or re-establish an image that enhances a coastal setting.

Re-establishment of a coastal culture may require a considerable amount of time and be characterized by a small-scale start and an incremental advance. Presumably, the support among property owners will come from the year-round residents, although this assumption has yet to be evaluated. Education programs for species can be enthusiastically received by hotel guests and property owners alike (Oertel *et al.* 1996), so it may be possible to target all stakeholders in the same campaign and combine concepts of ecotourism with concepts of environmental management and hazard reduction.

Education programs are required at all levels. The transition to sustainability is a long-term goal; support must come from parents who provide their children with direct contact with wild nature, so they will value it as they grow older (Grumbine 1994) and with restored nature, so they can assess its value and significance. Experience with students in 9th grade in the public school system in Hillsborough, New Jersey (Bogdon personal communication) indicates that students who visit the New Jersey coast rarely see the beach in a way other than as a recreation platform. They are often cognizant of the difference between the developed environment and the natural environment and the hazards associated with placing valued real estate in dynamic environments. They hear about beach erosion; they link waves with beach change in this erosion context; and they realize that changes in beach volume and shape can occur under these conditions. They may also be cognizant of the environmental losses associated with large-scale landscape conversions, especially loss of high-order faunal species. They may not link landforms with vegetation; link the types of vegetation with the dynamism of coastal landscapes; or link present-day landforms and vegetation with coastal landscapes of the past. They often do not comprehend what is truly natural in the landscape; realize the need to restore natural environmental values; or understand the roles people can and should play in the restoration process. For example, students accept a flat, groomed, unvegetated beach as a natural environment and think that this is the way a beach should be. Students in coastal communities may have a better conception of the dynamics of the coast, but may have no better conception of past characteristics or the potential for future restoration efforts.

Summary conclusions

The future of beaches and dunes as unique and diverse geomorphological and biological environments depends on the perceived value and use of these landforms by humans. Less developed coastal regions must be conserved to provide templates for restoration of degraded natural environments in developed areas. Self-regulation in sand dune systems requires availability of sediment and space to allow natural transfers to occur. Restoration of these systems should include consideration of off-site activities that effect sediment transport and supply.

Two key problems of managing beaches and dunes in developed areas are that: (1) the natural instability of the coast that results in such great natural diversity of landforms and biota is perceived as a hazard to shorefront development; and (2) natural values now play a minor role in recreational use of the developed coast. An important issue in coastal zone management is finding a way to retain natural and human values while providing a degree of stability acceptable to municipal managers and residents.

Compromise must occur among factions advocating natural and human values in regions characterized by a high level of development. There is great potential for restoring landforms and the processes of landform evolution on the local scale. Small and incremental changes on the level of municipalities and individual lots can allow the growth of dunes that can serve as templates for neighbors and neighboring towns. Complete reversion to dynamic systems is not likely where restoration poses threats to human investment or recreational needs, but greater landform diversity and mobility is possible to achieve. National and state authorities can offer financial and legislative support for these local restoration initiatives.

Environmental education that focusses on the ecology of the coast and its value to tourism must figure prominently in restoration planning along with hazard management. Dune systems make excellent environmental and science education subjects. Attention must be paid to the attitudes of tourists and local residents toward the beach. Visitors to beaches are increasingly coming to expect featureless sand platforms and commercial development of the backbeach as the normal image of a beach. Supporters of coastal tourism at national, state and local levels must reconsider the image of the beach that they present to visitors. At the same, time bare-sand and sand-drift areas in undeveloped dunes have become more acceptable to both beach visitors and scientists. Coastal managers must be aware of and responsive to these changes in attitude and perception and the potential that can be achieved by allowing for more dynamic geomorphic systems.

9

Directions for geomorphological research

Overview

Developed coasts evolve as a result of direct and indirect human actions that are occurring at larger scale and higher frequency through time due to improvements in technology and increases in the economic value of coastal property. Coastal landforms are eliminated, reshaped, remobilized, stabilized or entirely re-created as artifacts. The resulting landforms differ from natural landforms internally and externally and are generally: (1) less dynamic; (2) less diverse in vegetation cover; (3) smaller in area; and (4) subject to cycles of evolution that correspond more closely to human processes than natural processes (at least in the depositional phase). If they are enhanced or restored by human efforts, the alterations are usually designed to provide a small number of utility functions.

Existing human actions and regulations cannot ensure that landforms in the future will have the size, dynamism, topographic and species diversity to provide the number and variety of resource options available in the naturally functioning coastal landscapes that are being lost through incompatible development. Actions can be made more compatible with natural processes, but this will involve compromise solutions that are likely to be viewed as undesirable by many stakeholders.

The preceding chapters provide a review of the state of knowledge of the processes affecting beaches and dunes on developed coasts and the resulting landform characteristics, along with suggestions for ways that natural values can be maintained while accommodating human use. The state of knowledge of human-altered landforms is still primitive, in part because so few geomorphological investigations have been conducted of these landforms, particularly in a holistic, objective, basic-research context. Past reluctance on the part of geomorphologists to apply their expertise to assessing human-altered landforms and to adopt an objective approach to human-altered landforms has created a need to develop a research paradigm specifically directed toward evaluation of human-altered systems.

This chapter evaluates the kinds of research required to fill gaps in the state of knowledge of human-altered landforms, with special emphasis on dynamic management approaches that will increase their future value. Recognition that humans have become intrinsic agents in evolution of coastal landscapes is significant in that it places the problem of restoring the value of these landscapes squarely on human action, requiring management approaches that work with, rather than against, natural processes.

Space constraints prevent specification of all geomorphological research needs, but representative studies are identified for the themes represented in chapters 2 through 8. Proper understanding of human-altered coasts requires comprehensive research programs beyond the subdiscipline of geomorphology, including social science, ethics, ecology, economics and education. It is not possible to fully represent these themes, but examples of related research needs are presented to provide perspective on what a holistic approach to assessment of landforms on developed coasts entails.

This text is designed as a reference document for conducting basic and applied research on landforms on developed coasts. It is not designed as a "how to" management manual, but a management context is required to provide a process explanation for human-altered landforms and to make suggestions for how coasts can evolve under different scenarios of human input. Without this evolutionary context, it is difficult to imagine development of management guidelines that view landforms as more than human use structures, divorced from their natural counterparts.

Scales of investigation

Large-scale programs

The emphasis in this book has been on the characteristics, genesis and evolution of individual landforms, rather than on large-scale changes to the shoreline, but the increasing scale of human alterations increases the need to provide models for assessing change at greater spatial and temporal scales (Stive *et al.* 1990). Examples of human activities that will require large-scale investigations include alterations at inlets, creation of offshore mounds, perched beaches and reefs, closing of estuaries or tidal basins and changes in response to accelerated sea level rise. Many of these assessments will be undertaken as adjuncts to construction projects, where investigations are often limited to evaluation of whether design criteria have been achieved. Hopefully, the scope of these investigations will be increased to evaluate the broader implications of the project on landforms and land use within and outside project boundaries.

Some organizations and programs are already in place to evaluate large-scale landscape changes associated with shore protection strategies at the scale of longshore transport cells or reaches. One example is the project NOURTEC for Innovative Nourishment Techniques Evaluation funded jointly by the Commission of the European Communities Directorate for Science, Research and Development, the Dutch Rijkswaterstaat, the German Coastal Research Station Norderney and the Danish Coastal Authority (Mulder *et al.* 1994). Other initiatives include the European Commission LIFE project entitled "Biodiversity and Dune Protection (Favennec 1996) and the Scientific–Industrial Association for Seacoast Protection that calls for harmonic interaction between coastal processes and coastal changes and revival and subsequent control of coastal processes along with optimization of coastal development (Kiknadze 1993). This program confirms that optimal use of natural resources must have a theoretical basis in environmental science and be conducted by a scientific-research organization that is financially independent from those who exploit natural resources.

The need for small-scale projects

It is likely that most studies of human-altered systems will be conducted outside the framework of these large-scale investigations. Many human alterations have been made at the scale of individual properties or municipalities and have been conducted without the benefit of scientific input. Many suggestions for future research made in this chapter are directed at actions at this scale.

One of the most exciting and (to some) surprising facts is that nearly every stop on field excursions to developed coasts reveals something different about how humans use or alter the shoreline, and case studies of human alterations are almost unlimited in number and type (Walker 1985). Unraveling the rationale and history of human actions and attempting to reconstruct past natural environments and future evolution in some of these areas (e.g., Figure 1.2) can be a fascinating exercise.

Some potential investigations of small-scale uses of beaches and dunes may seem inconsequential in their usefulness (e.g., studying the effects of "sand art" on the beach or the effect of throwing stones on gravel beaches), but these phenomena are interesting if not of great economic concern. Many studies of what are now localized or small-scale effects may have increased value for application in the future. For example, studies of changes to sediment characteristics and landforms associated with waste-disposal operations and erosion of landfills provide perspective on both the direct effects of this dumping and potential future effects of using these kinds of materials as beachfill, when existing stocks of suitable material have been depleted. The effects of extraction on subsidence can be used in a management context to determine the local effect on flooding and wave energy concentration at the shoreline, but these studies will also have added value to other locations by providing insight to effects of accelerated sea level rise.

Many small-scale and temporary human-created landforms can be readily examined in low cost investigations. Examples of these kinds of projects include manipulation of dunes on individual properties and the performance of low-cost geotextile systems that are being increasingly used (Pilarczyk 1996). Many geotextile structures are designed as temporary or backup protection, and, as a result, have drawn little attention from scientists. It is likely that increased attention will be devoted to assessing effects of these projects as they become more conspicuous in the coastal landscape. Often no records are kept of municipal actions, and the ease with which management is practiced at this scale has raised little formal interest in the causes of erosion or in the necessity for carrying out evaluations of project performance (Anthony and Cohen 1995). The effects of harvesting beach vegetation and removing litter have also received little formal interest. The state of knowledge of effects of many of these alterations is so primitive that reconaissance-level investigations will be substantial contributions.

The need for a new research paradigm

Changing the culture of geomorphological investigations

The human environment has not been systematically studied in a way that compares in depth to the natural environment. True progress in the study of coastal evolution requires the development of conceptual and predictive models of landform dynamics that involves an orderly, objective and scientifically informed intellectual process, identifying gaps in understanding and a rational plan of study, followed by a concerted research effort backed by ample resources (Sherman and Bauer 1993). At present, there is no groundswell movement to mount such a program for human-altered shorelines.

Recent research on the geomorphology of developed coasts has provided ample evidence of divergences in the form, surface cover and rate of change of human-altered landforms relative to natural landforms, but has largely ignored the conclusions that natural landscapes are a myth, that human agency is not an intrusion but a part of the coastal environment and that human-altered landscapes can and should be modeled as a generic system (Nordstrom 1994b). It is difficult to obtain the proper level of objectivity when static natural systems are considered the standard by which human-altered systems are evaluated. Terms used to describe human modifications to shore-lines include "intervention," "disturbance," "alteration" and "bias." It would appear that the term "alteration" is the most appropriate of these terms if humans are considered intrinsic agents.

The traditional view of human-altered systems as an aberration is under-standable. Detailed monitoring of developed beaches or human-altered fore-dunes in a geomorphological context may be avoided because human alteration both hinders and promotes beach recovery and complicates inter-pretation of natural processes that may be the focus of the investigation (Arens and Wiersma 1994; Morton *et al.* 1994). Many coastal scientists may not think that it is in their interest to underscore the lack of naturalness of the landscape they subject to rigorous objective analysis because this documenta-tion conveys an impression that their results lack universal applicability, that they accept the human-altered condition as the norm, or that they accept the human-altered system as a blueprint for future coastal evolution (Nordstrom 1994b).

Increased importance of human-altered landforms

Many human-altered landforms may have escaped notice because they were considered too strange, too small, too temporary, too site specific or too artificial. These constraints are disappearing through time as the degree of human involvement increases. Landforms that now appear strange (e.g., bull-dozed mounds that serve as platforms for lifeguard towers) may become accepted as they become a part of the beach scene in subsequent years. Landforms that are small or temporary take on increased importance when they are ubiquitous or recurring (e.g., wind blown sand patches on cultural surfaces) or where they are allowed to survive because of artificial protection from wave erosion (Figure 6.4). Human-created landforms that are artifacts become accepted parts of the landscape whether they remain artifacts, such as platforms created on the backshore to accommodate tourist use (Figure 2.5), or whether they are reshaped by natural processes to more closely resemble natural landforms. All of these landforms will take on added interest as they increase in spatial coverage through time as more of the natural landscape is converted.

The need for an interdisciplinary approach

Few planners have an earth science background and few earth scientists have a planning background. Greater co-ordination and co-operation between scientists and planners is required to ensure that the dynamic nature of the coast is properly reflected in planning guidelines (Lee 1993). This co-operation may reduce the number of problems produced when scientists identify sound options that are not accommodated within prevailing policies (Gares 1989; Bray *et al.* 1997).

Co-operation is also required to determine the reasons managers create or reshape landforms and the means they use. Natural scientists may find profit in analyzing human process variables in addition to physical process variables to explain landform assemblages and facilitate finding procedures for incorporating natural ecosystem characteristics and values into public and private capital investment decisions and environmental praxis. Documenting reasons for human alterations to landscape features and rates and volumes of sediment moved may involve interviews with government personnel and shorefront residents and search of government correspondence. This kind of research may be second nature to social scientists, but it is not a common approach used by geomorphologists and sedimentologists.

Modeling human-altered geomorphological systems is not an easy task. Models that evaluate humans as an intrinsic process require a greater number of assumptions and more explicit assumptions about human actions than models that consider humans as an extrinsic process. Geomorphological factors must be combined with human factors such as preference and precedent, rights of property owners, legal implications, economic motivation, environmental interests and constraints of government programs (Nordstrom 1994b). Many activities conducted at the local level (e.g., grading), are often done on an *ad-hoc* basis, with little technical input. Study of the effects of these activities will require greater collaboration and co-operation between geomorphologists and managers at the local level. This kind of collaboration should be extended to actions taken by residents and municipal managers to remove or accommodate landforms that form landward of the beach.

Classifying and modeling human-altered landforms

Many human alterations are incompatible with the way coastal environments change through natural processes, but it is not axiomatic that classification systems or models of coastal evolution formulated for natural systems are the best to use to determine strategies for shorelines that have

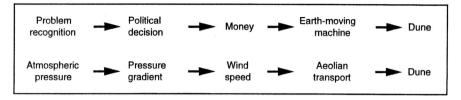

Figure 9.1 Comparison of human and natural processes that create a coastal dune. The outcome of both chains of events is a landform that may have a similar function according to restrictive human-use criteria (e.g., protection against flooding and overwash), but the characteristics of the two landforms, their subsequent evolution and their suitability for other functions may differ greatly.

passed the critical threshold beyond which return to a naturally functioning system is not an acceptable management option. Models of coastal evolution based on natural processes may be appropriate to anticipate future effects on undeveloped shorelines and shorelines where the scale of human activities is too small to overcome natural changes, but there are many developed shorelines where these models would not apply, and there may be fewer of these locations in the future (Nordstrom 1994b).

A process-response classification of human-altered landforms is rendered difficult by: (1) the great differences that can occur between natural and human process controls (Figure 9.1); (2) the great variety of processes that humans can use to create a landform; and (3) the great variety of subsequent changes that can be made to a landform that is created (due either to natural or human processes). The direct chain of events leading from human decisions to geomorphological response (presented simplistically in Figure 9.1) is difficult to trace, but these assessments are critical, given the inexorable transformation of the coast to a human artifact. Differences in the events proceeding from problem recognition to raising money for building a dune (Figure 9.1) could lead to selection of vegetation plantings rather than bulldozing, creating a dune with enhanced environmental value.

It is difficult to identify how human-altered landforms should be defined relative to their natural counterparts, if, indeed, representative natural landforms can be found under similar controls. Human-altered landforms are not at the opposite end of the spectrum from natural landforms according to most of the criteria identified in chapter 5, in that they can be smaller or larger, more mobile or less mobile, more densely vegetated or less densely vegetated. The ranking procedures using degree of naturalness (Figures 5.8 and 8.1) provide perspective on the extent to which human alterations allow natural processes to create landforms, but they do not provide objective, process-based systems for evaluating landforms by criteria other than degree of naturalness. Arens and Wiersma employ the conception of degree of dominance of certain types of management activities and degree of naturalness in their classification system. They also include presence of signs of aeolian activity, an important variable in identifying evidence of landform dynamism. It would be profitable to develop classification systems that predict landform outcomes based on human process inputs, rather than degree of naturalness, although the latter criterion may have greater usefulness in a management context.

A fruitful area of research is finding ways to present natural and human variables in the same matrix. Quantitative analyses that have included variables on human-induced changes have established a priori classes of shore types based on presence of human-induced landform characteristics (Anthony 1994; Dal Cin and Simeoni 1994) or have established landform classes based on the variables measured (Jackson *et al.* in press). The latter study reveals that geomorphological characteristics can account for a large percentage of the variance in shoreline characteristics, well after a shoreline is developed and protected. In their case, this resulted from beach nourishment and inlet dredging that changed the location of the ebb channel of a nearby inlet. The dominance of these two nature-engineering actions underscores the difficulty of using quantitative analysis to distinguish natural from human-induced geomorphological characteristics. Research on developed coastal landforms is still too preliminary to provide the basis for a definitive classification based on a single set of criteria or even a single methodology.

Assessing temporal scales of investigation

The recent past (150 years) has been described as a research gap lying between flourishing studies of long-term change and explorations of contemporary environmental dynamics (Cooke 1992). This time frame coincides with the period of greatest human alteration of the coast (Figure 1.3). There are some locations where the database is relatively long and human activities have covered a sufficient time span to draw meaningful conclusions about past effects (Nordstrom 1994a; Leidersdorf *et al.* 1993), but the database for most human-altered coasts spans such a short period and is so patchy in spatial coverage that meaningful extrapolation into the future is not possible. The lack of predictive power to determine landform evolution at temporal scales corresponding to human inputs inhibits design of a sustainable coastal system, leading to the protectionist strategy of weak sustainability (Pethick 1996).

The discussion of periodicities of cycles of change and relationships between size and longevity of landforms presented in chapter 6 is only one way of illustrating the differences in time–space relationships between natural and human altered landforms. This discussion was limited to only a few representative features. There is a need to specify the temporal and spatial scales of the full spectrum of human-altered landforms and their role in the evolution of landscapes.

Determining the proper temporal scale of investigation is important in evaluating the success of shore protection projects. Roelse *et al.* (1991) distinguish between system monitoring, project monitoring and project evaluation without data gathering. They point out that system monitoring has a longer period of execution, greater uniformity in measuring and handling data and increased potential for using the data for studies of large-scale coastal behavior. Project monitoring, in contrast, has a time span that (if done at all) lasts only a few years (Roelse *et al.* 1991). Project evaluation, without data gathering, lasts only during the life of the project. Human construction projects are often conducted as a series of activities, and monitoring of a single project does not reveal the complete picture of landscape evolution.

Natural erosion rates may not provide the best standard by which human actions should be evaluated. Many geomorphologists recommend using long-term records of shoreline change, such as >100 years (Paskoff 1992; Galgano and Leatherman 1991), but short-term trends that encompass shore protection activities may be better than long-term trends for predicting future erosion rates in intensively developed systems (Morton 1991; Byrnes *et al.* 1993). It is easy to introduce a bias into assessments of the significance of human alterations by selecting an inappropriate portion of the record of shoreline change.

The temporal scale of post-storm analyses of developed coasts (normally conducted within a few months of storm passage) should be extended to include post-reconstruction efforts (lasting several years) in order to place the long-term geomorphological impacts of storms into proper perspective (Nordstrom and Jackson 1995). This longer term, more holistic view of storm-induced changes, viewing humans as intrinsic to coastal evolution, will: (1) clarify changes that are not explainable under theories of natural evolution (dunes of a size or location that appear out of equilibrium with natural processes); (2) place shore protection strategies in perspective (by identifying their cumulative long-term effect on beach changes, not just their local, short-term effect); (3) identify the reason why successive storms cannot restore natural conditions; (4) provide guidelines for regulating construction of buildings that considers buildings integral to the coastal system; and (5) stimulate new ways to re-establish natural or naturally functioning components in a human-altered landscape (Nordstrom and Jackson 1995).

The potential differences in rate or scale of landscape changes resulting from future climatic changes should provide many interesting challenges for predicting future coastal characteristics and devising long-term management alternatives. It is likely that human actions will play a major role in evolution of these landscapes, requiring complex models that consider humans intrinsic geomorphological agents.

Providing bases for policy decisions

The preceding chapters indicate how humans are converting coastal landscapes into artifacts. Many management frameworks, even in intensively developed areas, do not directly treat the issue of whether humans are or should be the dominant agent of landscape evolution or whether human needs or actions should determine the characteristics of the resulting landforms. The nature of land use controls for an eroding coast will vary dramatically depending on whether the prescribed policy is one of no retreat, managed retreat or "muddle through." Policies that seem overtly environmentally friendly and adopt a stance that natural processes are or should be dominant may actually lead to destruction of natural environments. Policies that designate undeveloped shoreline enclaves for preservation as natural areas, while allowing protection structures in adjacent locations, may lead to sediment starvation in these enclaves, accompanied by accelerated erosion of desired environments. Policies that establish setback zones to accommodate erosion but fail to consider that long-term erosion will bring the shoreline to the location of fixed structures through time will also result in loss of beach and dune environments. In contrast, a coastal defense policy of no retreat may have greater potential for establishment and maintenance of naturally functioning landforms if the shoreline position is maintained using beach nourishment and if natural processes are allowed to rework the new sediments.

Geomorphological investigations are required to identify the coastal landscapes that could be created under alternative scenarios with humans recognized as intrinsic agents at the landscape scale so that planners can make rational choices about major long-term actions, such as those for sea level rise. For example, an interesting scenario would be to predict the shoreline of an eroding developed barrier island on the northeast coast of the USA if managed according to the Dutch policy of no retreat, using beach nourishment and nature engineering to create a naturalistic landscape. Creative solutions must be found for these eroding coasts where present development landward of the beach or dune restricts the amount of space available for landform migration.

Developing environmentally compatible alternatives

Increasing the number of management options

It is likely that the impetus for funding geomorphological investigations on developed coasts still will be in an applied-research mode, investigating impacts of new buildings or protection strategies. A creative task for geomorphologists is to provide guidelines to help maintain beach resources in a way that preserves as many elements of the natural system as possible within the existing management framework by maintaining the natural sediment budgets and preserving the mobility of landforms and their tendency to grow and be altered (Nordstrom 1990). Many coastal landforms have been modified into a form that accommodates only a limited number of uses. The most conspicuous examples are beaches that have been graded flat and raked for beach recreation and dunes that have been bulldozed into a single, linear, unvegetated ridge to provide emergency storm protection. A useful task for geomorphologists and other coastal resource specialists, is to identify ways of achieving a greater number of values and uses for these landforms consistent with views of the stakeholders and to change, where necessary, the perceptions of stakeholders in order to achieve these objectives.

There has been a perceptible shift from a position in which scientists offer their work for society and place it in an applied context to one in which science responds to societal goals in order to improve the environment (Trudgill and Richards 1997). One of these societal goals is the need to integrate nature-oriented management into coastal defense management as more of the coast becomes developed. Beach nourishment can be considered a form of "managed mobility" that can form the basis for improving natural characteristics. Louisse and van der Meulen (1991) present five options for a more dynamic coastal defense management based on "soft" defense measures that are applicable to the Dutch coast, where space–time scales of evolution can be 100s of hectares and 100s of years. The Dutch coastal defense policy of dynamic preservation offers new opportunities for the restoration and development of nature using a more integrated management approach; this policy is still in its infancy, and there is still much to be learned about strategies appropriate to this new policy (Koster and Hillen 1995). Finding appropriate strategies will be even more challenging for coasts where funding levels for shore protection are lower, where there is less room between buildings and the shoreline and where existing legal frameworks give stakeholders greater freedom to use beaches and dunes.

Making landscapes more dynamic

Managing sand dunes for conservation is more difficult than directly protecting dunes from destructive human activities because management requires a complete understanding of the implications of remedial action rather than simply requiring that the damaging development does not take place (Doody 1989). With change to more dynamic methods of management, it is increasingly important to distinguish between progressive deterioration and non-threatening system adjustments to changing inputs, requiring a carefully constructed research base capable of defining the limits of tolerances to change (Pethick 1996). The difficulty of predicting future evolution under alternative strategies (including specification of periods of recovery) for some of the dynamic approaches advocated by scientists (e.g., Wanders 1989) will greatly challenge geomorphologists working in both basic and applied research modes.

There is a need to anticipate the effects of allowing formerly managed landforms to revert to naturally functioning landforms (e.g., in formerly aforested dunes or where sand fences are no longer considered a management option). The temporal and spatial scales of the associated changes and the impact of these changes on future management decisions may be of long or short term, depending on the dominant natural process. Equilibration of dunes to a naturally functioning role following suspension of sand fencing programs has been predicted at upwards of 100 years on the outer Banks of North Carolina, where overwash is perceived to be dominant over dune building (DeKimpe *et al.* 1991). Equilibration has been predicted to be on the order of decades where aeolian transport and dune building play a greater role than overwash in maintaining the presence of a viable foredune (Gares and Nordstrom 1991, 1995).

In many locations, it may be unclear whether suspension of current human management practices will reinstate the former characteristics of the natural system or whether the imprint of human alterations will remain to provide a new blueprint for future evolution. Identification of the likelihood that natural evolution will occur and specification of additional management actions to assure this natural trajectory will provide interesting challenges in these locations.

Restoring environmental characteristics at specific sites

Each restoration site presents a unique set of problems and opportunities, and there is great need for the development of flexible implementation protocols (Tippets and Jorgensen 1991). A new role for geomorphologists is to provide the terrain parameters required to reshape coastal landforms in a way that is compatible in appearance and function if not in genesis (van der Meulen and Jungerius 1989b). Research projects are needed to evaluate the potential for re-establishing natural habitats at scales appropriate to municipalities and individual lots as well as at the scale of public parks and preserves. The complexities involved in dealing with so many stakeholders with different perceptions of the value of coastal resources will require an applied-research program that involves inputs from physical, biological and social scientists in ways that geomorphologists are not used to examining environmental problems.

Components of a research program for establishing habitats at the municipal level, as suggested in chapter 8, could include: (1) classifying the types of natural enclaves or zones that presently occur or could occur according to size, location, position relative to the beach and human structures; (2) identifying the mobility of landforms in these enclaves, their linkages to each other and their significance in providing species diversity and maintaining viable population sizes; (3) determining the value of these habitats to municipal managers, residents, commercial interests and beach users and their willingness to allow new habitats to be created; (4) evaluating the resources available within communities to implement new strategies; (5) establishing guidelines for municipal officials and local residents; (6) creating demonstration sites; and (7) establishing public information programs for residents and tourists. Establishing a working relationship with representative community officials and residents may be a requirement of these more creative or controversial projects. Table 9.1 identifies some of the research questions that might be evaluated in examining the feasibility of changing a management strategy for dunes from one of providing protection and recreation to one of providing recreation with enhanced (natural) values compatible with a developed shoreline (i.e. proceeding from conditions in Figure 8.6B to 8.6C). These research questions reveal the multidisciplinary nature of projects involving re-establishment of natural values on coasts where the landscape already has been transformed and maintained to accommodate a limited number of alternative recreational uses.

Investigating "traditional" problems

Effects of human actions on individual landforms

The number of meaningful geomorphological or sedimentological investigations that would provide valuable data and insight into the origin,

Table 9.1 *Questions for evaluating potential for restoring values of beaches and dunes in developed areas.*

General

How do values and uses for coastal dunes in developed areas differ from those in natural areas?

Which actions by municipal officials, private residents and commercial enterprises are non-compatible and should be suspended; which actions can be tolerated or expanded?

How can the concept of "coastal image" for a relevant segment of coast be defined?

How should the concept of ecotourism be framed on a developed coast?

How can coastal tourism include environmental values in a developed area?

What volunteer programs and labor are available to accomplish restoration goals?

How do we demonstrate that natural features in an urban setting provide a better recreational experience than alternative cultural features?

Is manual removal of cultural litter a feasible alternative to mechanical removal of all litter?

What are the best guidelines for coastal landscaping?

How should an environmental education project be formulated (forums, target audience, demonstration sites)?

Ecological

To what extent can the environmental gradient in the coastal setting be compressed?

Which species should be the objectives of management efforts?

How do vegetation and fauna differ on raked and unraked beaches?

How should the minimum critical size of a sand dune ecosystem be determined?

Which species can be accommodated on a foredune managed primarily to provide protection?

What is the nature of the relationship of aeolian transport and species interaction between the dunes on the beach and dune exclaves separated from the beach and foredune?

To what extent is mobile sand necessary to accommodate exclusively coastal species?

To what extent do dune exclaves landward of the beach provide habitat for dune flora and fauna?

Which species thrive best on dunes landward of shore-parallel protection structures?

How important is cultural litter to ecological values or evolution of incipient dune forms?

Should exotic vegetation be accepted?

Geomorphological

How do morphology, evolution and duration of landforms differ on raked and unraked beaches?

Is preservation of the upper litter line sufficient to regenerate useful incipient dune forms?

How can aeolian transport and dune mobility be accommodated without significantly increasing the degree of risk?

Will topographic variations lead to concentrations of wind flow and to blowouts?

Will elimination of sand fences on landward sides of dunes increase the hazard of blown sand?

What distance alongshore is required for the no-rake zone to ensure the viability of a naturally functioning dune?

Social science

What are norms and expectations of coastal residents and visitors regarding the appearance and accessibility of the beach?

What are the perceived values of dunes on private property by owners of the property?

Table 9.1 (*cont.*)

Can public appreciation of the inherent mobility of dune systems increase public awareness and understanding of coastal hazards?

What is the attitude of managers, residents and users of beaches as to the value of natural coastal features?

Does the public perceive dunes as more than structures to mitigate flooding and erosion?

Are dunes perceived and valued differently on a developed coast than on an undeveloped coast?

By what process do private coastal property owners choose between natural and exotic plants for use as landscaping agents?

What is the attitude of municipal officials toward symbolic fences?

What is the effect of natural debris on the perception of beach quality and cleanliness?

What is the relationship between knowledge and preference in the coastal setting?

Is development accepted as a legitimate part of the coastal landscape?

Is the absence of natural landforms on a raked beach viewed as unnatural or a legitimate version of the beach?

How does residential location *vis-à-vis* the shore influence environmental perception of beaches?

What is the relationship between relief (levelness) and preference in the coastal landscape?

How does the apparent degree of access affect landscape preference?

What is the arrangement of landscape features that allows the beach to be perceived as natural?

How does the spatial arrangement of dunes and paths affect landscape preference?

Which spatial arrangements are preferred in scenes that include dunes and the built environment?

How important are unoccluded views of the sea?

What is the influence of landform size on the perception that the landform belongs where it is or should remain where it is?

Source: Modified from Nordstrom *et al.* (1999).

evolution and function of individual landforms would number in the thousands. Many of the alterations to landforms identified in chapter 2 would have effects that are so site specific it would be impossible to provide a meaningful summary of research needs. Accordingly, only a few areas of investigation are mentioned here for perspective. The landforms discussed at greatest length in chapter 2 (and the other chapters) are those found in only a few key locations, largely western Europe and the USA. There is a need to identify the many types of landforms that occur in other areas to specify the full spectrum of forms and their temporal and spatial scales.

Little space was devoted in chapter 2 to landforms that are eliminated to facilitate construction of buildings and support infrastructure. These losses are frequently mentioned in the geomorphological literature, and it is important to continue to evaluate these losses to provide a context for evaluating effects of new development and for suggesting mitigating measures. However, more attention should be given to the landforms that remain and to ways that these landforms can be made more compatible with natural processes.

Many of the studies of landforms that are altered through use (e.g., trampling) also focus on degree of departure from (or return to) the previous condition without objective evaluation of the morphology and evolution of the new "disturbed" features that are created. Studies of changes to vegetation cover in coastal dunes by trampling and grazing have been well represented in the ecological literature, but geomorphological observations have been limited largely to statements of the dimensions of the area of bare sand. Future geomorphological studies should include insightful observations of types of changes, such as those provided by Mather and Ritchie (1977) and quantitative assessment of the form of these landforms, such as Bonner's (1988) assessment of walkovers, combined with instrumented studies of wind speed differences and sediment transport using traps (McCluskey et al. 1983). The value of many of these studies can be increased by expanding the context beyond the immediate needs of local managers. For example, greater understanding of the dynamics of gaps in dunes (now usually evaluated to select options for gap sealing) will take on increased importance in dune reserves that will be managed in the future as more dynamic systems where gaps will be allowed to evolve.

Reshaping by earth-moving equipment is one of the most ubiquitous ways landforms are altered by humans. Quantitative evaluations of these alterations are critical in both applied- and basic-research contexts. The applied studies are important in helping identify when and where beach grading and dune creation by mechanical means are advantageous or detrimental from the standpoint of hazard protection. Studies in a basic-research context are critical because mechanical transfer of sediment is the way many coastal landscapes can be expected to evolve in the future. There are far fewer studies of grading operations than there should be, considering the magnitude of change caused by this process.

There is a vast literature on use of vegetation for sand stabilization, but there are still important research needs, such as finding a solution for management of degenerated stands of *Ammophila* (van der Putten et al. 1993), and evaluating the success of dune stabilization programs in a long-term sense (Avis 1995). From the perspective of this book, many previous evaluations have been based on narrowly defined criteria, such as quantifying the area of dune that is stabilized and identifying rates of recovery of surface vegetation. It is also important to determine the significance of the loss of natural dynamism in the stabilized area and the role of the stabilized area in the broader coastal landscape. Many dune areas have been stabilized categorically without thorough investigations of alternatives to stabilization (Nordstrom and Lotstein 1989). Fruitful investigations can be conducted in locations where vegetation is removed to re-create blowouts (van Boxel *et al.* 1997), where sand fences are removed to create a more natural dune landscape (Gares and Nordstrom 1991; DeKimpe *et al.* 1991) and where aforested areas are allowed to revert to more dynamic dune landscapes (Sturgess 1992).

Beach and dune nourishment

There will always be a need for site-specific study of areas to be nourished to determine natural variability of profiles, identification of closure depth, choice of grain size for best retention and the cause of erosional hot spots (National Research Council 1995). There will also be a need for involvement of geomorphologists with engineers to assess the geomorphological impact of innovative nourishment designs as exemplified in the New Zealand experience (Healy *et al.* 1990).

Examination of past beach restoration and sand bypassing projects reveals a lack of monitoring programs and compilation of field data on project performance and biological impacts, with little standardization of format, content or reporting period to provide comparable data on different operations (Stauble and Nelson 1985; Roelse *et al.* 1991). Grosskopf and Kraus (1994) provide guidelines for surveying beach nourishment projects and Stauble and Nelson (1985) provide guidelines for monitoring borrow areas, fill placement, profile specifications, sediment analyses, shore processes and biota. Nelson (1993) presents a more detailed assessment of biological monitoring.

A large body of work has been performed by the US Army Corps of Engineers Dredging Research Program on creation and behavior of nearshore berms (Richardson 1994). Design criteria for the location, shape and dimensions of offshore nourishment projects are hard to define, and there is a need to monitor their morphological and sedimentological changes in time and space, along with the hydrodynamic processes operative on them (Hoekstra *et al.* 1994). Effects of bathymetry on variability in beachfill response is another important area of research (Stauble and Kraus 1993b), but these studies are likely to be too ambitious for most geomorphologists seeking independent projects.

The principal measures now used to determine success of beach nourishment programs are the lifespan of the fill and how actual performance conforms to predicted performance, but there is a need to define other measures of success because programs may serve a variety of objectives (National

Research Council 1995). The great importance of beach nourishment as a form of shore protection and the great significance of nourishment to future evolution of human-altered landscapes makes this option an important area of geomorphological research, especially if the nourished beach is to be considered as a landform in its own right and as a source of sediment for evolution of other landforms landward of it and downdrift of it (rather than merely as an engineering structure or recreation platform).

Identifying the geomorphological context

Investigations of nourished beaches in a geomorphological context are rare. Geomorphologists, like many of their counterparts in other disciplines, have been more concerned with determining lifetimes of fill volumes and cost effectiveness than determining either the influence of the fill on existing landforms or the potential landforms that could occur, given alternative management practices. It is not surprising that discussions of beach nourishment operations end up as discussions of erosion rates and volumes, given the present narrow conception of beach nourishment as a "slab of sand." It is curious that so much attention is focussed on the sediment that is lost from a nourished area while the landward portion of the beach that remains to affect aeolian transport and evolution of landforms on the beach and landward of it receives so little attention. There is a pronounced difference between the back-beach of a fill area and a natural beach in terms of surface sediments, rates of aeolian transport and microtopographic diversity. Van der Wal (1998) provides a good example of a study that concentrates on the geomorphological and sedimentological aspects of nourishment in a context other than simple volume or erosion rate.

There are guidelines for nourishment design and for sediment compatibility, but no specific requirements for mineralogical composition of nourishment material (Davis et al. 1991) or the way sediment differences affect subsequent evolution other than in the context of rates of erosion. Recent investigations indicate that mineralogical composition affects aeolian transport (van der Wal 1998), biota (Milton et al. 1997) and aesthetics (Bodge and Olsen 1992). It is assumed that studies of these effects will be increasingly important as local sediment supplies are exhausted and exotic materials must be found.

Considerations for renourishment

Design concepts for multiple nourishment operations are still in their infancy, and uncertainties about effectiveness of nourishment projects over the long term require flexibility that can be achieved when combining shore nourishment with appropriate monitoring (Stive *et al.* 1991). Geomorphological investigations are required to evaluate the characteristics and duration of landforms created as a result of previous nourishment projects and their survivability relative to the frequency of subsequent projects. This information will be useful in avoiding the crisis approach, where funds are procured and sand emplaced only when little beach is left (Leonard *et al.* 1990).

Dunes and aeolian transport

Dunes created by direct deposition of fill are often composed of materials that could not be transported by aeolian processes and they do not have the appropriate internal stratification. They are also usually created to achieve only one utility function (protection) and are shaped accordingly. Geomorphological investigations could be conducted to determine how nourished dunes could be made more compatible with natural processes or how alternative methods for achieving a dune through aeolian transport (with appropriate use of fences or vegetation) could occur in ways that are compatible with the constraints of the project.

High rates of aeolian transport, resulting in formation of a surface lag layer and subsequent reduction of aeolian transport, are readily observable effects of nourishment operations, but little is known about the phases of aeolian activity caused by reactivation of the surface of the fill, including the effect of layers of shell within the body of the fill (van der Wal 1998). There is also a need to evaluate the broader significance of initially enhanced aeolian transport so it can be managed as a resource rather than as a nuisance.

Longshore transport out of nourished areas

Fill materials that move alongshore out of the nourished area have been regarded as losses by critics of nourishment operations. Counter arguments refer to the benefits of added sediment inputs to the downdrift locations, without specifying the significance of these benefits in terms of landform evolution. It would be useful to determine how the added sediment inputs change the characteristics of landforms and associated environments in downdrift areas and specify how far changes occur. An increase in the sediment budget downdrift should enhance the likelihood for landforms to survive or grow, and increase topographic diversity and viability of habitat in a way that is less obtrusive than by direct nourishment. There is probably no better way to help undeveloped enclaves adjacent to developed areas recover from sediment starvation than having the developed areas re-establish their former role as source areas. The value of nourishment sites as feeder beaches to downdrift developed areas can be maximized if the new sediment is considered a natural resource and not simply new sand that can be graded into a wider recreation platform.

Effects of structures

Protection structures

Most evaluations of shore protection structures are conducted to assess effects on sediment budgets and erosion rates rather than on geomorphological features. Geomorphologists often associate protection structures with destruction of landforms or with creation of landforms that are less interesting or important than natural landforms. These perceptions may be true of seawalls that truncate the beach profile, but are not true of many other protection structures. Landforms at these structures are interesting because they are so poorly understood, and they are increasingly important topics of study because they occupy a greater proportion of the shore through time.

The widespread use of protection structures in some areas requires placing their effects in the context of large scales of shoreline evolution. Examples of studies at spatial scales >100 km and temporal scales of decades are provided by Nersesian *et al.* (1992), Short (1992) and Kana (1995). It would also be interesting to see the extent to which the structures that are no longer considered functional from the standpoint of protection still affect processes and create small-scale landforms. This comment could be applied to all cultural remnants, including piers, houses, swimming pools and roadways.

Groins produce great local differences in effects of waves, currents and aeolian transport, resulting in complex and interesting landforms with great variability across small spatial scales. The landforms created by these structures also may rank high in degree of naturalness in classifications of human-altered landforms (Figure 5.8). It is curious that structures that provide such a wealth of interesting geomorphological features (in a landscape that would be less interesting without them) have been treated with disdain by geomorphologists. It is also interesting that groins that are more compatible with natural coastal dynamics, such as pile groins (Trampenau *et al.* 1996), are less interesting from the standpoint of landform variability and complexity than impermeable groins (Everts 1979; Nersesian *et al.* 1992).

There is a great need for identifying groin effects seasonally and over longer periods (to identify conditions for filling, bypassing and offshore jetting) through monitoring changes in shoreline plan and profile, sediment characteristics, waves and currents, and quantifying longshore and cross-shore sediment transport (Kraus *et al.* 1994). Study of effects of traditional groins should be supplemented by study of effects of groins being built to new designs that: (1) allow sediment bypass (pile or notched groins); (2) have different effects on offshore-directed currents and sand transport (T-groins); or (3) have additional functions (e.g., artificial headlands). It is important to establish criteria for removing groins now that this option is being increasingly practiced. There is also a need to study the effects of groins in combination with beach nourishment to identify to what extent groins have a positive benefit in reducing initial sand volumes and future maintenance (Truitt *et al.* 1993) or in increasing the usefulness of landforms within and outside the limits of nourishment.

The great variety in design and use of breakwaters, headlands, sills and reefs, and the few extant field investigations of these structures (especially small-scale and innovative designs) leave room for many profitable studies. It is likely that there will be more geomorphological investigations of these structures as they are built in more locations. Headlands should be especially interesting to geomorphologists because of the complex wave and current circulation patterns they create and the crenulate shoreline planform. Saito *et al.* (1996) present examples of assessments of topographic effects of headlands. These structures and the landforms they create are accessible from land, greatly simplifying field investigations. Offshore structures create a lower energy environment in their lee, and it would be profitable to investigate changes in beach profile, modal beach state and grain-size characteristics to see to what extent the sheltered beaches become more reflective and characterized by coarser surface sediments as expected in lower energy environments (Nordstrom 1992; Masselink and Short 1993).

Wave dynamics and sedimentation processes at seawalls are poorly documented in the field, particularly during storms (Plant and Griggs 1992; FitzGerald *et al.* 1994; Kraus and McDougal 1996). Plant and Griggs (1992) offer an example of a field study relating processes to beach change. The geomorphological effects of the seawall they monitored were site-specific and monitored at daily to weekly intervals, so their measurements did not always capture the most significant interactions. However, their results point to some of the differences in beach processes and responses at a seawall and on the adjacent beach, and the interesting and valuable insight that can be gained in field investigation using state of the art instruments and sampling methodologies.

Few investigations have been conducted on landforms on the landward side of seawalls and other shore protection structures (Figure 5.1A). The increasing use of beach nourishment in front of seawalls creates additional opportunities for landforms to be associated with these structures. There is increased likelihood of aeolian transport across the structure in these cases, as well as increased potential for intentional or unintentional burial of structures under dunes. Dikes are another structure that can replace natural landforms or interact with them. Greater input from geomorphologists in the design of these features in a nature-engineering context will help ensure that they are incorporated into the coastal landscape in a more environmentally compatible manner.

Jetties and protection structures built at inlets, and dredging projects conducted to maintain predictable channels, result in pronounced changes at the inlets and at considerable distances downdrift. Considerations for study of managed inlets are provided in Dean (1988b); Bodge (1994a), Pope *et al.* (1994) and Stauble and Cialone (1996). The complexity of inlets and the great cost and difficulty of mounting projects that can provide definitive conclusions about the effects of human alterations make this area of research one of the most important and challenging of all.

Piers and pilings

There is now great opportunity for assessing the effects of scour at pilings and the effect of piers on the shoreline using existing research facilities such as the CERC pier at Duck, North Carolina (Miller *et al.* 1983) and the Port and Harbor Research Institute Pier at Hazaki, Japan (Maeno *et al.* 1996). The spatial scale of studies can be extended alongshore on both sides of piers to see how existing landforms are affected as they propagate past them and whether piers generate their own bedforms that migrate downdrift, as suggested by Plant and Holman (1996). There is no reason why study of structures cannot be incorporated into the sophisticated basic-research programs now being conducted at these facilities (e.g., Birkemeier *et al.* 1996). The increasing number of occasions where pile-supported houses will end up on the beach (Figure 6.3) justifies study of pilings beyond consideration of piers alone.

Fences

Sand-trapping fences are inexpensive and easy to install, and they have thus caused major alterations to the landscape in developed areas and coastal reserves. Their use for crowd control in some locations has extended their geomorphological impact to places where landform alterations are not required or intended. The effects of traditional single-row sand fences on trapping sand and the resulting characteristics of the dune are well known, but studies of alternative types and configurations of sand fences are limited in number, especially studies that use quantitative data to identify the characteristics of multi-row fences (Hotta *et al.* 1991; Hotta and Horikawa 1996). Studies of alternatives to use of sand fences will be of value, including removal of fences or suspension of their use where managers wish to have a more dynamic dune environment (DeKimpe *et al.* 1991; Gares and Nordstrom 1991) or use of symbolic fences rather than sand-trapping fences for crowd control where managers wish to minimize interference with natural sediment transfers.

Buildings

There are few investigations of the effects of houses on swash processes on the beach and on wind flow and aeolian transport in the dunes. Studies that do exist (e.g., Nordstrom and McCluskey 1984, 1985; Nordstrom *et al.* 1986) provide little quantitative data on processes. Investigations of impacts of houses will take on increased importance if management programs are implemented to: (1) prevent use of shore protection structures on eroding shorelines (resulting in more houses on the beach); or (2) make buildings more compatible with natural processes (to take advantage of environmentally friendly management solutions). There is a need for more field studies of performance of building foundations interacting with waves and currents in order to provide criteria for embedment depths (Nicholls *et al.* 1995). It may be possible to combine these studies of effects of natural processes on buildings (in a hazard-reduction context) with studies of the effects of houses on natural processes (in an environmental–management context).

Large buildings, such as high-rise structures, are often regulated under site review laws, and planning criteria are needed to specify the optimum dimensions, locations and shapes of the landforms that are created or modified by these buildings to expedite permit decisions in the future (Nordstrom and Jackson 1998). Conversion to high-rise buildings is new for many municipalities, and many do not have policies on wind control (Aynsley 1989) or a database to place new projects in perspective. Large-scale field studies are best handled at the community or larger scale because of the cost effectiveness of the monitoring programs and the significance of the results to development of generic models. The need for these field monitoring projects increases as building densities increase, the size of buildings increases, the impact of structures increases individually and cumulatively and the cost and scale of mitigating measures increase (Nordstrom and Jackson 1998).

Residential infrastructure

It will be useful to separate the effects of houses from the effects of support infrastructure (swimming pools, roads, walkways, parking areas) and the actions residents take to enhance use of their properties, because separate land use controls and performance criteria can be applied to each of these different human modifications. Field documentation of effects of swimming pools is lacking, and current assessments are based on models of scour at sea-walls and piers (Nnaji *et al.* 1996; Yazdani *et al.* 1997). Many different materials and designs are used to construct these pools, but some are superior to others in terms of impact on landforms (Nnaji *et al.* 1996).

As is the case with buildings, the impact of infrastructure may be studied in terms of both the direct effects of the structure and the indirect effects associated with maintenance and use. Studies of the impact of roads, for example, may include evaluation of the significance of enhanced sediment transport across the impermeable hard surface by waves and winds during storms, whereas studies of maintenance may include evaluation of the form and evolution of sand deposited to the side of the road during post-storm cleanup operations.

Artificial islands

Artificial islands created to provide a base for pumping stations, break-waters or dikes may be viewed solely as engineering structures or as opportunities for creation of new geomorphological environments. Landforms that have developed on offshore islands in the Beaufort Sea (Leidersdorf *et al.* 1990) and at Neeltje Jans, the artificial barrier island created in The Netherlands (Watson and Finkl 1990), show what can be done to enhance environmental values on these anthropogenic features and how terrain parameters can be developed to have applicability at large scales. Future study of these artificial environments will provide an important perspective on the long-term success of efforts to restore coastal landscapes at the barrier island scale, an option that may receive greater scrutiny with the need to evaluate alternatives to cope with accelerated sea level rise.

Related studies

The discussion in this book does not include hardware scale models or analytical models that provide information that is both supplementary and complementary to data gathered in the field. Kraus and McDougal (1996) and McDougal *et al.* (1996) provide examples of how these different means of assessing human alterations can be integrated. Modelling methods may be the best way to investigate projects where complex processes interact over large spatial scales where site-specific measurements cannot provide a comprehensive explanation (e.g., at inlets) and where field investigations are too costly.

Reference to biological investigations have been made frequently throughout the text because of the close relationship between landforms and biota.

This book concentrates only on the geomorphological perspective, and the biological context must be greatly expanded to provide a complete explanation of the developed coastal landscape and make suggestions for management.

The importance of social science research is readily seen in the number of questions that require evaluation when considering the option of restoring natural values to beaches and dunes in developed communities (Table 9.1). It is likely that presenting environmental issues as economic opportunities will improve personal and corporate behavior (Collins and Barkdull 1995), so economic analyses (e.g., Ruiz and de Quirós 1994) also are needed to ensure that suggestions for improving coastal resources have a chance of being successfully implemented. Ways of incorporating natural values into the economy of tourism would be especially valuable.

There is likely to be a strong undercurrent of advocacy in any study of developed coasts. In this book, for example, I advocate topographic and vegetational diversity and landform mobility, along with a more open acceptance and study of the results of establishing more dynamic landforms. The conception of good and bad in coastal landscapes is, in part, determined by our disciplinary background. Modern debates in environmental ethics provide much-needed perspective on alternative views of the relationship between societal interests and natural values.

Public involvement and educational programs (outreach) are important components of any project that involves human-induced transformation of the landscape. These programs are important when: (1) implementing costly or large-scale protection strategies, such as beach nourishment (Gordon 1992; Kana 1993; Koster and Hillen 1995; National Research Council 1995); (2) trying to get the public to think in terms of long-term solutions (Smith 1991); (3) suggesting radical management programs, such as relocating residents (Gares 1989; Bray *et al.* 1997); (4) considering devegetating dunes (Geelen *et al.* 1995); (5) implementing projects that may interfere with aesthetic values or views of the sea (Ulrich 1993; Jones 1996a); (6) attempting to get residents and policy makers to accept environmentally compatible measures over hard structures (Smith and Jackson 1990; Townend and Fleming 1991; Goss and Gooderham 1996); (7) attempting to discourage incompatible activities, such as mining sand and gravel (Smith 1994); (8) enhancing awareness of the mission and values of coastal management units (Morgan *et al.* 1993); (9) enhancing awareness of endangered species (Oertel *et al.* 1996); (10) making setback programs understandable to the public (Houlahan 1989); (11) dramatizing the local relevance of state policy documents (Burby and Dalton 1993); and (12) helping visitors chose among a variety of recreation opportunities and match their preferences with the appropriate level of development (Morgan 1996).

Concluding statement

Study of beaches and dunes in the context of human actions has become a necessity, whether this study is conducted as basic or applied research. The distinction between these two modes of research is blurred on developed coasts, as is the relationship between what is natural and what is an artifact. It is regretable that so much of the natural system has been lost, but it does not appear that a return to a natural condition is possible for most locations that humans wish to use. Perhaps the greatest research challenge is determining the standard by which future human alterations will be evaluated, given the difficulty of determining what is natural and the difficulty of substantiating that return to a natural system would actually be achievable, given the nature of humans. Our ability to alter the processes, forms and functions of the coast is awesome. Hopefully, this power can be directed to activities that are more environmentally compatible and our definition of what is environmentally compatible can be properly formulated and be meaningful.

References

Adriaanse, L.A. and Choosen, J. 1991. Beach and dune nourishment and environmental aspects. *Coastal Engineering* 16: 129–46.

Ahrendt, K. and Köster, R. 1996. An artificial longshore bar at the west coast of the island of Sylt/German Bight – first experiences. *Journal of Coastal Research* 12: 354–67.

Ahrens, J.P. and Cox, J. 1990. Design and performance of reef breakwaters. *Journal of Coastal Research* Special Issue 7: 61–75.

Ahrens, J.P. and Heimbaugh, M.S. 1989. Dynamic stability of dumped riprap. *Coastal Zone 89*. New York: American Society of Civil Engineers, 3377–89.

Aibulatov, N.A. 1993. Geoecology of the Black Sea coastal zone. In Kos'yan, R. editor, *Coastlines of the Black Sea*. New York: American Society of Civil Engineers, 103–24.

Allison, M.C. and Pollock, C.B. 1993. Nearshore berms: an evaluation of prototype designs. *Coastal Zone 93*. New York: American Society of Civil Engineers, 2938–50.

Amos, W.H and Amos, S.H. 1985. *Atlantic and Gulf Coasts: Audubon Society Nature Guides*. New York: Alfred A. Knopf.

Anctil, F. and Ouellet, Y. 1990. Preliminary evaluation of impacts of sand extraction near Illes-de-la-Madeleine Archipelago, Québec, Canada. *Journal of Coastal Research* 6: 37–51.

Anders, F.J. and Leatherman, S.P. 1987. Effects of off-road vehicles on coastal foredunes at Fire Island, New York, USA. *Environmental Management* 11: 45–52.

Andersen, U.V. 1995a. Resistance of Danish coastal vegetation types to human trampling. *Biological Conservation* 71: 223–30.

Andersen, U.V. 1995b. Invasive aliens: a threat to the Danish coastal vegetation? In Healy, M.G. and Doody, J.P. editors, *Directions in European Coastal Management*. Cardigan, UK: Samara Publishing Ltd., 335–44.

Anderson, P. and Romeril, M.G. 1992. Mowing experiments to restore a species-rich sward on sand dunes in Jersey, Channel Islands, G.B. In Carter, R.W.G., Curtis, T.G.F. and Sheehy-Skeffington, M.J. editors, *Coastal Dunes: Geomorphology, Ecology and Management for Conservation*, Rotterdam: A.A. Balkema, 218–34.

Andrassy, C.L. 1991. Monitoring of a nearshore disposal mound at Silver Strand State Park. *Coastal Sediments 91*. New York: American Society of Civil Engineers, 1970–84.

Angus, S. and Elliott, M.M. 1992. Erosion in Scottish machair with particular reference to the Outer Hebrides. In Carter, R.W.G., Curtis, T.G.F. and Sheehy-Skeffington,

M.J. editors, *Coastal Dunes: Geomorphology, Ecology and Management for Conservation*, Rotterdam: A.A. Balkema, 93–112.

Anthony, E.J. 1994. Natural and artificial shores of the French Riviera: an analysis of their interrelationship. *Journal of Coastal Research* 10: 48–58.

Anthony, E.J. 1997. The status of beaches and shoreline development options on the French Riviera: a perspective and a prognosis. *Journal of Coastal Conservation* 3: 169–78.

Anthony, E.J. and Cohen, O. 1995. Nourishment solutions to the problem of beach erosion in France: the case of the French Riviera. In Healy, M.G. and Doody, J.P. editors, *Directions in European Coastal Management*. Cardigan, UK: Samara Publishing Ltd, 199–212.

Apostolova, I., Meshinev, T. and Petrova, A. 1996. Habitat diversity in the Veleka River mouth and Silistar protected areas in Bulgaria. In Jones, P.S., Healy, M.G. and Williams, A.T. editors, *Studies in European Coastal Management*. Cardigan, UK: Samara Publishing Ltd., 183–90.

Arens, S.M. and Wiersma, J. 1994. The Dutch foredunes: inventory and classification. *Journal of Coastal Research* 10: 189–202.

Armah, A.K. 1991. Coastal erosion in Ghana: causes, patterns, research needs and possible solutions. *Coastal Zone 91*. New York: American Society of Civil Engineers, 2463–73.

Armstrong, G. and R.E. Flick 1989. Storm damage assessment for the January 1988 storm along the southern California Shoreline. *Shore and Beach* 57 (4): 18–23.

Atherley, K.A., Nurse, L.A. and Toppin, Y.B. 1991. Facing management challenges on the Barbados coastline: the problem of coastline access. In Cambers, G. editor, *Coastlines of the Caribbean*. New York: American Society of Civil Engineers, 17–31.

Avis, A.M. 1995. An evaluation of the vegetation development after artificially stabilizing South African coastal dunes with indigenous species. *Journal of Coastal Conservation* 1: 41–50.

Awosika, L.F. and Ibe, A.C. 1993. Geomorphology and tourism related aspects of the Lekki barrier-lagoon coastline in Nigeria. In Wong, P.P. editor, *Tourism vs Environment: the Case for Coastal Areas*. Dordrecht: Kluwer Academic Publishers, 109–24.

Awosika, L.F., Ibe, A.C., Ibe, C.E. and Inegbedion 1991. Geomorphology of the Lekki barrier-lagoon coastline in Nigeria and its vulnerability to rising sea level. *Coastal Zone 91*. New York: American Society of Civil Engineers, 2380–93.

Awosika, L.F., Ibe, A.C. and Ibe, C.E. 1993. Anthropogenic activities affecting sediment load balance along the west African coastline. In Awosika, L.F., Ibe, A.C. and Shroader, P. editors, *Coastlines of Western Africa*. New York: American Society of Civil Engineers, 26–39.

Aynsley, R.M. 1989. Politics of pedestrian level urban wind control. *Building and Environment*. 24: 291–5.

Ballinger, R.C., Havard, M., Pettit, S. and Smith, H.D. 1996. Towards a more integrated management approach for the Welsh coastal zone. In Jones, P.S., Healy, M.G. and Williams, A.T. editors, *Studies in European Coastal Management*. Cardigan, UK: Samara Publishing Ltd., 35–44.

Bandeira, J.V., Araújo, L.C. and do Valle, A.B. 1990. Emergency situation in the shoreline reach of an offshore oilfield pipeline and remedial measures. *Coastal Engineering:*

Proceedings of the Twenty-second Coastal Engineering Conference. New York: American Society of Civil Engineers, 3171–82.

Barr, D. A. and McKenzie, J. B. 1976. Stabilization in Queensland, Australia, using vegetation and mulches. *International Journal of Biometeorology* 20: 1–8.

Barron, P. and Dalton, G. 1996. Direct seeding of native trees and shrubs in coastal environments. *Journal of Coastal Research* 12: 1006–8.

Basco, D. R. 1990. The effect of seawalls on long-term shoreline change rates for the southern Virginia ocean shoreline. *Coastal Engineering: Proceedings of the Twenty-second Coastal Engineering Conference*. New York: American Society of Civil Engineers, 1292–305.

Basco, D. R. and Shin, C. S. 1996. Dune damage curves and their use to estimate dune maintenance costs. *Coastal Engineering 1996: Proceedings of the Twenty-fifth International Conference*. New York: American Society of Civil Engineers, 2969–81.

Basco, D. R., Bellomo, D. A. and Pollock, C. 1992. Statistically significant beach profile change with and without the presence of seawalls. *Coastal Engineering: Proceedings of the Twenty-third Coastal Engineering Conference*. New York: American Society of Civil Engineers, 1924–37.

Basco, D. R., Bellomo, D. A., Hazelton, J. M. and Jones, B. N. 1997. The influence of seawalls on subaerial beach volumes with receding shorelines. *Coastal Engineering* 30: 203–33.

Baskaran, A. and Stathopoulos, T. 1989. Computational evaluation of wind effects on buildings. *Building and Environment* 24: 325–33.

Bauer, B. O., Allen, J. R., Nordstrom, K. F. and Sherman, D. J. 1991. Sediment redistribution in a groin embayment under shore-normal wave approach. *Zeitschrift für Geomorphologie* Supplementband 81: 135–48.

Baye, P. 1990. Ecological history of an artificial foredune ridge on a northeastern barrier spit. In Davidson-Arnott, R. G. D. editor, *Proceedings of the Symposium on Coastal Sand Dunes*, Ottawa: National Research Council Canada, 389–403.

Beachler, K. E. and Higgins, S. H. 1992. Hollywood/Hallandale building Florida's beaches in the 1990's. *Shore and Beach* 60 (3): 15–22.

Beachler, K. E. and Mann, D. W. 1996. Long range positive effects of the Delray beach nourishment program. *Coastal Engineering 1996: Proceedings of the Twenty-fifth International Conference*. New York: American Society of Civil Engineers, 4613–20.

Beatley, T., Brower, D. J. and Schwab, A. K. 1994. *An introduction to Coastal Zone Management*. Washington, DC: Island Press.

Beatley, T., Manter, S. and Platt, R. H. 1992. Erosion as a political hazard: Folly Beach after Hugo. In Platt, R. H., Miller, H. C., Beatley, T., Melville. J. and Mathenia, B. G. editors, *Coastal Erosion: Has Retreat Sounded?* Boulder, CO: University of Colorado Institute of Behavioral Science, 140–52.

Beeftink, A. 1985. Interactions between *Limonium vulgare* and *Plantago maritima* in the *Plantagini-Limonietum* on the Boschplaat, Terschelling, The Netherlands. *Vegetatio* 61: 33–44.

Bell, F. W. and Leeworthy, V. R. 1985. An economic analysis of saltwater recreational beaches in Florida 1984. *Shore and Beach* 53 (2): 16–21.

Berendsen, H. J. A. and Zagwijn, W. H. 1984. Some conclusions reached at the symposium on geological changes in the western Netherlands during the period 1000–1300 AD. *Geologie en Mijnbouw* 63: 225–9.

Binderup, M. 1997. Recent changes of the coastline and nearshore zone, Vejrø Island, Denmark: possible consequences for future development. *Journal of Coastal Research* 13: 417–20.

Bird, E.C.F. 1993. *Submerging Coasts: the Effects of a Rising Sea Level on Coastal Environments.* Chichester: John Wiley & Sons.

Bird, E.C.F. 1996. *Beach Management.* Chichester: John Wiley & Sons.

Bird, E.C.F. and Fabbri, P. 1987. Archaeological evidence of coastline changes illustrated with reference to Latium, Italy. *Colloques Internationaux C.N.R.S.: Déplacements des Lignes de Rivage en Méditérranée.* Paris: Editions du CNRS.

Birkemeier, W.A., Long, C.E. and Hathaway, K.K. 1996. DELILAH, DUCK 94 and Sandy Duck: three nearshore field experiments. *Coastal Engineering 1996: Proceedings of the Twenty-fifth International Conference.* New York: American Society of Civil Engineers, 4052–65.

Blackstock, T. 1985. Nature conservation within a conifer plantation on a coastal dune system, Newborough Warren, Anglesey. In Doody, P. editor, *Sand Dunes and their Management,* Peterborough: Nature Conservancy Council, 145–9.

Blair, C. and Rosenberg, E.L. 1987. Virginia's CZM Program. *Coastal Zone 87.* New York: American Society of Civil Engineers, 2983–98.

Bocamazo, L.M. 1991. Sea Bright to Manasquan, New Jersey beach erosion control projects. *Shore and Beach* 59 (3): 37–42.

Bodge, K.R. 1994a. The extent of inlet impacts upon adjacent shorelines. *Coastal Engineering: Proceedings of the Twenty-fourth Coastal Engineering Conference.* New York: American Society of Civil Engineers, 2943–57.

Bodge, K.R. 1994b. Performance of nearshore berm disposal at Port Canaveral, Florida. *Dredging '94.* New York: American Society of Civil Engineers, 1182–91.

Bodge, K.R. and E.J. Olsen. 1992. Aragonite beachfill at Fisher Island, Florida. *Shore and Beach* 60 (1): 3–8.

Bokuniewicz, H.C. 1990. Tailoring local responses to rising sea level: a suggestion for Long Island. *Shore and Beach* 58 (3): 22–5.

Boldyrev, V.L., Dimitriev, A.E. and Shelkov, I.A. 1996. New approaches to shoreline protection along the Baltic Sea coast of Kaliningrad region. In Jones, P.S., Healy, M.G. and Williams, A.T. editors, *Studies in European Coastal Management.* Cardigan, UK: Samara Publishing Ltd., 105–10.

Bondesan, M., Castiglioni, G.B., Elmi, C., Gabbianelli, G. Marocco, R., Pirazzoli, P.A. and Tomasin, A. 1995. Coastal areas at risk from storm surges and sea-level rise in northeastern Italy. *Journal of Coastal Research* 11: 1354–79.

Bonner, A.E. 1988. Pedestrian walkover form and eolian sediment movement at Fire Island, New York. *Shore and Beach* 56 (1): 23–7.

Boorman, L.A. 1989. The grazing of British sand dune vegetation. *Proceedings of the Royal Society of Edinburgh* 96B: 75–88.

Boorman, L.A. and Fuller, R.M. 1977. Studies on the impact of paths on the dune vegetation at Winterton, Norfolk, England. *Biological Conservation* 12: 203–16.

Born, S.M. and Miller, A.H. 1987. An assessment of the "soft" approach to CZM. *Coastal Zone 87.* New York: American Society of Civil Engineers, 2939–53.

Borówka, R.K. 1990. The Holocene development and present morphology of the Leba

Dunes, Baltic coast of Poland. In Nordstrom, K. F., Psuty, N. P. and Carter, R.W.G. editors, *Coastal Dunes: Form and Process*, Chichester: John Wiley & Sons, 289–313.

Bortz, B.M. 1991. Pre-storm and post-storm hurricane planning in Nags Head, North Carolina. *Coastal Zone 91*. New York: American Society of Civil Engineers, 1034–45.

Bosma, K. F. and Dalrymple, R.A. 1996. Beach profile analysis around Indian River Inlet, Delaware, USA. *Coastal Engineering 1996: Proceedings of the Twenty-fifth International Conference*. New York: American Society of Civil Engineers, 2829–42.

Bourman, R.P. 1990. Artificial beach progradation by quarry waste disposal at Rapid Bay, South Australia. *Journal of Coastal Research* Special Issue 6: 69–76.

Boyles, R. Jr. 1993. The economics of managing coastal erosion. *Coastal Zone 93*. New York: American Society of Civil Engineers, 791–7.

Branan, W.V., Hale, L.Z. and Srinian, K. 1991. What future for Phuket? *Coastal Zone 91*. New York: American Society of Civil Engineers, 1713–19.

Bray, M.J. and Hooke, J.M. 1995. Strategies for conserving dynamic coastal landforms. In Healy, M.G. and Doody, J.P. editors, *Directions in European Coastal Management*. Cardigan, UK: Samara Publishing Ltd., 275–90.

Bray, M., Hooke, J. and Carter, D. 1997. Planning for sea-level rise on the south coast of England: advising the decision-makers. *Transactions of the Institute of British Geographers* 22: 13–30.

Breton, F. and Esteban, P. 1995. The management and recuperation of beaches in Catalonia. In Healy, M.G. and Doody, J.P. editors, *Directions in European Coastal Management*. Cardigan, UK: Samara Publishing Ltd, 511–17.

Brindell, J.R. 1990. Florida coastal management moves to local government. *Journal of Coastal Research* 6: 727–33.

Bringas Rábago, N.L. 1993. Tourism development issues in the Tijuana-Ensenada corridor, Baja California, Mexico. In Fermán-Almada, J.L., Gómez-Morin, L. and Fischer, D.W. editors, *Coastal Zone Management in Mexico: the Baja California Experience*. New York: American Society of Civil Engineers, 24–9.

Browder, A.E., Dean, R.G. and Chen, R. 1996. Performance of a submerged breakwater for shore protection. *Coastal Engineering 1996: Proceedings of the Twenty-fifth International Conference*. New York: American Society of Civil Engineers, 2312–23.

Bruno, M.S., Herrington, D.O. and Rankin, K.L. 1996. The use of artificial reefs in erosion control: results of the New Jersey pilot reef project. In Tait, L.S. editor, *The Future of Beach Nourishment*, Tallahassee, FL: Florida Shore and Beach Preservation Association, 239–54.

Brunsden, D. 1996. Geomorphological events and landform change: the centenary lecture to the Department of Geography, University of Heidelberg. *Zeitschrift für Geomorphologie* 40: 273–88.

Bruun, P. 1993. Economic aspects of backpassing: new type profiling by new-type equipment. *Journal of Coastal Research* 9: 1106–9.

Bruun, P. 1995. The development of downdrift erosion. *Journal of Coastal Research* 11: 1242–57.

Bruun, P. and Willekes, G. 1992. Bypassing and backpassing at harbors, navigation channels, and tidal entrances: use of shallow-water draft hopper dredgers with pump-out capabilities. *Journal of Coastal Research* 8: 972–7.

Bryche, A., de Putter, B. and De Wolf, P. 1993. The French and Belgian coast from Dunkirk to De Panne: a case study of transborder cooperation in the framework of the Interreg Initiative of the European Community. In Hillen, R. and H.J. Verhagen editors, *Coastlines of the Southern North Sea*. New York: American Society of Civil Engineers, 336–43.

Buckley, M. and Rhodes, O.E. 1989. The Upton/Jones Amendment and the zone of imminent collapse. *Coastal Zone 89*. New York: American Society of Civil Engineers, 3999–4009.

Bull, C.F.J., Davis, A.M. and Jones, R. 1998. The influence of fish-tail groynes (or breakwaters) on the characteristics of the adjacent beach at Llandudno, North Wales. *Journal of Coastal Research* 14: 93–105.

Bunpapong, S. and Ausavajitanond, S. 1991. Saving what's left of tourism development at Patong Beach, Phuket, Thailand. *Coastal Zone 91*. New York: American Society of Civil Engineers, 1685–97.

Burby, R.J. and Dalton, L.C. 1993. State planning mandates and coastal management. *Coastal Zone 93*. New York: American Society of Civil Engineers, 1069–83.

Burger, J. 1989. Least tern populations in coastal New Jersey: monitoring and management of a regionally-endangered species. *Journal of Coastal Research* 5: 801–11.

Bush, D.M. 1991. Impact of Hurricane Hugo on the rocky coast of Puerto Rico. *Journal of Coastal Research* Special Issue 8: 49–67.

Butler, R.W. 1980. The concept of a tourist area cycle of evolution: implications for management of resources. *Canadian Geographer* 24: 5–12.

Byrnes, M.R., Hiland, M.W. and McBride, R.A. 1993. Historical shoreline position change for the mainland beach in Harrison County, Mississippi. *Coastal Zone 93*, New York: American Society of Civil Engineers, 1406–20.

Cambers, G. 1993a. Beach stability and coastal erosion in the eastern Caribbean islands – a regional program. *Coastal Zone 93*, New York: American Society of Civil Engineers, 870–82.

Cambers, G. 1993b. Coastal zone management in the smaller islands of the eastern Caribbean: an assessment and future perspectives. *Coastal Zone 93*, New York: American Society of Civil Engineers, 2343–53.

Camfield, F.E. 1993. Different views of beachfill performance. *Shore and Beach* 61 (4) 4–8.

Canning, D.J. 1993 Dunes management plan: Long Beach Peninsula, Washington. *Coastal Zone 93*. New York: American Society of Civil Engineers, 1938–50.

Capelli, M.H. 1991. Recreational impacts on coastal habitats: Ventura County fairgrounds, California. *Coastal Zone 91*. New York: American Society of Civil Engineers, 3331–51.

Caputo, C., D'Alessandro, G.B., LaMonica, B., Landini, B. and Palmieri, E.L. 1991. Present erosion and dynamics of Italian beaches. *Zeitschrift für Geomorphologie* Supplementband 81: 31–9.

Carbognin, L., Cipriani, M. and Marabini, F. 1989. First results of a new shore protection work installed along the eastern Italian coast. In Fabbri, P. editor, *Coastlines of Italy*. New York: American Society of Civil Engineers, 170–6.

Carlson, L.H. and Godfrey, P.J. 1989. Human impact management in a coastal recreation and natural area. *Biological Conservation* 49: 141–56.

Carter, R.W.G. 1980. Human activities and geomorphic processes: the example of recreation pressure on the Northern Ireland coast. *Zeitschrift für Geomorphologie* Supplementband 34:155–64.

Carter, R.W.G. 1985. Approaches to sand dune conservation in Ireland. In Doody, P. editor, *Sand Dunes and their Management*, Peterborough: Nature Conservancy Council, 29–41.

Carter, R.W.G. 1988. *Coastal Environments*. London: Academic Press.

Carter, R.W.G. 1992. Sea level changes: past, present and future. *Quaternary Proceedings* 2: 111–32.

Carter, R.W.G., Eastwood, D.A. and Pollard, H.J. 1993. Man's impact on the coast of Ireland. In Wong, P.P. editor, *Tourism vs Environment: the Case for Coastal Areas*. Dordrecht: Kluwer Academic Publishers, 211–25.

Cencini, C. and Varani, L. 1988. Italy. In Walker, H.J. editor, *Artificial Structures and Shorelines*. Dordrecht: Kluwer Academic Publishers, 193–206.

Cencini, C. and Varani, L. 1989. Degradation of coastal dune systems through anthropogenic action. In Fabbri, P. editor, *Coastlines of Italy*. New York: American Society of Civil Engineers, 55–69.

Cencini, C., Marchi, M., Torresani, S. and Varani, L. 1988. The impact of tourism on Italian deltaic coastlands. *Ocean and Shoreline Management* 11: 353–74.

Cervantes-Borja, J.F. and Meza-Sanchez, M. 1993. Geoecodynamic assessmant to improve the landscape tourist resources in Cancun, Yucatan Peninsula, Mexico. In Wong, P.P. editor, *Tourism vs Environment: the Case for Coastal Areas*. Dordrecht: Kluwer Academic Publishers, 55–65.

Chapman, D.M. 1989. *Coastal Dunes of New South Wales: Status and Management*, Sydney: University of Sydney Coastal Studies Unit Technical Report 89/3.

Chasten, M.A., McCormick, J.W. and Rosati, J.D. 1994. Using detached breakwaters for shoreline and wetland stabilization. *Shore and Beach* 62 (2): 17–22.

Chou, L.M. and Sudara, S. 1991. Implications of sea-level rise and climate change on two developing resort islands in Thailand and Singapore. *Coastal Zone 91*. New York: American Society of Civil Engineers, 2793–800.

Cialone, M.A. and Stauble, D.K. 1998. Historical findings on ebb shoal mining. *Journal of Coastal Research* 14: 537–63.

Ciavola, P. and Simeoni, U. 1995. A review of the coastal geomorphology of Karavasta Lagoon, (Albania): short term coastal change and implications for coastal conservation. In Healy, M.G. and Doody, J.P. editors, *Directions in European Coastal Management*. Cardigan, UK: Samara Publishing Ltd, 301–16.

City of Stirling. 1984. *Coastal Report: a Report on Coastal Management and Development in the Coastal Reserve for the City of Stirling Municipality in Western Australia*. Perth: Imperial Printing Company.

Clarke, D. and Kasul, R. 1994. Habitat value of offshore dredged material berms for fishery resources. *Dredging '94*. New York: American Society of Civil Engineers, 938–45.

Clarke, J.J. 1991. Regional coastal planning and regulation: the Cape Cod experience. *Coastal Zone 91*. New York: American Society of Civil Engineers, 2947–61.

Clausner, J.E., Gebert, J.A., Rambo, A.T. and Watson, K.D. 1991. Sand bypassing at Indian River Inlet, Delaware. *Coastal Sediments 91*. New York: American Society of Civil Engineers, 1177–91.

Clayton,. T.D. 1989. Artificial beach replenishment on the U.S. Pacific shore. *Coastal Zone 89*. New York: American Society of Civil Engineers, 2033–45.

Coastal Engineering Research Center (CERC) 1984. *Shore Protection Manual*, Ft. Belvoir, VA: U.S. Army Corps of Engineers.

Coch, N.K. 1994. Geologic effects of hurricanes. *Geomorphology*. 10: 37–63.

Collins, D. and Barkdull, J. 1995. Capitalism, environmentalism, and mediating structures: from Adam Smith to stakeholder panels. *Environmental Ethics* 17: 227–44.

Cooke, R.U. 1992. Common ground, shared inheritance: research imperatives for environmental geography. *Transactions, Institute of British Geographers* 17: 131–51.

Cooper, W.S. 1958. The coastal sand dunes of Oregon and Washington. *Geological Society of America Memoir 72*.

Corona, M.G., Vicente, A.M. and Novo, F.G. 1988. Long-term vegetation changes on the stabilized dunes of Doñana National Park (SW Spain). *Vegetatio* 75: 73–80.

Corre, J.J. 1989. Ecological and geomorphological problems of sand dunes along Mediterranean coast of France (Golfe du Lion). In van der Meulen, F., Jungerius, P.D. and DeGroot, R.S. editors, *Landscape Ecological Impact of Climate Change on Coastal Dunes in Europe: Discussion Report on Coastal Dunes*. The Hague: Dutch Ministry of the Environment, 1–19.

Cortright, R. 1987. Foredune management on a developed shoreline: Nedonna Beach, Oregon. *Coastal Zone 87*. New York: American Society of Civil Engineers, 1343–56.

Coughlan, P.M. and Robinson, D.A. 1990. The Gold Coast Seaway, Qeensland, Australia. *Shore and Beach* 58 (1): 9–16.

Cousins, K. 1991. Ecotourism – examples from United States coastal management programs and marine and estuarine parks. *Coastal Zone 91*. New York: American Society of Civil Engineers, 1603–10.

Cowell, C.M. 1993. Ecological restoration and environmental ethics. *Environmental Ethics* 15: 19–32.

Cowell, M., Leatherman, S.P. and Buckley, M.K. 1993. Shoreline change rate analysis: long term versus short term data. *Shore and Beach* 61 (2): 13–20.

Cowell, P.J. and Thom, B.G. 1994. Morphodynamics of coastal evolution. Pages 33–86 In Carter, R.W.G. and Woodroffe, C.D. editors, *Coastal Evolution*. Cambridge University Press.

Cunniff, S.E. 1985. Impacts of severe storms on beach vegetation. *Coastal Zone 85*. New York: American Society of Civil Engineers, 1022–37.

Curtis, W.R., Davis, J.E. and Turner, I.L. 1996. Evaluation of a beach dewatering system: Nantucket, USA. *Coastal Engineering 1996: Proceedings of the Twenty-fifth International Conference*. New York: American Society of Civil Engineers, 2677–90.

d'Angremond, K., van den Berg, E.J.F. and de Jager, J.H. 1992. Use and behavior of gabions in coastal protection. *Coastal Engineering: Proceedings of the Twenty-third Coastal Engineering Conference*. New York: American Society of Civil Engineers, 1748–57.

Dal Cin, R. and Simeoni, U. 1994. A model for determining the classification, vulnerability and risk in the southern coastal zone of the Marche (Italy). *Journal of Coastal Research* 10: 18–29.

Davies, D.J., Parker, S.J. and Smith, W.E. 1993. Geological characterization of selected offshore sand resources on the OCS, offshore Alabama, for beach nourishment. *Coastal Zone 93*, New York: American Society of Civil Engineers, 1173–87.

Davies, J.L. 1972. *Geographical Variation in Coastal Development*, New York: Hafner.

Davis, D.W. 1993. Cheniere Caminada and the hurricane of 1893. *Coastal Zone 93*. New York: American Society of Civil Engineers, 2256–69.

Davis, G.A., Hanslow, D.J., Hibbert, K. and Nielsen, P. 1992. Gravity drainage: a new method of beach stabilisation through drainage of the watertable. *Coastal Engineering: Proceedings of the Twenty-third Coastal Engineering Conference*. New York: American Society of Civil Engineers, 1129–41.

Davis, R.A. 1991. Performance of a beach nourishment project based on detailed multi-year monitoring: Redington Beach, FL. *Coastal Sediments 91*, New York: American Society of Civil Engineers, 2101–15.

Davis, R.A., Fox, W.T., Hayes, M.O. and J.C. Boothroyd. 1972. Comparison of ridge and runnel systems in tidal and non-tidal environments. *Journal of Sedimentary Petrology* 2:413–21.

Davis, R.A., Herrygers, R.F. and Hogue, R.C. 1991. Effect of shell on beach performance: examples from the west-central coast of Florida. *Coastal Zone 91*. New York: American Society of Civil Engineers, 525–33.

Davidson, N.C. and Stroud, D.A. 1996. Conserving international coastal habitat networks on migratory waterfowl flyways. *Journal of Coastal Conservation* 2: 41–54.

Davison, A.T. 1993. The National Flood Insurance Program and coastal hazards. *Coastal Zone 93*. New York: American Society of Civil Engineers, 1377–91.

Davison, A.T., Nicholls, R.J. and Leatherman, S.P. 1992. Beach nourishment as a coastal management tool: an annotated bibliography on developments associated with the artificial nourishment of beaches. *Journal of Coastal Research* 8: 984–1022.

De Moor, G. and Bloome, E. 1988. Belgium. In Walker, H.J. editor, *Artificial Structures and Shorelines*. Dordrecht: Kluwer Academic Publishers, 115–26.

De Raeve, F. 1989. Sand dune vegetation and management dynamics. In van der Meulen, F. Jungerius, P.D. and Visser, J.H. editors, *Perspectives in Coastal Dune Management*, The Hague: SPB Academic Publishing, 99–109.

de Ruig, J.H.M. 1995. The Dutch experience: four years of dynamic preservation of the coastline. In Healy, M.G. and Doody, J.P. editors, *Directions in European Coastal Management*. Cardigan, UK: Samara Publishing Ltd., 253–66.

de Ruig, J.H.M. and Roelse, P. 1992. A feasibility study of a perched beach concept in The Netherlands. *Coastal Engineering: Proceedings of the Twenty-third Coastal Engineering Conference*. New York: American Society of Civil Engineers, 2581–98.

De Wolf, P., Frasaer, D., van Sieleghem, J. and Houthuys, R. 1993. Morphological trends of the Belgian Coast shown by 10 years of remote-sensing based surveying. In Hillen, R. and H.J. Verhagen editors, *Coastlines of the Southern North Sea*. New York: American Society of Civil Engineers, 245–57.

Dean, R.G. 1986 Coastal armoring: effects, principles and mitigation. *Coastal Engineering: Proceedings of the Twentieth Coastal Engineering Conference*. New York: American Society of Civil Engineers, 1843–57.

Dean, R.G. 1988a. Recommended modifications in benefit/cost sand management methodology. *Shore and Beach* 56 (4): 13–19.

Dean, R.G. 1988b. Sediment interaction at modified coastal inlets: processes and policies. In Aubrey, D.G. and Weishar, L. editors, *Hydrodynamics and Sediment Dynamics of Tidal Inlets*, New York: Springer-Verlag, 412–39.

Dean, R.G. 1991a. Equilibrium beach profiles: characteristics and applications. *Journal of Coastal Research* 7: 53–84.

Dean, R.G. 1991b. Beach erosion in the town of Jupiter Island: causes and remedial measures. *Coastal Zone 91*. New York: American Society of Civil Engineers, 1904–21.

Dean, R.G. 1997. Models for barrier island restoration. *Journal of Coastal Research* 13: 694–703.

Dean, R.G. and Yoo, C-H. 1994. Beach nourishment in presence of seawall. *Journal of Waterway, Port, Coastal, and Ocean Engineering* 120: 302–16.

Dean, R.G., Armstrong, G.A. and Sitar, N. 1984. *California Coastal Erosion and Storm Damage During the Winter of 1982–1983*. Washington, DC: National Academy Press.

Dean, R.G., Chen, R. and Browder, A.E. 1997. Full scale monitoring study of a submerged breakwater, Palm Beach, Florida, USA. *Coastal Engineering* 29: 291–315.

DeKimpe, N.M., Dolan, R. and Hayden, B.P. 1991. Predicted dune recession on the Outer Banks of North Carolina, USA. *Journal of Coastal Research* 7: 451–63.

Demirayak, F. 1995. A management plan for Belek Beach. In Healy, M.G. and Doody, J.P. editors, *Directions in European Coastal Management*. Cardigan, UK: Samara Publishing Ltd., 113–17.

Demos, C.J. 1991. Success of dune restoration after removal of UXO. *Coastal Zone 91*. New York: American Society of Civil Engineers, 2863–76.

Denevan, W.M. 1992. The pristine myth: the landscape of the Americas in 1492. *Annals of the Association of American Geographers* 82: 369–85.

Denison, P.S. 1998. Beach nourishment/groin field construction project: Bald Head Island, North Carolina. *Shore and Beach* 66 (1): 2–9.

Dette, H.-H. and Raudkivi, A.J. 1994. Beach nourishment and dune protection. *Coastal Engineering: Proceedings of the Twenty-fourth Coastal Engineering Conference*. New York: American Society of Civil Engineers, 1934–45.

Dette, H.-H., Führböter, A. and Raudkivi, A.J. 1994. Interdependence of beach fill volumes and repetition intervals. *Journal of Waterway, Port, Coastal, and Ocean Engineering* 120: 580–93.

Dettmer, A. and Cave, N. 1993. Permit enforcement, the Achilles heel of coastal protection: strategies for effective coastal regulation. *Coastal Zone 91*, New York: American Society of Civil Engineers, 3063–75.

Dias, J.M.A. and Neal, W.J. 1992. Sea cliff retreat in southern Portugal: profiles, processes, and problems. *Journal of Coastal Research* 8: 641–54.

Dickinson, W.W. and Woolfe, K.J. 1997. An in-situ transgressive barrier model for the Nelson Boulder Bank, New Zealand. *Journal of Coastal Research* 13: 937–52.

Doing, H. 1989. Introduction to the landscape ecology of southern Texel. In van der Meulen, F., Jungerius, P.D. and Visser, J.H. editors, *Perspectives in Coastal Dune Management*, The Hague: SPB Academic Publishing, 279–85.

Dolan, R. 1987. The Ash Wednesday Storm of 1962: 25 years later. *Journal of Coastal Research* 3 (2): ii-v.

Dolan, R. and Davis, R.E. 1994. Coastal storm hazards. *Journal of Coastal Research* SI12: 103–14.

Dolan, R. and Godfrey, P. 1973. Effects of Hurricane Ginger on the barrier islands of North Carolina. *Geological Society of America Bulletin* 84: 1329–34.

Dolan, R., Hayden, B. and Heywood, J. 1978. Analysis of coastal erosion and storm surge hazards. *Coastal Engineering* 2: 41–54.

Domroes, M. 1993. Maldivian tourist resorts and their environmental impact. In Wong,

P. P. editor, *Tourism vs Environment: the Case for Coastal Areas*. Dordrecht: Kluwer Academic Publishers, 69–82.

Domurat, G.W. 1987. Beach nourishment – a working solution. *Shore and Beach* 55 (3–4): 92–5.

Doody, J. P. 1989. Conservation and development of the coastal dunes in Great Britain. In van der Meulen, F., Jungerius, P. D. and Visser, J. H. editors, *Perspectives in Coastal Dune Management*, The Hague: SPB Academic Publishing, 53–67.

Doody, J. P. 1996. Management and use of dynamic estuarine shorelines. In Nordstrom, K. F. and Roman, C. T. editors, *Estuarine Shores: Evolution, Environments and Human Alterations*. Chichester: John Wiley & Sons, 421–34.

Dornhelm, R. B. 1995. The Coney Island public beach and boardwalk improvement of 1923. *Shore and Beach* 63 (1): 7–11.

Douglass, S. L. and Hinesley, P. 1993. Dauphin Island, Alabama beaches: real decisions in the real world. In Laska, S. and Puffer, A. editors, *Coastlines of the Gulf of Mexico*. New York: American Society of Civil Engineers, 172–85.

Douglass, S. L. and Weggel, J. R. 1987. Performance of a perched beach – Slaughter Beach, Delaware. *Coastal Sediments 87*. New York: American Society of Civil Engineers, 1385–98.

Draga, M. 1983. Eolian activity as a consequence of beach nourishment observations at Westerland (Sylt), German North Sea coast. *Zeitschrift für Geomorphologie* Supplement-Band 45: 303–19.

Druery, B. M. and Nielsen, A. F. 1980. Mechanisms operating at a jettied river entrance. *Coastal Engineering: Proceedings of the Seventeenth Coastal Engineering Conference*. New York: American Society of Civil Engineers, 2607–26.

Dustin, D. L. and McAvoy, L. H. 1982. Decline and fall of quality recreation opportunities and environments. *Environmental Ethics* 4: 49–57.

Dzhaoshvili, Sh.V. and Papashvili, I.G. 1993. Development and modern dynamics of alluvial-accumulative coasts of the eastern Black Sea. In Kos'yan, R. editor, *Coastlines of the Black Sea*. New York: American Society of Civil Engineers, 224–33.

Eastwood, D. A. and Carter, R.W.G. 1981. The Irish dune consumer. *Journal of Leisure Research* 13: 273–81.

Edge, B. L., Dowd, M., Dean, R. G. and Johnson, P. 1994. The reconstruction of Folly Beach. *Coastal Engineering: Proceedings of the Twenty-fourth Coastal Engineering Conference*. New York: American Society of Civil Engineers, 3491–506.

Edmondson, S. E. and Gateley, P. S. 1996. Dune heath on the Sefton Coast sand dune system, Merseyside, UK. In Jones, P.S., Healy, M.G. and Williams, A.T. editors, *Studies in European Coastal Management*. Cardigan, UK: Samara Publishing Ltd., 207–20.

Ehlers, J. and Kunz, H. 1993. Morphology of the Wadden Sea: natural processes and human interference. In Hillen, R. and H. J. Verhagen editors, *Coastlines of the Southern North Sea*. New York: American Society of Civil Engineers, 65–84.

Eitner, V. 1996. Morphological and sedimentological development of a tidal inlet and its catchment area (Otzumer Balje, Southern North Sea) *Journal of Coastal Research* 12: 271–93.

Espejel, I. 1993. Conservation and management of dry coastal vegetation. In Fermán-Almada, J.L., Gómez-Morin, L. and Fischer, D.W. editors, *Coastal Zone Management in Mexico: the Baja California Experience.* New York: American Society of Civil Engineers, 119–36.

Essig, S. 1993. Political coalitions to conserve coastal barriers. *Coastal Zone 93.* New York: American Society of Civil Engineers, 3024–35.

Everts, C.H. 1979. Beach behavior in the vicinity of groins – two New Jersey field examples. *Coastal Structures 79.* New York: American Society of Civil Engineers, 853–67.

Everts, C.H. and Czerniak, M.T. 1977. Spatial and temporal changes in New Jersey beaches. *Coastal Sediments 77.* New York: American Society of Civil Engineers, 444–59.

Fabbri, P. 1985a. Coastline variations in the Po Delta since 2500 B.P. *Zeitschrift für Geomorphologie,* Supplementband 57: 155–67.

Fabbri, P. 1985b. Coastal planning in the Mediterranean Sea: lines for scientific co-operation. In *Coastal Planning: Realities and Perspectives,* Genoa: Comune di Genova, 163–84.

Fabbri, P. 1989. Italians and their coastal zone: an introduction to some peculiarities. In Fabbri, P. editor, *Coastlines of Italy.* New York: American Society of Civil Engineers, v–viii.

Fabbri, P. 1990. Introduction. In Fabbri, P. editor, *Recreational Uses of Coastal Areas.* Dordrecht, Kluwer Academic Publishers, vii–xviii.

Fabbri, P. 1996. Present and future directions for EUCC – the European Union for Coastal Conservation. In Jones, P.S., Healy, M.G. and Williams, A.T. editors, *Studies in European Coastal Management.* Cardigan, UK: Samara Publishing Ltd., 1–2.

Fanos, A.M., Khafagy, A.A. and Dean, R.G. 1995. Protective works on the Nile Delta coast. *Journal of Coastal Research* 11: 516–28.

Favennec, J. 1996. Coastal management by the French National Forestry Service in Aquitaine, France. In Jones, P.S., Healy, M.G. and Williams, A.T. editors, *Studies in European Coastal Management.* Cardigan, UK: Samara Publishing Ltd., 191–6.

Fermán-Almada, J.L. and Gómez-Morin, L. 1993. Legal and regulatory aspects to support a Mexican coastal zone management program. In Fermán-Almada, J.L., Gómez-Morin, L. and Fischer, D.W. editors, *Coastal Zone Management in Mexico: the Baja California Experience.* New York: American Society of Civil Engineers, 7–13.

Fermán-Almada, J.L., Gómez-Morin, L. and Fischer, D.W. 1993. Coastal zone management in Mexico: the Baja California experience. In Fermán-Almada, J.L., Gómez-Morin, L. and Fischer, D.W. editors, *Coastal Zone Management in Mexico: the Baja California Experience.* New York: American Society of Civil Engineers, 1–6.

Fernandez-Rañada, J.C. 1989. Conditioning of Estepona Beach, Malaga, Spain. *Shore and Beach* 57 (2): 10–19.

Ferrante, A., Franco, L. and Boer, S. 1992. Modelling and monitoring of a perched beach at Lido di Ostia (Rome). *Coastal Engineering: Proceedings of the Twenty-third Coastal Engineering Conference.* New York: American Society of Civil Engineers, 3305–18.

Finkl, C.W. Jnr. 1996a. What might happen to America's shorelines if artificial beach replenishment is curtailed: a prognosis for southeastern Florida and other sandy regions along regressive coasts. *Journal of Coastal Research* 12: iiii–ix.

Finkl,. C.W. Jnr. 1996b. Beach fill from recycled glass: a new technology for mitigation of localized erosional 'hot spots' in Florida. In Tait, L.S. editor, *The Future of Beach Nourishment,* Tallahassee, FL: Florida Shore and Beach Preservation Association, 174–5.

Finkl, C.W. Jnr. editor. 1994. Coastal hazards: perception, susceptibility and mitigation. *Journal of Coastal Research* Special Issue 12.

Finkl, C.W. Jnr. and Pilkey, O.H. editors. 1991. Impacts of Hurricane Hugo: September 10–22, 1989. *Journal of Coastal Research* Special Issue 8.

Fischer, D.L. 1985. Beach erosion control: public issues in beach stabilization decisions, Florida. *Journal of Coastal Research* 2: 51–9.

Fischer, D.L. 1989. Response to coastal storm hazard: short-term recovery versus long-term planning. *Ocean and Shoreline Management* 12: 295–308.

Fischer, D.L., Rivas, V. and Cendrero, 1995. Local government planning for coastal protection: a case study of Cantabrian municipalities, Spain. *Journal of Coastal Research* 11: 858–74.

FitzGerald, D.M., Hubbard, D.K. and Nummedal, D. 1978. Shoreline changes associated with tidal inlets along the South Carolina coast. *Coastal Zone 78*. New York: American Society of Civil Engineers, 1973–94.

FitzGerald, D.M., Penland, S. and Nummedal. D. 1984. Control of barrier island shape by inlet sediment bypassing: East Frisian Islands. *Marine Geology* 60: 355–76.

FitzGerald, D.M. van Heteren, S. and Montello, T.M. 1994. Shoreline processes and damage resulting from the Halloween Eve storm of 1991 along the north and south shores of Massachusetts Bay, USA. *Journal of Coastal Research* 10: 113–32.

Fletcher, C.H., Mullane, R.A. and Richmond, B.M. 1997. Beach loss along armored shorelines on Oahu, Hawaiian Islands. *Journal of Coastal Research* 13: 209–15.

Fletcher, C.H., Richmond, B.M., Barnes, G.M. and Schroeder, T.A. 1995. Marine flooding on the coast of Kaua'i during Hurricane Iniki: hindcasting inundation components and delineating washover. *Journal of Coastal Research* 11: 188–204.

Flick, R.E. 1993. The myth and reality of southern California beaches. *Shore and Beach* 61 (3): 3–13.

Flick, R.E., Armstrong, G.A. and Sterrett, E.H. 1991. Shoreline erosion assessment and atlas of the San Diego region. In Domurat, G.W. and Wakeman, T.H. editors, *The California Coastal Experience*. New York: American Society of Civil Engineers, 160–71.

Foster, G.A., T.R. Healy and DeLange, W.P. 1996. Presaging beach renourishment from a nearshore dredge dump mound, Mt Maunganui Beach, New Zealand. *Journal of Coastal Research* 12: 395–405.

Fowler, J.E. Chasten, M.A. and Chu, Y-h. 1993. Federal damage assessments in New England after the Halloween '91 northeaster. *Coastal Zone 93*. New York: American Society of Civil Engineers, 2842–52.

Franco, L. and Tomasicchio, G.R. 1992. Hydraulic and mathematical modelling of historical and modern seawalls for the defence of Venice Lagoon. *Coastal Engineering: Proceedings of the Twenty-third Coastal Engineering Conference*. New York: American Society of Civil Engineers, 1879–93.

Freestone, D. 1991. Problems of coastal zone management in Antigua and Barbuda. In Cambers, G. editor, *Coastlines of the Caribbean*. New York: American Society of Civil Engineers, 61–9.

Frihy, O.E. and Komar, P.D. 1996. Shoreline erosion and human impacts along the Damietta Promontory of the Nile Delta, Egypt. *Shore and Beach* 64 (3): 40–7.

Funnell, C.E. 1975. *By the Beautiful Sea*. New York: Alfred Knopf Inc.

Gadd, P. E. and Leidersdorf, M. 1990. Recent performance of linked concrete mat armor under wave and ice impact. *Coastal Engineering: Proceedings of the Twenty-second Coastal Engineering Conference*. New York: American Society of Civil Engineers, 2768–81.

Gale, T., Botterill, D. and Morgan, N. 1995. Cultural change and the British seaside resort – implications for the quality and integrity of resort environments. In Healy, M.G. and Doody, J.P. editors, *Directions in European Coastal Management*. Cardigan, UK: Samara Publishing Ltd., 129–33.

Galgano, F.A. and Leatherman, S.P. 1991. Shoreline change analysis: a case study. *Coastal Sediments 91*. New York: American Society of Civil Engineers, 1043–53.

Galvin, C. 1983. Sea level rise and shoreline recession. *Coastal Zone 83*. New York: American Society of Civil Engineers, 684–705.

Galvin, C.J. 1991. Native sand beaches: a disappearing research resource. *SEPM News*, 3 (3): 3.

Galvin, C.J. 1995. Coastal processes at Shark River Inlet, Monmouth County, New Jersey. *Shore and Beach* 63 (3): 33.

Gardner, L.R., Michener, W.K., Kjerfve, B. and Karinshak, D.A. 1991. The geomorphic effects of Hurricane Hugo on an undeveloped coastal landscape at North Inlet, South Carolina. *Journal of Coastal Research* Special Issue 8: 181–6.

Gares, P.A. 1989. Geographers and public policy making: lessons learned from the failure of the New Jersey Dune Management Plan. *Professional Geographer* 41: 20–9.

Gares, P.A. 1990. Eolian processes and dune changes at developed and undeveloped sites, Island Beach, New Jersey. In Nordstrom, K.F., Psuty, N.P. and Carter, R.W.G. editors, *Coastal Dunes: Form and Process*, Chichester: John Wiley & Sons, 361–80.

Gares, P.A. and Nordstrom, K.F. 1988. Creation of dune depressions by foredune accretion. *Geographical Review* 78: 194–204.

Gares, P.A. and Nordstrom, K.F. 1991. Coastal dune blowouts – dynamics and management implications. *Coastal Zone 91*. New York: American Society of Civil Engineers, 2851–62.

Gares, P.A. and Nordstrom, K.F. 1995. A cyclic model of foredune blowout evolution for a leeward coast: Island Beach, New Jersey. *Annals of the Association of American Geographers* 85: 1–20.

Gares P.A., Nordstrom, K.F. and Psuty, N.P. 1980. Delineation and implementation of a dune management district. *Coastal Zone 80*. New York: American Society of Civil Engineers, 1269–88.

Garniel, A. and Mierwald, U. 1996. Changes in the morphology and vegetation along the human-altered shoreline of the lower Elbe. In Nordstrom, K.F. and Roman, C.T. editors, *Estuarine Shores: Evolution, Environments and Human Alterations*. Chichester: John Wiley & Sons, 375–96.

Garson, P. 1985. Rabbit grazing and the dune slack flora of Holy Island, Lindisfarne N.N.R. In Doody, P. editor, *Sand Dunes and their Management*, Peterborough: Nature Conservancy Council, 205–16.

Gayes, P.T. 1991. Post-Hurricane Hugo nearshore side scan sonar survey: Myrtle Beach to Folly Beach, South Carolina. *Journal of Coastal Research* Special Issue 8: 95–111.

Geelen, L.H.W., Cousin, E.F.H. and Schoon, C.F. 1995. Regeneration of dune slacks in the Amsterdam Waterworks dunes. In Healy, M.G. and Doody, J.P. editors, *Directions in European Coastal Management*. Cardigan, UK: Samara Publishing Ltd, 525–32.

Gibson, D.J. and Looney, P.B. 1994. Vegetation colonization of dredge spoil on Perdido Key, Florida. *Journal of Coastal Research* 10: 133–43.

Gibson, D.J., Ely, J.S. and Looney, P.B. 1997. A Markovian approach to modeling succession on a coastal barrier island following beach nourishment. *Journal of Coastal Research* 13: 831–41.

Godfrey, P.J. 1987 A successful local program for preserving and maintaining dunes on a developed barrier island. In Platt, R.H. Pelczarski, S.G. and Burbank, B.K.R. editor, *Cities on the Beach*. Chicago: University of Chicago Department of Geography Research paper 224, 163–9.

Godfrey, P.J. and Godfrey, M.M. 1973. Comparison of ecological and geomorphic interactions between altered and unaltered barrier island systems in North Carolina. In Coates, D.R. editor, *Coastal Geomorphology*, Binghamton, NY: State University of New York, 239–58.

Godfrey, P.J. and Godfrey, M.M. 1981. Ecological effects of off-road vehicles on Cape Cod. *Oceanus* 23: 56–67.

Godfrey, P.J., Leatherman, S.P and Zaremba, R. 1979. A geobotanical approach to classification of barrier beach systems. In Leatherman, S.P. editor, *Barrier Islands from the Gulf of St. Lawrence to the Gulf of Mexico*. New York: Academic Press, 99–126.

Godschalk, D.R. 1992. Implementing coastal zone management: 1972–1990. *Coastal Management* 20: 93–116.

Godschalk, D.R. Brower, D.J. and Beatley, T. 1989. *Catastrophic Coastal Storms*. Durham, NC: Duke University Press.

Goldsmith, V. 1989. Coastal sand dunes as geomorphological systems. *Proceedings of the Royal Society of Edinburgh*. 96B: 1–15.

Golick, A., Rosen, D.S., Golan, A., Shoshany, M., DiCastro, D. and Harari, P. 1996. Ashdod Port's effect on the shoreline, seabed and sediment. *Coastal Engineering 1996: Proceedings of the Twenty-fifth International Conference*. New York: American Society of Civil Engineers, 4376–89.

Gómez-Pina, G. and Ramírez, J.L. 1994. The complementary interaction between beach nourishment and harbour management: four cases in Spain. *Coastal Engineering: Proceedings of the Twenty-fourth Coastal Engineering Conference*. New York: American Society of Civil Engineers, 3507–21.

González-Yajimovich, O.E. and Escofet, A. 1991. Ecological and geomorphic impact of the destruction of a coastal sand dune system in a sand spit. *Coastal Zone 91*. New York: American Society of Civil Engineers, 2877–82.

Good, J.W. 1994. Shore protection policy and practices in Oregon: an evaluation of implementation success. *Coastal Management* 22: 325–52.

Gordon, A.D. 1992. The restoration of Bate Bay, Australia – plugging the sink. *Coastal Engineering: Proceedings of the Twenty-third Coastal Engineering Conference*. New York: American Society of Civil Engineers, 3319–30.

Gornitz, V. 1995. Sea level rise: a review of recent past and near-future trends. *Earth Surface Processes and Landforms* 20: 7–20.

Goss, C. and Gooderham, K. 1996. The importance of citizen involvement in developing a local beach management plan. In Tait, L.S. editor, *The Future of Beach Nourishment*, Tallahassee, FL: Florida Shore and Beach Preservation Association, 339–49.

Granja, H.M. 1996. Some examples of inappropriate coastal management practice in northwestern Portugal. In Jones, P.S., Healy, M.G. and Williams, A.T. editors, *Studies in European Coastal Management*. Cardigan, UK: Samara Publishing Ltd., 121–8.

Granja, H.M. and Carvalho, G.S. 1996. Is the coastline "protection" of Portugal by hard engineering structures effective? *Journal of Coastal Research* 11: 1229–41.

Gray, D.H., Hryciw, R.D. and Ghiassian, H. 1996. Protection of coastal sand dunes with anchored geonets. In Tait, L.S. editor, *The Future of Beach Nourishment*, Tallahassee, FL: Florida Shore and Beach Preservation Association, 255–70.

Gray, T. 1909. *The Buried City of Kenfig*. London: Unwin.

Green, K.M. and Cambers, G. 1991. The economic and environmental considerations of beach sand mining in St. Lucia, West Indies. In Cambers, G. editor, *Coastlines of the Caribbean*. New York: American Society of Civil Engineers, 124–35.

Grechischev, E.K, Rybak, O.L. and Yaroslavtzev, N.A. 1993. Specific features of shoreline development in Sochi region and engineering methods of coast protection. In Kos'yan, R. editor, *Coastlines of the Black Sea*. New York: American Society of Civil Engineers, 303–15.

Griggs, G.B. 1994. California's coastal hazards. *Journal of Coastal Research* Special Issue 12: 1–15.

Griggs, G.B. 1995. Relocation or reconstruction of threatened coastal structures: a second look. *Shore and Beach* 63 (2): 31–7.

Griggs, G.B. and Johnson, R.E. 1983. Impact of 1983 storms on the coastline. *California Geology*, 163–74.

Griggs, G.B., Tait, J.F., Scott, K. and Plant, N. 1991a. The interaction of seawalls and beaches: four years of field monitoring, Monterey Bay, California. *Coastal Sediments 91*. New York: American Society of Civil Engineers, 1871–85.

Griggs, G.B., Pepper, J.E. and Jordan, M.E. 1991b. California's coastal hazards policies: a critique. In Domurat, G.W. and Wakeman, T.H. editors, *The California Coastal Experience*. New York: American Society of Civil Engineers, 89–107.

Grosskopf, W.G. and Behnke, D.L. 1993. An emergency remedial beach fill design for Ocean City, Maryland. *Shore and Beach* 61 (1): 8–12.

Grosskopf, W.G. and Kraus, N.C. 1994. Guidelines for surveying beach nourishment projects. *Shore and Beach* 62 (2): 9–16.

Grumbine, R.E. 1994. Wildness, wise use, and sustainable development. *Environmental Ethics* 16: 241–9.

Guilcher, A. 1985. Problems of shore and foreshore management exemplified by the French coasts. In *Coastal Planning: Realities and Perspectives*, Genoa: Comune di Genova, 91–124.

Guilcher, A. and Hallégouët, B. 1991. Coastal dunes in Brittany and their management. *Journal of Coastal Research* 7: 517–33.

Günbak, A.R., Gökçe, K.T. and Güler, I. 1992. Sedimentation and erosion problems of Yakakent fishery harbor. *Coastal Engineering: Proceedings of the Twenty-third Coastal Engineering Conference*. New York: American Society of Civil Engineers, 3081–92.

Gundlach, E.R. and Siah, S.J. 1987. Cause and elimination of the deflation zones along the Atlantic City (New Jersey) shoreline. *Coastal Zone 87*. New York: American Society of Civil Engineers, 1357–69.

Gusmão, L.A., Cassar, J.C. and Neves, C.F. 1993. The northern coast of the state of Rio de Janiero. *Coastal Zone 93*, New York: American Society of Civil Engineers, 106–20.

Haeseler, V. 1989. The situation of the invertebrate fauna of coastal dunes and sandy coasts in the western Mediterranean (France, Spain). In van der Meulen, F., Jungerius, P.D. and Visser, J.H. editors, *Perspectives in Coastal Dune Management*, The Hague: SPB Academic Publishing, 125–31.

Hales, L. 1995. Accomplishments of the Corps of Engineers Dredging Research Program. *Journal of Coastal Research* 11: 68–88.

Hall, M.J. and Halsey, S.D. 1991. Comparison of overwash penetration from Hurricane Hugo and pre-storm erosion rates for Myrtle Beach and North Myrtle Beach, South Carolina, U.S.A. *Journal of Coastal Research* Special Issue 8: 229–35.

Hall, M.J. and Pilkey, O.H. 1991. Effects of hard stabilization on dry beach width for New Jersey. *Journal of Coastal Research* 7: 771–85.

Hallermeier, R.J. 1987. Applying large scale replicas of shore erosion by storms. *Coastal Sediments 87*. New York: American Society of Civil Engineers, 1415–29.

Hamilton, R.P., Ramsey, J.S. and Aubrey, D.G. 1996. Numerical predictions of erosional "hot spots" at Jupiter Island, Florida. In Tait, L.S. editor, *The Future of Beach Nourishment*, Tallahassee, FL: Florida Shore and Beach Preservation Association, 75–90.

Hands, E.B. and Allison, M.C. 1991. Mound migration in deeper water and methods of categorizing active and stable depths. *Coastal Sediments '91*. New York: American Society of Civil Engineers, 1985–99.

Hands, E.B. and Resio, D.T. 1994. Empirical guidance for siting berms to promote stability or nourishment benefits. *Dredging '94*. New York: American Society of Civil Engineers, 220–8.

Hansen, M. and Knowles, S.C. 1988. Ebb-tidal delta response to jetty construction at three South Carolina inlets, In Aubrey, D.G. and Weishar, L. editors, *Hydrodynamics and Sediment Dynamics of Tidal Inlets*. New York: Springer-Verlag, 364–81.

Hanson, H. and Kraus, N.C. 1993. Optimization of beach fill transitions. In Stauble. D.K. and Kraus, N.C. editors, *Beach Nourishment: Engineering and Management Considerations*. New York: American Society of Civil Engineers, 103–17.

Harvey, H.J. 1996. The National Trust: 100 years of coastal habitat and species conservation in Britain. In Jones, P.S., Healy, M.G. and Williams, A.T. editors, *Studies in European Coastal Management*. Cardigan, UK: Samara Publishing Ltd., 173–81.

Haward, M. and Bergin, A. 1991. Australian intergovernmental relations and coastal zone policy. *Coastal Zone 91*. New York: American Society of Civil Engineers, 737–51.

Hawkins, R. 1996. Green Globe: a new approach to environmentally responsible tourism. In Jones, P.S., Healy, M.G. and Williams, A.T. editors, *Studies in European Coastal Management*. Cardigan, UK: Samara Publishing Ltd., 75–80.

Hayes, M.O. 1967. Hurricanes as geological agents, south Texas coast. *American Association of Petroleum Geologists Bulletin* 51: 937–42.

Hazelton, J.M., Basco, D.R., Bellomo, D. and Williams, G. 1994. Statistical variations in beach parameter change rates for walled and non-walled profiles at Sand Bridge, VA. *Coastal Engineering: Proceedings of the Twenty-fourth Coastal Engineering Conference*. New York: American Society of Civil Engineers, 1812–26.

Healy, R.G. and Zinn, J.A. 1985. Environment and development conflicts in coastal zone management. *Journal of the American Planning Association* 51: 299–311.

Healy, T.R., Kirk, R.M. and de Lange, W.P. 1990. Beach nourishment in New Zealand. *Journal of Coastal Research* Special Issue 6: 77–90.

Hearon, G.E., McDougal, W.G. and Komar, P.D. 1996. Long-term beach response to shore stabilization structures on the Oregon coast. *Coastal Engineering 1996: Proceedings of the Twenty-fifth International Conference*. New York: American Society of Civil Engineers, 2718–31.

Heilman, D.J. and Edge, B.L. 1996. Interaction of the Colorado River Project, Texas, with longshore sediment transport. *Coastal Engineering 1996: Proceedings of the Twenty-fifth International Conference*. New York: American Society of Civil Engineers, 3309–22.

Helewaut, M. and Malherbe, B. 1993. Design and execution of beach nourishments in Belgium. In Hillen, R. and H.J. Verhagen editors, *Coastlines of the Southern North Sea*. New York: American Society of Civil Engineers, 258–66.

Hellström, G.B. and Lubke, R.A. 1993. Recent changes to a climbing-falling dune system on the Robberg Peninsula Southern Cape coast, South Africa. *Journal of Coastal Research* 9: 647–53.

Hesp, P.A. and Hilton, M.J. 1996. Nearshore-surfzone system limits and the impacts of sand extraction. *Journal of Coastal Research* 12: 726–47.

Hewett, D.G. 1985. Grazing and mowing as management tools on dunes. *Vegetatio* 62: 441–7.

Heywood, V.H. 1989. Patterns, extents and modes of invasions by terrestrial plants. In Drake, J.A. *et al.* (eds.) *Biological Invasions: a Global Perspective*, Chichester: John Wiley, 31–55.

Higgins, L.S. 1933. An investigation into the problem of the sand dune areas on the south Wales coast. *Archaeologia Cambrensis* 88: 26–67.

Hillen, R. and Roelse, P. 1995. Dynamic preservation of the coastline in The Netherlands. *Journal of Coastal Conservation* 1: 17–28.

Hillyer, T.M., Stakhiv, and Sudar, R.A. 1997. An evaluation of the economic performance of the U.S. Army Corps of Engineers Shore Protection Program. *Journal of Coastal Research* 13: 8–22.

Hoekstra, P., Houwman, K.T., Kroon, A., van Vessem, P. and Ruessink, B.G. 1994. The Nourtec experiment of Terschelling: process-oriented monitoring of a shoreface nourishment. *Coastal Dynamics '94*. New York: American Society of Civil Engineers, 402–16.

Hoekstra, P., Houwman, K.T., Kroon, A., Ruessink, B.G., Roelvink, J.A. and Spanhoff, R. 1996. Morphological develoment of the Terschelling shoreface nourishment in response to hydrodynamic and sediment transport processes. *Coastal Engineering 1996: Proceedings of the Twenty-fifth International Conference*. New York: American Society of Civil Engineers, 2897–910.

Hoffman, P.R. 1992. Tourism and language in Mexico's Los Cabos. *Journal of Cultural Geography* 12: 77–92.

Holgate-Pollard, D. 1996. Coastal management in England: the policy context. In Jones, P.S., Healy, M.G. and Williams, A.T. editors, *Studies in European Coastal Management*. Cardigan, UK: Samara Publishing Ltd., 29–33.

Hoogeboom, K.R. 1989. Restoration and development guidelines for ocean beach recreation areas. *Coastal Zone 89*. New York: American Society of Civil Engineers, 3120–34.

Hooke, J.M. and Bray, M.J. 1995. Coastal groups, littoral cells, policies and plans in the UK. *Area* 27: 358–68.

Hosier, P. E. and Cleary, W. J. 1977. Cyclic geomorphic patterns of washover on a barrier island in southeastern North Carolina. *Environmental Geology* 2: 23–31.

Hotta, S. and Horikawa, K. 1996. Countermeasures against wind-blown sand on beaches. *Coastal Engineering 1996: Proceedings of the Twenty-fifth International Conference*. New York: American Society of Civil Engineers, 4188–99.

Hotta, S., Kraus, N. C. and Horikawa, K. 1987. Function of sand fences in controlling wind-blown sand. *Coastal Sediments 87*. New York: American Society of Civil Engineers, 772–87.

Hotta, S., Kraus, N. C. and Horikawa, K. 1991. Functioning of multi-row sand fences in forming foredunes. *Coastal Sediments 91*. New York: American Society of Civil Engineers, 261–75.

Hotten, R. D. 1988. Sand mining on Mission Beach, San Diego, California. *Shore and Beach* 56 (2): 18–21.

Houlahan, J. M. 1989. Comparison of state construction setbacks to manage development in coastal hazard areas. *Coastal Management* 17: 219–28.

Houston, J. R. 1991. Beachfill performance. *Shore and Beach* 59 (3): 15–24.

Houston, J. R. 1995. Beach nourishment. *Shore and Beach* 63 (1): 21–4.

Houston, J. R. 1996a. International tourism and U.S. beaches. *Shore and Beach* 64 (2): 3–4.

Houston, J. R. 1996b. Simplified Dean's method for beach-fill design. *Journal of Waterway, Port, Coastal and Ocean Engineering*. 122: 143–6.

Houston, J. R. 1996c. Engineering practice for beach-fill designs. *Shore and Beach* 64 (3): 27–35.

Huber, R. M. Jr. and Meganck, R. 1990. The management challenge of Grande Anse Beach erosion, Grenada, West Indies. *Ocean and Shoreline Management* 13: 99–109.

Hughes, P. and Brundrit, G. B. 1995. Sea level rise and coastal planning: a call for stricter controls in river mouths. *Journal of Coastal Research* 11: 887–98.

Humphries, L. P. and Scott, W. B. 1991. A study of the impact of the dumping of spoil on beach processes. *Coastal Zone 91*. New York: American Society of Civil Engineers, 2246–59.

Ingram, S. J. and Chapman, D. M. 1993. DUNECARE: healing the coast. *Coastal Zone 93*. New York: American Society of Civil Engineers, 2739–52.

Inman, D. L. 1974. Ancient and modern harbors: a repeating phylogeny. *14th Coastal Engineering Conference*. New York: American Society of Civil Engineers, 2049–67.

Inman, D. J., Masters, P. M. and Stone, K. E. 1991. Induced subsidence: environmental and legal implications. *Coastal Zone 91*. New York: American Society of Civil Engineers, 16–27.

Innocenti, L and Pranzini, E. 1993. Geomorphological evolution and sedimentology of the Ombrone River delta, Italy. *Journal of Coastal Research* 9: 481–93.

Irish, J. L., Lillycrop, W. J. and Parson, L. E. 1996. Accuracy of sand volumes as a function of survey density. *Coastal Engineering 1996: Proceedings of the Twenty-fifth International Conference*. New York: American Society of Civil Engineers, 3736–49.

Iwagaki, Y. 1994. The present and future of coastal engineering in Japan. *Coastal Engineering: Proceedings of the Twenty-fourth Coastal Engineering Conference*. New York: American Society of Civil Engineers, 1–9.

Jackson, L. A. and Tomlinson, R. B. 1990. Nearshore nourishment implementation, monitoring and model studies of 1.5 m³ at Kirra Beach. *Coastal Engineering: Proceedings of the Twenty-second Coastal Engineering Conference*. New York: American Society of Civil Engineers, 2241–54.

Jackson, M.E. and O'Donnell, M. 1993. Responsible government structures: coastal management in the United States, Great Britain, and Australia. *Coastal Zone 93*. New York: American Society of Civil Engineers, 773–90.

Jackson, N.L., Nordstrom, K.F., Bruno, M.S. and Spalding, V.L. in press. Classification of spatial and temporal changes to a developed barrier island, Seven Mile Beach, New Jersey, USA. In Slaymaker, O. editor, *Geomorphology and Global Environmental Change*. London: John Wiley & Sons.

Janssen, M.P. 1995. Coastal management: restoration of natural processes in foredunes. In Healy, M.G. and Doody, J.P. editors, *Directions in European Coastal Management*. Cardigan, UK: Samara Publishing Ltd., 195–8.

Jensen, F. A 1995. Long term management plan for the Skaw Spit. In Healy, M.G. and Doody, J.P. editors, *Directions in European Coastal Management*. Cardigan, UK: Samara Publishing Ltd., 137–42.

Jones, D.K.C. 1993. Global warming and geomorphology. *The Geographical Journal* 159: 124–30.

Jones, S. 1996a. Kenfig National Nature Reserve: a profile of a British west coast dune system. In Jones, P.S., Healy, M.G. and Williams, A.T. editors, *Studies in European Coastal Management*. Cardigan, UK: Samara Publishing Ltd., 255–67.

Jones, S. 1996b. A comparative analysis of coastal zone management plans in England and The Netherlands. In Jones, P.S., Healy, M.G. and Williams, A.T. editors, *Studies in European Coastal Management*. Cardigan, UK: Samara Publishing Ltd., 45–56.

Judd, F.W., Lonard, R.I., Everitt, J.H. and Villarreal, R. 1989. Effects of vehicular traffic in the secondary dunes and vegetated flats of South Padre Island, Texas. *Coastal Zone 89*. New York: American Society of Civil Engineers, 4634–45.

Judd, F.W., Lonard, R.I., Everitt, J.H., Escobar, D.E. and Davis, R. 1991. Resilience of seacoast bluestem barrier island communities. *Coastal Zone 91*. New York: American Society of Civil Engineers, 3513–24.

Julien, B. 1996. Integrated management for the European coastal zone. In Jones, P.S., Healy, M.G. and Williams, A.T. editors, *Studies in European Coastal Management*. Cardigan, UK: Samara Publishing Ltd., 5–22.

Kana, T.W. 1983. Soft engineering alternatives for shore protection. *Coastal Zone 83*. New York: American Society of Civil Engineers, 912–29.

Kana, T.W. 1989. Erosion and beach restoration at Seabrook Island, South Carolina. *Shore and Beach* 57 (3): 3–18.

Kana, T.W. 1991. The South Carolina coast II: development and beach management. In Stauble, D.K. editor, *Barrier Islands: Process and Management*. New York: American Society of Civil Engineers, 274–83.

Kana, T.W. 1993. The profile volume approach to beach nourishment. In Stauble. D.K. and Kraus, N.C. editors, *Beach Nourishment: Engineering and Management Considerations*. New York: American Society of Civil Engineers, 176–90.

Kana, T.W. 1995. A mesoscale sediment budget for Long Island, New York. *Marine Geology* 126: 87–110.

Kana, T.W., Stevens, F.D. and Lennon, G. 1991. Post-Hugo beach restoration in South

Cariolina. *Coastal Sediments 91*. New York: American Society of Civil Engineers, 1697–711.

Kaplin, P.A., Porotov, A.V., Selivanov, A.O. and Yesin, N.V. 1993. The north Black Sea and the Sea of Azov coasts under a possible greenhouse-induced global sea-level rise. In Kos'yan, R. editor, *Coastlines of the Black Sea*. New York: American Society of Civil Engineers, 316–54.

Katoh, K. and Yanagishima, S-I. 1996. Field experiment on the effect of gravity drainage system on beach sedimentation. *Coastal Engineering 1996: Proceedings of the Twenty-fifth International Conference*. New York: American Society of Civil Engineers, 2654–65.

Katuna, M.P. 1991. The effects of Hurricane Hugo on the Isle of Palms, South Carolina: from destruction to recovery. *Journal of Coastal Research* Special Issue 8: 229–35.

Katz, E. 1992. The big lie: human restoration of nature. *Research in Philosophy and Technology: Technology and the Environment* JAI Press Inc., 231–41.

Kawaguchi, T., Takaki, N and Oka, S. 1991. Examples of the coast environment improvement works at fishing port areas in Japan. In Nagao, Y. editor, *Coastlines of Japan*. New York: American Society of Civil Engineers, 352–63.

Kawaguchi, T., Hashimoto, O., Mizumoto, T. and Kamata, A. 1994. Construction of offshore fishing port for prevention of coastal erosion. *Coastal Engineering: Proceedings of the Twenty-fourth Coastal Engineering Conference*. New York: American Society of Civil Engineers, 1197–291.

Kay, R.C. and Alder, J. 1999. *Coastal planning and management*. London: Routledge.

Kelletat, D. 1993. Coastal geomorphology and tourism on the German North Sea coast. In Wong, P.P. editor, *Tourism vs Environment: the Case for Coastal Areas*. Dordrecht: Kluwer Academic Publishers, 139–65.

Kieslich, J.M. 1981. *Tidal Inlet Response to Jetty Construction*. GITI Report 19. Vicksburg, MI: US Army Engineer Waterway Experiment Station.

Kieslich, J.M. and Brunt D.H. III. 1989. Assessment of a two-layer beachfill at Corpus Christi Beach, TX. *Coastal Zone 89*. New York: American Society of Civil Engineers, 3975–3984.

Kiknadze, A.G. 1993. Scientific basis of regulation of coastal processes (case study: coast of the Black Sea). In Kos'yan, R. editor, *Coastlines of the Black Sea*. New York: American Society of Civil Engineers, 201–13.

Klarin, P. and Hershman, M. 1990. Response of coastal management programs to sea level rise. *Coastal Management* 18: 143–65.

Klijn, J.A. 1990. The younger dunes in The Netherlands: chronology and causation. *Catena* Supplement 18: 89–100.

Klomp, W.H. 1989 Het Zwanenwater: a Dutch dune wetland reserve. In van der Meulen, F. Jungerius, P.D. and Visser, J.H. editors, *Perspectives in Coastal Dune Management*, The Hague: SPB Academic Publishing, 305–12.

Knecht, R.W., Cicin-Sain, B. and Fisk, G.W. 1997. Perceptions of the performance of state coastal zone management programs in the United States. II regional and state comparisons. *Coastal Management* 25: 325–43.

Knight, D., Mitchell, B. and Wall, G. 1997. Bali: sustainable development, tourism and coastal management. *Ambio* 26: 90–6.

Koarai, M., Saito, T., Watanabe, T. and Izumi, Y. 1993. Method for understanding the characteristics of coastal zones from historical and geographical viewpoints. *Coastal Zone 93*. New York: American Society of Civil Engineers, 1243–57.

Koedel, C. R. 1983. History of New Jersey's barrier islands. In *New Jersey's Barrier Islands: an Ever-Changing Public Resource*. New Brunswick, NJ: Rutgers University Center for Coastal and Environmental Studies, 17–28.

Koehler, H. H., Harder, H., Meyerdirks, J. and Voigt, A. 1996. The effects of trampling on the microarthropod fauna of dune sediments: a case study from Jutland, Denmark. In Jones, P. S., Healy, M. G. and Williams, A. T. editors, *Studies in European Coastal Management*. Cardigan, UK: Samara Publishing Ltd., 221–31.

Koike, K. 1988. Japan. In Walker, H. J. editor, *Artificial Structures and Shorelines*. Dordrecht: Kluwer Academic Publishers, 317–30.

Kooijman, A. M. and de Haan, M. W. A. 1995. Grazing as a measure against grass encroachment in Dutch dry dune grassland: effects on vegetation and soil. *Journal of Coastal Conservation* 1:127–34.

Koltasz Smith and Partners. 1994. *Mandurah Coastal Strategy*. Perth: State of Western Australia Department of Planning and Urban Development.

Komar, P. D. 1979. *Physical Processes and Geologic Hazards on the Oregon Coast*. Newport, Oregon: Oregon Coastal Zone Management Association, Inc.

Komar, P. D. 1998. *Beach Processes and Sedimentation*, 2nd Edition. Upper Saddle River, NJ: Prentice Hall.

Komar, P. D. and Shih, S-M. 1991. Sea-cliff erosion along the Oregon coast. *Coastal Sediments 91*. New York: American Society of Civil Engineers, 1558–70.

Koske, R. and Gemma, J. 1997. Mycorrhyzae and succession in plantings of beach grass in sand dunes. *American Journal of Botany* 84: 118–30.

Koster, M. J. and Hillen, R. 1995. Combat erosion by law: coastal defense policy for The Netherlands. *Journal of Coastal Research* 11: 1221–8.

Kos'yan, R. D. and Magoon, O. T. 1993. Man on the Black Sea coast. In Kos'yan, R. editor, *Coastlines of the Black Sea*. New York: American Society of Civil Engineers, 1–13.

Kraus, N. C. 1987. The effects of seawalls on the beach: a literature review. *Coastal Sediments 87*. New York: American Society of Civil Engineers, 945–60.

Kraus, N. C. 1988. The effects of seawalls on the beach: an extended literature review. In Kraus, N. C. and Pilkey, O. H. editors, Effects of Seawalls on the Beach. *Journal of Coastal Research* Special Issue 4, 1–28.

Kraus, N. C. 1993. The January 4, 1992 storm at Ocean City, Maryland. *Shore and Beach* 61 (1): 2.

Kraus, N. C. and McDougal, W. G. 1996. The effects of seawalls on the beach: Part I: an updated literature review. *Journal of Coastal Research* 12: 691–702.

Kraus, N. C. and Pilkey, O. H. editors. 1988. Effects of Seawalls on the Beach. *Journal of Coastal Research* Special Issue. 4.

Kraus, N. C., Hanson, H. and Blomgren, S. H. 1994. Modern functional design of groin systems. *Coastal Engineering: Proceedings of the Twenty-fourth Coastal Engineering Conference*. New York: American Society of Civil Engineers, 1327–42.

Kroon, A., Hoekstra, P., Houwman, K. and Ruessink, G. 1994. Morphological monitoring of a shoreface nourishment: NOURTEC experiment at Terschelling, The Netherlands. *Coastal Engineering: Proceedings of the Twenty-fourth Coastal Engineering Conference*. New York: American Society of Civil Engineers, 2222–36.

Kuhn, G. G. and Shepard, F. P. 1980. Coastal erosion in San Diego County, California. *Coastal Zone 80*. New York: American Society of Civil Engineers, 1899–918.

Kunz, H. 1990. Artificial beach nourishment on Norderney, a case study. *Coastal Engineering:*

Proceedings of the Twenty-second Coastal Engineering Conference. New York: American Society of Civil Engineers, 3254–67.

Kunz, H. 1993a. Sand losses from an artificially nourished beach stabilized by groynes. In Stauble. D.K. and Kraus, N.C. editors, *Beach Nourishment: Engineering and Management Considerations*. New York: American Society of Civil Engineers, 191–205.

Kunz, H. 1993b. Coastal protection in the past; coastal zone management in the future? Case study for the Ems-Weser area, Germany. In Hillen, R. and H.J. Verhagen editors, *Coastlines of the Southern North Sea*. New York: American Society of Civil Engineers, 314–35.

Kunz, H. 1996. Groynes on the East Frisian Islands: history and experiences. *Coastal Engineering 1996: Proceedings of the Twenty-fifth International Conference*. New York: American Society of Civil Engineers, 2128–41.

Lahousse, B., Clabout, P., Chamley, H. and van der Valk, L. 1993. Morphology of the Southern North Sea Coast from Cape Blanc-Nez (F) to Den Helder (NL). In Hillen, R. and H.J. Verhagen editors, *Coastlines of the Southern North Sea*. New York: American Society of Civil Engineers, 85–95.

Lamb, F.H. 1898. Sand-dune reclamation on the Pacific Coast. *The Forester* 4: 141–2.

Lamberti, A. and Mancinelli, A. 1996. Italian experience on submerged barriers as beach defence structures. *Coastal Engineering 1996: Proceedings of the Twenty-fifth International Conference*. New York: American Society of Civil Engineers, 2352–65.

Larson, M. and Kraus, N.C. 1991. Mathematical modeling of the fate of beach fill. *Coastal Engineering*. 16: 83–114.

Laubscher, W.I., Swart, D.H., Schoonees, J.S., Pfaff, W.M. and Davis, A.B. 1990. The Durban beach restoration scheme after 30 years. *Coastal Engineering: Proceedings of the Twenty-second Coastal Engineering Conference*. New York: American Society of Civil Engineers, 3227–38.

Laustrup, C. 1993. Coastal erosion management of sandy beaches in Denmark. In Hillen, R. and H.J. Verhagen editors, *Coastlines of the Southern North Sea*. New York: American Society of Civil Engineers, 267–76.

Lawrenz-Miller, S. 1991. Grunion spawning versus beach nourishment: nursery or burial ground. *Coastal Zone 91*. New York: American Society of Civil Engineers, 2197–208.

Leach, S. 1985. The problem of setting grazing levels on dune-heath Earlshall Muir – Fife. In Doody, P. editor, *Sand Dunes and their Management*. Peterborough: Nature Conservancy Council, 135–43.

Leadon, M.E. 1996. Hurricane Opal: damage to Florida's beaches, dunes and coastal structures. In Tait, L.S. editor, *The Future of Beach Nourishment*, Tallahassee, FL: Florida Shore and Beach Preservation Association, 313–28.

Leatherman, S.P. 1984. Shoreline evolution of North Assateague Island, Maryland. *Shore and Beach* 52: 3–10.

Ledesma Vasquez, J. and Huerta Santana, D.M. 1993. Slides and slumps on the coastal zone between San Miguel and El Descanso, Baja California, Mexico. In Fermán-Almada, J.L., Gómez-Morin, L. and Fischer, D.W. editors, *Coastal Zone Management in Mexico: the Baja California Experience*. New York: American Society of Civil Engineers, 30–42.

Lee, E.M. 1993. The political ecology of coastal planning and management in England and

Wales: policy responses to the implications of sea-level-rise. *The Geographical Journal* 159: 169–78.

Lee, E.M. 1995. Coastal cliff recession in Great Britain: the significance for sustainable coastal management. In Healy, M.G. and Doody, J.P. editors, *Directions in European Coastal Management*. Cardigan, UK: Samara Publishing Ltd, 185–93.

Lee, L.J. 1980. Sea cliff erosion in southern California. *Coastal Zone 80*. New York: American Society of Civil Engineers, 1919–38.

Lee, L.J. and Crampton, W. 1980. Sunset Cliffs stabilization San Diego, California. *Coastal Zone 80*. New York: American Society of Civil Engineers, 2271–90.

Leidersdorf, C.B., Gadd, P.E. and McDougal, W.G. 1990. Arctic slope protection methods. *Coastal Engineering: Proceedings of the Twenty-second Coastal Engineering Conference*. New York: American Society of Civil Engineers, 1687–701.

Leidersdorf, C.B., Hollar, R.C. and Woodell, G. 1993. Beach enhancement through nourishment and compartmentalization: the recent history of Santa Monica Bay. In Stauble. D.K. and Kraus, N.C. editors, *Beach Nourishment: Engineering and Management Considerations*. New York: American Society of Civil Engineers, 71–85.

Leidersdorf, C.B., Hollar, R.C. and Woodell, G. 1994. Human intervention with the beaches of Santa Monica Bay, California. Shore and Beach 62 (3): 29–38.

Lennon, G. 1991. The nature and causes of hurricane-induced ebb scour channels on a developed shoreline. *Journal of Coastal Research* Special Issue 8: 237–46.

Leonard, L.A., Dixon, K.L. and Pilkey, O.H. 1990. A comparison of beach replenishment on the U.S Atlantic, Pacific and Gulf coasts. *Journal of Coastal Research* Special Issue 6: 127–40.

Lewis, D. 1992. The sands of time: Cornwall's Hayle to Gwithian Towans. In Carter, R.W.G., Curtis, T.G.F. and Sheehy-Skeffington, M.J. editors, *Coastal Dunes: Geomorphology, Ecology and Management for Conservation*, Rotterdam: A.A. Balkema, 463–73.

Liddle, M.J. and Greig-Smith, P. 1975a. A survey of tracks and paths in a sand dune ecosystem. I. Soils. *Journal of Applied Ecology* 12: 893–908.

Liddle, M.J. and Greig-Smith, P. 1975b. A survey of tracks and paths in a sand dune ecosystem. II. Vegetation. *Journal of Applied Ecology* 12: 909–30.

Light, A. and Higgs, E.S. 1996. The politics of ecological restoration. *Environmental Ethics* 18: 227–47.

Lin, J-C. 1997. Coastal modification due to human influence in south-western Taiwan. *Quaternary Science Reviews* 15: 895–900.

Lin, P. C-P., Hansen, I. and Sasso, R.H. 1996. Combined sand bypassing and navigation improvements at Hillsboro Inlet, Broward County, Florida: the importance of a regional approach. In Tait, L.S. editor, *The Future of Beach Nourishment*, Tallahassee, FL: Florida Shore and Beach Preservation Association, 43–59.

Lindsay, B.E. and Tupper, H.C. 1990. Coastal beach regulation in sand dune areas for Maine and New Hampshire: institutional approach and beach user attitudes. *Shore and Beach* 58 (3): 19–21.

Llamas, M.R. 1990. Geohydrology of the eolian sands of the Doñana National Park (Spain). *Catena* Supplement 18: 145–54.

Lloyd, R.J. 1980. Noosa beach restoration scheme. *Coastal Engineering: Proceedings of the Seventeenth Coastal Engineering Conference*. New York: American Society of Civil Engineers, 1619–35.

Löffler, M. and Coosen, J. 1995. Ecological impact of sand replenishment. In Healy, M.G.

and Doody, J. P. editors, *Directions in European Coastal Management*. Cardigan, UK: Samara Publishing Ltd, 291–9.

Looney, P. B. and Gibson, D. J. 1993. Vegetation monitoring of beach nourishment. In Stauble. D. K. and Kraus, N. C. editors, *Beach Nourishment: Engineering and Management Considerations*. New York: American Society of Civil Engineers, 226–41.

Louisse, C. J. and van der Meulen, F. 1991. Future coastal defence in The Netherlands: strategies for protection and sustainable development. *Journal of Coastal Research 7*: 1027–41.

Louters, T., Mulder, J. P. M., Postma, R. and Hallie, F. P. 1991. Changes in coastal morphological processes due to the closure of tidal inlets in the SW Netherlands. *Journal of Coastal Research 7*: 635–52.

Maeda, M., Kajiwara, Y., Ishii, M. and Yabe, Y. 1991. Long-term vision of development, utilization and conservation of Tokyo metropolitan coastal zone. In Y. Nagao editor, *Coastlines of Japan*. New York: American Society of Civil Engineers, 6–20.

Maeno, Y., Ishikawa, M., Bando, K., Akiyama, Y. and Yabe, K. 1996. Sediment movement and stress conditions in sea bed. *Coastal Engineering 1996: Proceedings of the Twenty-fifth International Conference*. New York: American Society of Civil Engineers, 3860–73.

Mann, D.W. 1996. Beach nourishment benefit estimates: past, present and future? In Tait, L. S. editor, *The Future of Beach Nourishment*, Tallahassee, FL: Florida Shore and Beach Preservation Association, 146–56.

Manoha, B. and Teisson, C. 1990. To retreat in order to better fight: littoral protection of shingle beaches in the north of France. *Coastal Engineering: Proceedings of the Twenty-second Coastal Engineering Conference*. New York: American Society of Civil Engineers, 2542–51.

Marabani, F. and Veggiani, A. 1993. The evolution of the northern Adriatic coastal zone (Italy) and the environment. In Fabbri, P. editor, *Coastlines of the Mediterranean*. New York: American Society of Civil Engineers, 1–15.

Marchi, E. 1992. Coastal engineering in Venice. *Coastal Engineering: Proceedings of the Twenty-third Coastal Engineering Conference*. New York: American Society of Civil Engineers, 4–39.

Marcomini, S. C. and López, R. A. 1993. Coastal protection effects at Buenos Aires, Argentina. *Coastal Zone 93*, New York: American Society of Civil Engineers, 2724–38.

Marqués, M. A. and Julià, R. 1987. Geomorphological mapping of Mediterranean coastal features, northeast Spain. *Journal of Coastal Research 3*: 29–36.

Marra, J. P. 1993. Sand management planning in Oregon. *Coastal Zone 93*. New York: American Society of Civil Engineers, 1913–24.

Marsh, G. P. 1885. *Earth as Modified by Human Action*. New York: Charles Scribner's Sons.

Martí, J. L. J., Hernández, C. G. and Montero, G. G. 1996. Researches and measures for beach preservation: the case of Varadero Beach, Cuba. In Tait, L. S. editor, *The Future of Beach Nourishment*, Tallahassee, FL: Florida Shore and Beach Preservation Association, 297–312.

Masselink, G. and Short, A. D. 1993. The effect of tidal range on beach morphodynamics and morphology: a conceptual beach model. *Journal of Coastal Research 9*: 785–800.

Mather, A.S. and Ritchie, W. 1977. *The Beaches of the Highlands and Islands of Scotland*. Perth: Countryside Commission for Scotland.

Matondo, J.I. 1991. Coastline management: a challenge for developing countries. *Coastal Zone 91*. New York: American Society of Civil Engineers, 2763–6.

Maun, M. 1993. Adaptations enhancing survival and establishment of seedlings on coastal dune systems. *Vegetatio* 111: 59–70.

Mauriello, M.N. 1989. Dune maintenance and enhancement: a New Jersey example. *Coastal Zone 89*. New York: American Society of Civil Engineers, 1023–37.

Mauriello, M.N. 1991. Beach nourishment and dredging: New Jersey's policies. *Shore and Beach* 59 (3): 25–8.

Mauriello, M.N. and Halsey, S.D. 1987. Dune building on a developed coast. *Coastal Zone 87*. New York: American Society of Civil Engineers, 1313–27.

McCluskey, J.M., Nordstrom, K.F. and Rosen, P.S. 1983. An eolian sediment budget for the south shore of Long Island, N.Y. National Park Service/Rutgers University Cooperative Research Unit Technical Report.

McComb, D.G. 1986. *Galveston: a History*. Austin: University of Texas Press.

McDougal, W.G., Kraus, N.C. and Ajiwibowo, H. 1996. The effects of seawalls on the beach: Part II, numerical modeling of SUPERTANK seawall tests. *Journal of Coastal Research* 12: 702–13.

McDowell, A.J., Carter, R.W.G. and Pollard, H.J. 1993. The impact of man on the shoreline environment of the Costa del Sol, southern Spain. In Wong, P.P. editor, *Tourism vs Environment: the Case for Coastal Areas*. Dordrecht: Kluwer Academic Publishers, 189–209.

McLachlan, A. and Burns, M. 1992. Headland bypass dunes on the South African coast: 100 years of (mis)management. In Carter, R.W.G., Curtis, T.G.F. and Sheehy-Skeffington, M.J. editors, *Coastal Dunes: Geomorphology, Ecology and Management for Conservation*. Rotterdam: A.A. Balkema, 71–9.

McLellan, T.N. 1990. Nearshore mound construction using dredged material. *Journal of Coastal Research* Special Issue 7: 99–107.

McLellan, T.N. and Kraus, N.C. 1991. Design guidance for nearshore berm construction. *Coastal Sediments '91*. New York: American Society of Civil Engineers, 2000–11.

McLouth, M.E., Lapolla, J. and Bodge, K. 1994. Port Authority's role in inlet management and beneficial use of dredged material. *Dredging '94*. New York: American Society of Civil Engineers, 971–980.

McNinch, J.E. and Wells, J.T. 1992. Effectiveness of beach scraping as a method of erosion control. *Shore and Beach* 60 (1): 13–20.

Medina, J.R. 1992. A robust armor design to face uncertainties. *Coastal Engineering: Proceedings of the Twenty-third Coastal Engineering Conference*. New York: American Society of Civil Engineers, 1371–84.

Melvin, S.M., Griffin, C.R. and MacIvor, L.H. 1991. Recovery strategies for piping plovers in managed coastal landscapes. *Coastal Management* 19: 21–34.

Mendelssohn, I.A., Hester, M.W., Monteferrante, F.J. and Talbot, F. 1991. Experimental dune building and vegetative stabilization in a sand-deficient barrier island setting on the Louisiana coast, USA. *Journal of Coastal Research* 7: 137–49.

Mesa, C. 1996. Nearshore berm performance at Newport Beach, California. *Coastal Engineering 1996: Proceedings of the Twenty-fifth International Conference*. New York: American Society of Civil Engineers, 4636–49.

Meur-Ferec, C. 1995. The role of territorial departments in the conservation of French coastal sites. In Healy, M.G. and Doody, J.P. editors, *Directions in European Coastal Management*. Cardigan, UK: Samara Publishing Ltd, 383–90.

Meyer-Arendt, K.J. 1985. The Grand Isle, Louisiana resort cycle. *Annals of Tourism Research* 12: 449–65.

Meyer-Arendt, K.J. 1990. Recreational business districts in Gulf of Mexico seaside resorts. *Journal of Cultural Geography* 11: 39–55.

Meyer-Arendt, K.J. 1991. Tourism development on the north Yucatán coast: human response to shoreline erosion and hurricanes. *Geojournal* 23: 327–36.

Meyer-Arendt, K.J. 1992. Historical coastal environmental changes: human response to shoreline erosion. In Dilsaver, L. and Colten, C. editors, *The American Environment: Interpretation of Past Geographies*, Savage, MD: Rowman and Littlefield, 217–33.

Meyer-Arendt, K.J. 1993a. Geomorphic impacts of resort evolution along the Gulf of Mexico coast; applicability of resort cycle models. In Wong, P.P. editor, *Tourism vs Environment: the Case for Coastal Areas*. Dordrecht: Kluwer Academic Publishers, 125–38.

Meyer-Arendt, K.J. 1993b. Shoreline changes along the north Yucatán coast. In Laska, S. and Puffer, A. editors, *Coastlines of the Gulf of Mexico*. New York: American Society of Civil Engineers, 103–17.

Michigan Sea Grant. 1982. *Bluff Slumping: Proceedings of the 1982 Workshop*. Ann Arbor, MI: Michigan Sea Grant Office.

Miles, J.R., Russell, P.E. and Huntley, D.A. 1996. Sediment transport and wave reflection near a seawall. *Coastal Engineering 1996: Proceedings of the Twenty-fifth International Conference*. New York: American Society of Civil Engineers, 2612–24.

Miller, H.C., Birkemeier, W.A. and DeWall, A.E. 1983. Effects of CERC research pier on nearshore processes. *Coastal Structures 83*. New York: American Society of Civil Engineers, 769–84.

Milton, S.L., Schulman, A.A. and Lutz, P.L. 1997. The effect of beach nourishment with aragonite versus silicate sand on beach temperature and loggerhead sea turtle nesting success. *Journal of Coastal Research* 13: 904–15.

Mimura, N. and Nunn, P.D. 1998. Trends of beach erosion and shoreline protection in rural Fiji. *Journal of Coastal Research* 14: 37–46.

Ministry of Transport and Public Works 1990. *A New Coastal Defence Policy for The Netherlands*. The Hague: Rijkswaterstaat.

Miossec, A. 1988. The physical consequences of touristic development on the coastal zone as exemplified by the Atlantic coast of France between Girone and Finistère. *Ocean and Shoreline Management* 11: 303–18.

Miossec, A. 1993. Tourist development and coastal conservation in France. In Wong, P.P. editor, *Tourism vs Environment: the Case for Coastal Areas*. Dordrecht: Kluwer Academic Publishers, 167–87.

Mitchell, J.K. 1987. A management-oriented, regional classification of developed coastal barriers. In Platt, R.H., Pelczarski, S.G. and Burbank, B.K.R. editors, *Cities on the Beach*, Chicago: University of Chicago Department of Geography Research paper 224, 143–54.

Møller, J.T. 1990. Artificial beach nourishment on the Danish North Sea coast. *Journal of Coastal Research* Special Issue 6: 1–9.

Monadier, P., Ropert, F., Bellessort, B. and Viguier, J. 1992. New types of shore protection.

Possibilities of application along the coast of France. *Coastal Engineering: Proceedings of the Twenty-third Coastal Engineering Conference*. New York: American Society of Civil Engineers, 1414–19.

Morang, A. 1992. Inlet migration and hydraulic processes at East Pass, Florida. *Journal of Coastal Research* 8: 457–81.

Morang, A., Irish, J. L. and Pope, J. 1996. Hurricane Opal morphodynamic impacts on East Pass, Florida: preliminary findings. In Tait, L. S. editor, *The Future of Beach Nourishment*, Tallahassee, FL: Florida Shore and Beach Preservation Association, 192–208.

Morgan, J. M. 1996. Resources, recreationists, and revenues: a policy dilemma for today's state park systems. *Environmental Ethics* 18: 279–90.

Morgan, R., Jones, T. C. and Williams, A. T. 1993. Opinions and perceptions of England and Wales Heritage Coast beach users: some management implications from the Glamorgan Heritage Coast, Wales. *Journal of Coastal Research* 9: 1083–93.

Morton, R. A. 1976. Effects of Hurricane Eloise on beach and coastal structures, Florida Panhandle. *Geology* 4: 277–80.

Morton, R. A. 1979. Temporal and spatial variations in shoreline changes and their implications, examples from the Texas gulf coast. *Journal of Sedimentary Petrology* 49: 1101–12.

Morton, R. A. 1991. Geologic framework and depositional history of the southeastern Texas coast. *Coastal Sediments '91*. New York: American Society of Civil Engineers, 1069–81.

Morton, R. A., Paine, J. G. and Gibeaut, J. C. 1994. Stages and durations of post-storm beach recovery, southeastern Texas coast. *Journal of Coastal Research* 10: 884–908.

Moutsouris, C. I. 1995. Shoreline structures for coastal defence. In Healy, M. G. and Doody, J. P. editors, *Directions in European Coastal Management*. Cardigan, UK: Samara Publishing Ltd., 153–9.

Moutzouris C. I. and Maroukian, H. 1988. Greece. In Walker, H. J. editor, *Artificial Structures and Shorelines*. Dordrecht: Kluwer Academic Publishers, 207–15.

Mulder, J. P. M., van de Kreeke, J. and van Vessem, P. 1994. Experimental shoreface nourishment, Terschelling (NL). *Coastal Engineering: Proceedings of the Twenty-fourth Coastal Engineering Conference*. New York: American Society of Civil Engineers, 2886–99.

Mullard, J. 1995. Gower: a case study in integrated coastal management initiatives in the UK. In Healy, M. G. and Doody, J. P. editors, *Directions in European Coastal Management*. Cardigan, UK: Samara Publishing Ltd., 471–6.

Mullarda, J., Atkins, J., Hughes, M. R., Sell, S. and Winder, M. 1996. The Gower Peninsula, Wales, UK. In Jones, P. S., Healy, M. G. and Williams, A. T. editors, *Studies in European Coastal Management*. Cardigan, UK: Samara Publishing Ltd., 269–74.

Mura, P. M. 1989. Coastal planning and conservation policies in Italy. In Fabbri, P. editor, *Coastlines of Italy*. New York: American Society of Civil Engineers, 1–15.

Nakashima, L. D. 1989. Shoreline responses to Hurricane Bonnie in southwestern Louisiana. *Journal of Coastal Research* 5: 127–36.

Nagao, Y. editor, 1991. *Coastlines of Japan*. New York: American Society of Civil Engineers.

Nagao, Y. and Fujii, T. 1991. Construction of man-made island and preservation of coastal zone. In Nagao, Y. editor, *Coastlines of Japan*. New York: American Society of Civil Engineers, 212–26.

Nakashima, L.D. 1989. Shoreline responses to Hurricane Bonnie in southwestern Louisiana. *Journal of Coastal Research* 5: 127–36.

Namikas, S.L. 1992. *Beach Nourishment Performance at South Beach, Sandy Hook Unit, Gateway National Recreation Area*. Technical Report 99–22. Rutgers University Institute of Marine and Coastal Sciences, New Brunswick, NJ.

National Research Council. 1990. *Managing Coastal Erosion*. Washington, DC: National Academy Press.

National Research Council. 1995. *Beach Nourishment and Protection*. Washington, DC: National Academy Press.

Nelson, D.D. 1991. Factors effecting beach morphology changes caused by Hurricane Hugo, northern South Carolina. *Journal of Coastal Research* Special Issue 8: 229–35.

Nelson, W.G. 1993. Beach restoration in the southeastern US: environmental effects and biological monitoring. *Ocean and Coastal Management*. 19: 157–82.

Nersesian, G.K., Kraus, N.C. and Carson, F.C. 1992. Functioning of groins at Westhampton Beach, Long Island, New York. *Coastal Engineering: Proceedings of the Twenty-third Coastal Engineering Conference*. New York: American Society of Civil Engineers, 3357–70.

New Jersey Department of Environmental Protection (NJDEP). 1996. *CAFRA: a Guide for Single-Family and Duplex Homeowners*. Trenton, NJ: Office of Permit Information and Assistance.

Nicholls, R.J., and Leatherman, S.P. 1996. Adapting to sea-level rise: relative sea-level trends to 2100 for the United States. *Coastal Management* 24: 301–24.

Nicholls, R.J., Davison, A.T. and Gambel, J. 1993. Vertical erosion on beaches and the National Flood Insurance Program. *Coastal Zone 93*. New York: American Society of Civil Engineers, 1362–76.

Nichols, R.J., Davison, A.T. and Gambel, J. 1995. Erosion in coastal settings and pile foundations. *Shore and Beach* 63 (4): 11–17.

Nielsen, A.F., Hesp, P.A. and Lord, D.B. 1991. Marine dredging and aggregate extraction. *Proceedings of the 10th Australasian Conference on Coastal and Ocean Engineering*, 67–72.

Niemeyer, H.D. 1994. Long-term morphodynamical development of the East Frisian Islands and coast. *Coastal Engineering: Proceedings of the Twenty-fourth Coastal Engineering Conference*. New York: American Society of Civil Engineers, 2417–33.

Niemeyer, H.D., Kaiser, R. and Knaak, K. 1996. Effectiveness of a combined beach and shoreface nourishment on the island of Norderney/East Frisia, Germany. *Coastal Engineering 1996: Proceedings of the Twenty-fifth International Conference*. New York: American Society of Civil Engineers, 4621–34.

Nir, Y. 1982. Offshore artificial structures and their influence on the Israel and Sinai Mediterranean beaches. *Coastal Engineering: Proceedings of the Eighteenth Coastal Engineering Conference*. New York: American Society of Civil Engineers, 1837–56.

Nnaji, S., Yazdani, N. and Rambo-Rodenberry, M. 1996. Scour impact of coastal swimming pools on beach systems. *Journal of Coastal Research* 12: 186–91.

Noble, R.M. 1978. Coastal structures' effects on shorelines. *Coastal Engineering: Proceedings of the Sixteenth Coastal Engineering Conference*. New York: American Society of Civil Engineers, 2069–85.

Nobuoka, H., Irie, I., Kato, H. and Mimura, N. 1996. Regulation of nearshore circulation by submerged breakwater for shore protection. *Coastal Engineering 1996: Proceedings of the Twenty-fifth International Conference.* New York: American Society of Civil Engineers, 2391–403.

Nordberg, L. 1995. Coastal conservation in selected European states. In Healy, M.G. and Doody, J.P. editors, *Directions in European Coastal Management.* Cardigan, UK: Samara Publishing Ltd., 47–50.

Nordstrom, K.F. 1987a. Shoreline changes on developed coastal barriers. In Platt, R.H., Pelczarski, S.G. and Burbank, B.K.R. editors. *Cities on the Beach,* Chicago: University of Chicago Department of Geography Research paper 224, 65–79.

Nordstrom, K.F. 1987b. Management of tidal inlets on barrier island shorelines. *Journal of Shoreline Management* 3: 169–90.

Nordstrom, K.F. 1988a. Effects of shore protection and dredging projects on beach configuration near tidal inlets in New Jersey. In Aubrey, D.G. and Weishar, L. editors, *Hydrodynamics and Sediment Dynamics of Tidal Inlets.* New York: Springer-Verlag, 440–54.

Nordstrom, K.F. 1988b. Dune grading along the Oregon coast, USA: a changing environmental policy. *Applied Geography* 8: 101–16.

Nordstrom, K.F. 1990. The concept of intrinsic value and depositional coastal landforms. *Geographical Review* 80: 68–81.

Nordstrom, K.F. 1992. *Estuarine Beaches.* London: Elsevier Science Publishers.

Nordstrom, K.F. 1993. Intrinsic value and landscape evaluation. *Geographical Review* 83: 473–6.

Nordstrom, K.F. 1994a. Developed coasts. In Carter, R.W.G. and Woodroffe, C. editors, *Coastal Evolution.* Cambridge: Cambridge University Press, 477–509.

Nordstrom, K.F. 1994b. Beaches and dunes of human altered coasts. *Progress in Physical Geography* 18: 497–516.

Nordstrom, K.F. 1998. *The evolution and value of landforms on human altered coasts.* In Ewing, L. and Sherman, D.J. editors, *California's coastal natural hazards.* Los Angeles: University of Southern California Sea Grant, 7–27.

Nordstrom, K.F. and Arens, S.M. 1998. The role of human actions in evolution and management of foredunes in The Netherlands and New Jersey, USA. *Journal of Coastal Conservation* 4: 169–80.

Nordstrom, K.F. and Jackson, N.L. 1995. Temporal scales of landscape change following storms on a human-altered coast, New Jersey, USA. *Journal of Coastal Conservation* 1:51–62.

Nordstrom, K.F. and Jackson, N.L. 1998. Effects of high rise buildings on wind flow and beach characteristics at Atlantic City, New Jersey, USA. *Ocean and Coastal Management* 39: 245–63.

Nordstrom, K.F. and Lotstein, E.L. 1989. Perspectives on resource use of dynamic coastal dunes. *Geographical Review* 79: 1–12.

Nordstrom, K.F. and McCluskey, J.M. 1984. Considerations for the control of house construction in coastal dunes. *Coastal Zone Management Journal* 12: 385–402.

Nordstrom, K.F. and McCluskey, J.M. 1985. The effects of houses and sand fences on the eolian sediment budget at Fire Island, New York. *Journal of Coastal Research* 1: 39–46.

Nordstrom, K.F. and Psuty, N.P. 1983. The value of coastal dunes as a form of shore protection in California, USA. *Coastal Zone 83.* New York: American Society of Civil Engineers, 873–85.

Nordstrom, K. F. and Renwick, W. H. 1984. A coastal cliff management district for protection of eroding high relief coasts. *Environmental Management* 8: 197–202.

Nordstrom, K. F., Lampe, R. and Vandemark, L. M. 1999. Re-establishing naturally-functioning dunes on developed coasts. *Environmental Management*, in press.

Nordstrom, K. F., McCluskey, J. M. and Rosen, P. S. 1986. Aeolian processes and dune characteristics of a developed shoreline. In Nickling, W. G. editor, *Aeolian Geomorphology*. Boston: Allen and Unwin, 131–47.

Nugent, N. K. and Hart, T. F. 1989. Protecting New York's scenic coast. *Coastal Zone 89*. New York: American Society of Civil Engineers, 3569–80.

Oertel, G. F., Foster, W. G. and Graham, W. R. 1996. Elements of a successful beach management plan, Sea Island, Georgia. In Tait, L. S. editor, *The Future of Beach Nourishment*, Tallahassee, FL: Florida Shore and Beach Preservation Association, 27–42.

Olsauskas, A. 1995. Influence of recreation on flora stability on the Lithuanian coastal dunes. In Healy, M. G. and Doody, J. P. editors, *Directions in European Coastal Management*. Cardigan, UK: Samara Publishing Ltd., 103–5.

Olsen, E. J. 1982. South Seas Plantation beach improvement project. *Shore and Beach* 50 (1): 6–10.

Olsen, E. J. 1996. Tybee Island South Beach stabilization project. In Tait, L. S. editor, *The Future of Beach Nourishment*, Tallahassee, FL: Florida Shore and Beach Preservation Association, 5–16.

Oosterveld, P. 1985. Grazing in dune areas: the objectives of nature conservation and the aims of research for nature conservation management. In Doody, P. editor, *Sand Dunes and their Management*, Peterborough: Nature Conservancy Council, 187–203.

Orford, J. D., Carter, R. W. G., Forbes, D. L. and Taylor, R. B. 1988. Overwash occurrence consequent on morphodynamic changes following lagoon outlet closure on a coarse clastic barrier. *Earth Surface Processes and Landforms* 13: 27–35.

Orford, J. D., Carter, R. W. G., Jennings, S. C. and Hinton, A. C. 1995. Processes and time scales by which a coastal gravel-dominated barrier responds geomorphically to sea-level rise: Story Head Barrier, Nova Scotia. *Earth Surface Processes and Landforms* 20: 21–37.

Orme, A. R. 1980. Energy–sediment interaction around a groin. *Zeitschrift für Geomorphologie* Supplementband 34: 111–28.

Ortman, D. E. 1987. Washington's CZMP – the first shall be last. *Coastal Zone 87*. New York: American Society of Civil Engineers, 2968–82.

Otvos, E. G. 1993. Mississippi-Alabama: natural and man-made shores a study in contrasts. *Coastal Zone 93*. New York: American Society of Civil Engineers, 2600–15.

Owens, D. W. 1985. Coastal Management in North Carolina: building a regional consensus. *Journal of the American Planning Association* 51: 322–9.

Pacini, M., Pranzini, E. and Sirito, G. 1997. Beach nourishment with angular gravel at Cala Gonone (eastern Sardinia, Italy). *Proceedings of the Third International Conference on the Mediterranean Coastal Environment, MEDCOAST*, 1043–58.

Packham, J. R., Harmes, P. A. and Spiers, A. 1995. Development of a shingle community related to a specific sea defence structure. In Healy, M. G. and Doody, J. P. editors, *Directions in European Coastal Management*. Cardigan, UK: Samara Publishing Ltd., 369–71.

Palmetto Dunes Resort. 1990. Practical aspects of beach scraping. *Journal of Coastal Research* Special Issue 9: 877–82.

Parks, J. 1991. Pumping in and pumping out: case histories of fluidized sand bypassing for channels and beachface dewatering for beaches. *Coastal Zone 91*. New York: American Society of Civil Engineers, 193–203.

Paskoff, R. 1987. Archaeological data on shoreline displacements in the Mediterranean since antiquity. *Journal of Coastal Research* 3: 107–11.

Paskoff, R. 1992. Eroding Tunisian beaches: causes and mitigation. *Bollettino di Oceanologia Teorica ed Applicata* 1: 85–91.

Paskoff, R. and Kelletat, D. 1991. Introduction: review of coastal problems. *Zeitschrift für Geomorphologie* Supplementband. 81: 1–13.

Paskoff, R. and Oueslati, A. 1991. Modifications of coastal conditions in the Gulf of Gabes (southern Tunisia) since classical antiquity. *Zeitschrift für Geomorphologie* Supplementband 81: 149–62.

Paskoff, R. and Petiot, R. 1990. Coastal progradation as a by-product of human activity: an example from Chañaral Bay, Atacama Desert, Chile. *Journal of Coastal Research* Special Issue 6: 91–102.

Paternoster, R. J. 1991. Public access evolution: Long Beach, California coastline. *Coastal Zone 91*. New York: American Society of Civil Engineers, 3477–82.

Patterson, D. R., Bisher, D. R. and Brodeen, M. R. 1991. The Oceanside experimental sand bypass: the next step. *Coastal Sediments 91*. New York: American Society of Civil Engineers, 1165–76.

Pearsall, S. 1993. Terrestrial coastal environments and tourism in Western Samoa. In Wong, P. P. editor, *Tourism vs Environment: the Case for Coastal Areas*. Dordrecht: Kluwer Academic Publishers, 33–53.

Peña, C., Carrion, V. and Castañeda, A. 1992. Projects, works and monitoring at Barcelona coast. *Coastal Engineering: Proceedings of the Twenty-third Coastal Engineering Conference*. New York: American Society of Civil Engineers, 3385–98.

Penland, S., Nummedal, D. and Schramm, W. E. 1980. Hurricane impact at Dauphin Island, Alabama. In *Coastal Zone 80*. New York: American Society of Civil Engineers, 1425–49.

Peshkov, V. M. 1993. Artificial gravel beaches in the coastal protection. In Kos'yan, R. editor, *Coastlines of the Black Sea*. New York: American Society of Civil Engineers, 82–102.

Pethick, J. 1996. The sustainable use of coasts: monitoring, modelling and management. In Jones, P. S., Healy, M. G. and Williams, A. T. editors, *Studies in European Coastal Management*. Cardigan, UK: Samara Publishing Ltd., 83–92.

Philip, N. A. and Whaite, P. H. 1980. The beach improvement programme: New South Wales. *Coastal Engineering: Proceedings of the Seventeenth Coastal Engineering Conference*. New York: American Society of Civil Engineers, 1669–79.

Phillips, J. D. 1991. The human role in earth surface systems: some theoretical considerations. *Geographical Analysis* 23: 316–31.

Pilarczyk, K. W. 1996. Geotextile systems in coastal engineering – an overview. *Coastal Engineering 1996: Proceedings of the Twenty-fifth International Conference*. New York: American Society of Civil Engineers, 2114–27.

Pilkey, O. H. 1981. Geologists, engineers, and a rising sea level. *Northeastern Geology* 3/4: 150–8.

Pilkey, O. H. 1990. Editorial: A time to look back at beach replenishment. *Journal of Coastal Research* 6: iii–vii.

Pilkey, O. H. 1992. Another view of beachfill performance. *Shore and Beach* 60 (2): 20–5.

Pilkey, O. H. and Clayton, T. D. 1987. Beach replenishment: the national solution? *Coastal Zone 87*. New York: American Society of Civil Engineers, 1408–19.

Pilkey, O. H. and Wright III, H. L. 1988. Seawalls versus beaches. *Journal of Coastal Research* Special Issue 4: 41–64.

Pinto, C., Silovsky, E., Henley, F., Rich, L., Parcell, J. and Boyer, D. 1972. *The Oregon Dunes NPA Resource Inventory*. Portland, OR: US Department of Agriculture, Forest Service, Pacific Northwest Region.

Piotrowska, H. 1989. Natural and anthropogenic changes in sand-dunes and their vegetation on the southern Baltic coast. In van der Meulen, F., Jungerius, P. D. and Visser, J. H. editors, *Perspectives in Coastal Dune Management*, The Hague: SPB Academic Publishing, 33–40.

Plant, N. and Griggs, G. B. 1991. The impact of the October 17, 1989 Loma Prieta earthquake on coastal bluffs and implications for land use planning. *Coastal Sediments 91*. New York: American Society of Civil Engineers, 827–41.

Plant, N. G. and Griggs, G. B. 1992. Interactions between nearshore processes and beach morphology near a seawall. *Journal of Coastal Research* 8: 183–200.

Plant, N. and Holman, R. 1996. Interannual shoreline variations at Duck, NC, USA. *Coastal Engineering 1996: Proceedings of the Twenty-fifth International Conference*. New York: American Society of Civil Engineers, 3521–33.

Platt, R. H. 1985. Congress and the coast. *Environment* 27: 12–17; 34–40.

Platt, R. H. 1994. Evolution of coastal hazards policies in the United States. *Coastal Management* 22: 265–84.

Pluijm, M., van der Lem, J. C. and de Ruig, J. H. W. 1994. Offshore breakwaters versus beach nourishments: a comparison. *Coastal Engineering: Proceedings of the Twenty-fourth Coastal Engineering Conference*. New York: American Society of Civil Engineers, 3208–22.

Podufaly, E. T. 1964. Operation Five High. *Shore and Beach* 30 (2): 9–18.

Pope, J. 1991. Ebb delta and shoreline response to inlet stabilization, examples from the southeast Atlantic coast. *Coastal Zone 91*. New York: American Society of Civil Engineers, 643–54.

Pope, J. 1997. Responding to coastal erosion and flooding damages. *Journal of Coastal Research* 13: 704–10.

Pope, J., Smith, J. B. and Gorman, L. T. 1994. History of entrance channel dredging and disposal operations: St. Marys, Georgia/Florida. *Dredging '94*. New York: American Society of Civil Engineers, 906–17.

Postma, R. 1989. Erosional trends along the cuspate river-mouths in the Adriatic. In Fabbri, P. editor, *Coastlines of Italy*. New York: American Society of Civil Engineers, 84–97.

Postolache, I. and Diaconeasa, D. 1991. Coastal erosion problems in Romania. *Coastal Zone 91*. New York: American Society of Civil Engineers, 2260–70.

Premaratne, A. 1991. Difficulties of coastal resources management in developing countries: Sri Lankan experiences. *Coastal Zone 91*. New York: American Society of Civil Engineers, 3026–41.

Prestedge, G.K. 1992. Sharp Rock pier and submerged groin. *Shore and Beach* 60 (3): 6–14.

Pronio, M.A. 1989. Distribution, community ecology and local spread of *Carex cobomugi* (Japanese sedge) at Island Beach State Park, New Jersey. Unpublished MS thesis, New Brunswick, NJ: Rutgers University.

Psuty, N.P. 1986. Impacts of impending sea-level rise scenarios: the New Jersey barrier island responses. *Bulletin of the New Jersey Academy of Sciences* 31: 29–36.

Psuty, N.P. 1987. Dune management: planning for change. In Platt, R.H., Pelczarski, S.G. and Burbank, B.K.R. editors, *Cities on the Beach*, Chicago: University of Chicago Department of Geography Research paper 224, 55–161.

Psuty, N.P. 1989. An application of science to the management of coastal dunes along the Atlantic coast of the U.S.A. *Proceedings of the Royal Society of Edinburgh* 96B: 289–307.

Psuty, N.P. 1992. Spatial variation in coastal foredune development. In Carter, R.W.G., Curtis, T.G.F. and Sheehy-Skeffington, M.J. editors, *Coastal Dunes: Geomorphology, Ecology and Management for Conservation*, Rotterdam: A.A. Balkema, 3–13.

Psuty, N.P. and Moreira, M.E.S.A., 1992. Characteristics and longevity of beach nourishment at Praja da Rocha, Portugal. *Journal of Coastal Research* Special Issue 8: 660–76.

Purnell, R. 1996. Recent developments in coastal defence policy and guidance in England. *Coastal Engineering 1996: Proceedings of the Twenty-fifth International Conference*. New York: American Society of Civil Engineers, 4253–60.

Pye, K. 1990. Physical and human influences on coastal dune development between the Ribble and Mersey estuaries, northwest England. In Nordstrom, K.F., Psuty, N.P. and Carter, R.W.G. editors, *Coastal Dunes: Form and Process*, Chichester: John Wiley & Sons, 339–59.

Qu, M., Liu, Y. and Xia, D. 1993. Coastal erosion and protection in China. *Coastal Zone 93*. New York: American Society of Civil Engineers, 121–35.

Rakocinski, C.F., Heard, R.W., LeCroy, S.E., McLelland, J.A. and Simons, T. 1996. Responses by macrobenthic assemblages to extensive beach restoration at Perdido Key, Florida, U.S.A. *Journal of Coastal Research* 12: 326–53.

Randall, R.E. 1983. Management for survival – a review of the plant ecology and protection of the "machair" beaches of northwest Scotland. In McLachlan, A. and Erasmus, T. editors, *Sandy Beaches as Ecosystems*, Boston: Dr. W. Junk, 733–40.

Randall, R.E. and Doody, J.P. 1995. Habitat inventories and the European Habitats Directive: the example of shingle beaches. In Healy, M.G. and Doody, J.P. editors, *Directions in European Coastal Management*. Cardigan, UK: Samara Publishing Ltd., 19–36.

Ranwell, D.S. 1959. Newborough Warren, Anglesey I. The dune system and dune slack habitat. *Journal of Ecology* 47: 571–601.

Ranwell, D.S. 1972. *Ecology of Salt Marshes and Sand Dunes*. London: Chapman and Hall.

Reynolds, W.J. 1987. Coastal structures and long term shore migration. *Coastal Zone 87*. New York: American Society of Civil Engineers, 414–26.

Richardson, T.W. 1994. Technical Area 5 – Dredging Research Program. *Dredging '94*. New York: American Society of Civil Engineers, 1–10.

Ridolfi, G. 1989. The use of the sea for leisure activities: the case of the Ligurian Sea. In Fabbri, P. editor, *Coastlines of Italy*. New York: American Society of Civil Engineers, 111–25.

Rijkswaterstaat. 1990. *A New Coastal Defence Policy for The Netherlands.* 's-Gravenhage: Ministry of Transport and Public Works.

Ritchie, W. and Gimingham, C.H. 1989. Restoration of coastal dunes breached by pipeline landfalls in north-east Scotland. *Proceedings of the Royal Society of Edinburgh* 96B: 231–45.

Ritchie, W. and Penland, S. 1988. Rapid dune changes associated with overwash processes on the deltaic coast of south Louisiana. *Marine Geology* 81: 97–122.

Rizzo, G. 1989. Marinas and minor harbors in the northern Adriatic. In Fabbri, P. editor, *Coastlines of Italy.* New York: American Society of Civil Engineers, 70–83.

Roberts, N. 1989. *The Holocene: an Environmental History.* New York: Basil Blackwell.

Rodríguez, R.W., Webb, R.M.T. and Bush, D.M. 1994. Another look at the impact of Hurricane Hugo on the shelf and coastal resources of Puerto Rico, U.S.A. *Journal of Coastal Research* 10: 278–96.

Roelse, P. 1990. Beach and dune nourishment in The Netherlands. *Coastal Engineering: Proceedings of the Twenty-second Coastal Engineering Conference.* New York: American Society of Civil Engineers, 1984–97.

Roelse, P., Coosen, J. and Minneboo, F.A.J. 1991. Beach nourishment and monitoring program. *Coastal Engineering* 16: 43–59.

Rogers, S.M. 1987. Artificial seaweed for erosion control. *Shore and Beach* 55 (1): 19–29.

Rogers, S.M 1991a. Foundations and breakaway walls of small coastal buildings in Hurricane Hugo. *Coastal Zone 91.* New York: American Society of Civil Engineers, 1220–30.

Rogers, S.M. 1991b. Flood insurance construction standards: can they work on the coast? *Coastal Zone 91.* New York: American Society of Civil Engineers, 1064–78.

Rogers, S.M. 1993. Relocating erosion-threatened buildings: a study of North Carolina housemoving. *Coastal Zone 93.* New York: American Society of Civil Engineers, 1392–405.

Roman, C.T. and Nordstrom, K.F. 1988. The effect of erosion rate on vegetation patterns of an east coast barrier island. *Estuarine, Coastal and Shelf Science* 29: 233–42.

Rothwell, P. 1985. Ainsdale sand dunes – National Nature Reserve. In: Doody, P. editor, *Sand Dunes and their Management,* Peterborough: Nature Conservancy Council, 151–8.

Rouch, F. and Bellessort, B. 1990. Man-made beaches more than 20 years on. *Coastal Engineering: Proceedings of the Twenty-second Coastal Engineering Conference.* New York: American Society of Civil Engineers, 2394–401.

Ruiz, L.F.V. and de Quirós, F.B. 1994. Cost–benefit analysis of shore protection investments. *Coastal Engineering: Proceedings of the Twenty-fourth Coastal Engineering Conference.* New York: American Society of Civil Engineers, 3237–50.

Saffir, H.S. 1991. Hurricane Hugo and implications for design professionals and code-writing authorities. *Journal of Coastal Research* Special Issue 8: 25–32.

Saito, K., Uda, T., Ohara, S., Kawanakajima, Y. and Uchida, K. 1996. Observations of nearshore currents and beach changes around headlands built on the Kashimanada coast, Japan. *Coastal Engineering 1996: Proceedings of the Twenty-fifth International Conference.* New York: American Society of Civil Engineers, 4000–13.

San Diego Association of Governments (SANDAG) 1995. Shoreline preservation strategy for the San Diego region. *Shore and Beach* 63 (2): 17–30.

Sanjaume, E. 1988. The dunes of Saler, Valencia, Spain. *Journal of Coastal Research* Special Issue 3: 63–9.

Sanjaume, E. and Pardo, J. 1992. The dunes of the Valencian coast (Spain): past and present. In Carter, R.W.G., Curtis, T.G.F. and Sheehy-Skeffington, M.J. editors, *Coastal Dunes: Geomorphology, Ecology and Management for Conservation*. Rotterdam: A.A. Balkema, 475–86.

Sato, S. and Tanaka, N. 1980. Artificial resort beach protected by offshore breakwaters and groins. *Coastal Engineering: Proceedings of the Seventeenth Coastal Engineering Conference*. New York: American Society of Civil Engineers, 2003–22.

Savov, B. and Borissova, P. 1993. Bulgarian coastal management. In Fabbri, P. editor, *Coastlines of the Mediterranean*. New York: American Society of Civil Engineers, 16–21.

Sawaragi, T. 1988. Current shore protection works in Japan. *Journal of Coastal Research* 4: 531–41.

Schmahl, G.P. and Conklin, E.J. 1991. Beach erosion in Florida: a challenge for planning and management. *Coastal Zone 91*. New York: American Society of Civil Engineers, 261–71.

Scholl, S.F. 1991. Regulating visual quality of the California coastline. In Domurat, G.W. and Wakeman, T.H. editors, *The California Coastal Experience*. New York: American Society of Civil Engineers, 172–86.

Schwartz, M.L., Marti, J.L.J., Foyo-Herrera, J. and Montero, G.G. 1991. Artificial nourishment at Varadero Beach, Cuba. *Coastal Sediments '91*. New York: American Society of Civil Engineers, 2081–8.

Scott, J.M. 1982. Federal policy and coastal communities. *Shore and Beach* 50: 8–10.

Sea Isle City. 1982. *Sea Isle City Centennial 1882–1982*. Sea Isle City, NJ: City Hall.

Seabloom, E.W. and Wiedemann, A.M. 1994. Distribution and effects of *Ammophila breviligulata* Fern. (American beachgrass) on the foredunes of the Washington coast. *Journal of Coastal Research* 10: 178–88.

Sexton, W.J. 1995. The post-storm Hurricane Hugo recovery of the undeveloped beaches along the South Carolina coast, "Capers Island to the Santee Delta" *Journal of Coastal Research* 11: 1020–5.

Sexton, W.J. and Hayes, M.O. 1991. The geologic impact of Hurricane Hugo and post-storm recovery along the undeveloped coastline of South Carolina, Dewees Island to the Santee delta. *Journal of Coastal Research* Special Issue 8: 275–90.

Shah, N.J., Linden, O. Lundin, C.G. and Johnstone, R. 1997. Coastal management in eastern Africa: status and future. *Ambio* 26: 227–34.

Sherman, D.J. in press. Coastal dunes of California. In Flick, R. editor, *California's Changing Coastline*. Berkeley: University of California Press.

Sherman, D.J. and Bauer, B.O. 1993. Coastal geomorphology through the looking glass. *Geomorphology* 7: 225–49.

Sherman, D.J. and Nordstrom, K.F. 1994. Hazards of wind blown sand and sand drift. *Journal of Coastal Research* Special Issue 12: 263–75.

Sherman, D.J., Bauer, B.O., Nordstrom, K.F. and Allen, J.R. 1990. A tracer study of sediment transport in the vicinity of a groin. *Journal of Coastal Research* 6: 427–38.

Shipman, H. 1993. Potential application of the Coastal Barrier Resources Act to Washington state. *Coastal Zone 93*, New York: American Society of Civil Engineers, 2243–51.

Shiraishi, N., Ohhama, H., Endo, T. and Peña-Santana, P.G. 1994. Stability and management of an artificial beach. *Coastal Engineering: Proceedings of the Twenty-fourth Coastal Engineering Conference*. New York: American Society of Civil Engineers, 2395–405.

Short, A. D. 1992. Beach systems of the central Netherlands coast: processes, morphology and structural impacts in a storm driven multi-bar system. *Marine Geology* 107: 103–37.

Shuisky, Y. D. and Schwartz, M. L. 1988. Human impact and rates of shoreline retreat along the Black Sea coast. *Journal of Coastal Research* 4: 405–16.

Silvester, R. and Hsu, J. R. C. 1993. *Coastal Stabilization: Innovative Concepts*. Englewood Cliffs, NJ: Prentice Hall.

Simison, E. J., Leslie, K. C. and Noble, R. M. 1978. Potential shoreline impacts from proposed structures at Point Conception, California. *Coastal Zone 78*. New York: American Society of Civil Engineers, 1639–52.

Skarregaard, P. 1989. Stabilisation of coastal dunes in Denmark. In van der Meulen, F., Jungerius, P. D. and Visser, J. H. editors, *Perspectives in Coastal Dune Management*, The Hague: SPB Academic Publishing, 151–61.

Smith, A. 1994. Coastal protection in the Pacific Island regions: issues and needs. *Ocean and Coastal Management* 25: 53–61.

Smith, A. W. and Jackson, L. A. 1990. The timing of beach nourishment placements. *Shore and Beach* 58 (1): 17–24.

Smith, A. W. and Piggott, T. L. 1989. An estimate of the value of a beach in terms of beach users. *Shore and Beach* 57 (2): 32–6.

Smith, G. G., Mocke, G. P. and Swart, D. H. 1994. Modelling and analysis techniques to aid mining operations on the Namibian coastline. *Coastal Engineering: Proceedings of the Twenty-fourth Coastal Engineering Conference*. New York: American Society of Civil Engineers, 3335–49.

Smith, K. J. 1991. Beach politics – the importance of informed, local support for beach restoration projects. *Coastal Zone 91*. New York: American Society of Civil Engineers, 56–61.

Smith, M., Rhind, P. and Richards, A. 1995. The Welsh coastal zone: the EC Habitats Directive set in a European context. In Healy, M. G. and Doody, J. P. editors, *Directions in European Coastal Management*. Cardigan, UK: Samara Publishing Ltd., 59–68.

Smith, W. G., Rosati, J. D. and Lemire, A. 1993. Revere Beach and Point of Pines, Massachusetts, shore front study. In Stauble. D. K. and Kraus, N. C. editors, *Beach Nourishment: Engineering and Management Considerations*. New York: American Society of Civil Engineers, 118–32.

Snyder, M. R. and Pinet, P. R. 1981. Dune construction using two multiple sand-fence configurations: implications regarding protection of eastern Long Islands south shore. *Northeastern Geology* 3: 225–9.

Sorensen, J. 1997. National and international efforts at integrated coastal management: definitions, achievements, and lessons. *Coastal Management* 25: 3–41.

Sorensen, R. M. 1990. Beach behavior and effect of coastal structures: Bradley Beach, New Jersey. *Shore and Beach* 58 (1): 25–9.

Sorensen, R. M. and Brandani, A. 1987. A comparative assessment of coastal area management initiatives in Latin America. *Coastal Zone 87*. New York: American Society of Civil Engineers, 3065–84.

Sorensen, R.M. and Schmeltz, E.J. 1982. Closure of the breach at Moriches Inlet. *Shore and Beach* 50 (4): 33–40.

Springman, R. and Born, S.M. 1979. *Wisconsin's shore erosion plan: an appraisal of options and strategies*. Madison, WI: Wisconsin Department of Administration.

Stansfield, C.A. 1990. Cape May: selling history by the sea. *Journal of Cultural Geography* 11: 25–37.

Stansfield, C.A. and Rickert, J.E. 1970. The recreational business district. *Journal of Leisure Research* 2: 213–25.

Stauble, D.K. 1991. Coastal impacts of Hurricane Hugo. *Coastal Sediments 91*. New York: American Society of Civil Engineers, 1666–80.

Stauble, D.K. and Cialone, M.A. 1996. Ebb shoal evolution and sediment management techniques, Ocean City, Maryland. In Tait, L.S. editor, *The Future of Beach Nourishment*, Tallahassee, FL: Florida Shore and Beach Preservation Association, 209–24.

Stauble, D.K. and Holem, G.W. 1991. Long term assessment of beach nourishment project performance. *Coastal Zone 91*. New York: American Society of Civil Engineers, 510–24.

Stauble. D.K. and Kraus, N.C. 1993a. *Beach Nourishment: Engineering and Management Considerations*. New York: American Society of Civil Engineers.

Stauble. D.K. and Kraus, N.C. 1993b. Project performance: Ocean City, Maryland beach nourishment. In Stauble. D.K. and Kraus, N.C. editors, *Beach Nourishment: Engineering and Management Considerations*. New York: American Society of Civil Engineers, 1–15.

Stauble. D.K. and Nelson, W.G. 1985. Guidelines for beach nourishment: a necessity for project management. *Coastal Zone 85*. New York: American Society of Civil Engineers, 1002–21.

Stauble, D.K., Seabergh, W.C. and Hales, L.Z. 1991. Effects of Hurricane Hugo on the South Carolina coast. *Journal of Coastal Research* Special Issue 8: 129–62.

Steller, D.L., Good, B., Clark, C., Bahlinger, K., Rasi, J. and Steyer, G. 1993. Coastal restoration: Louisiana's saving grace. *Coastal Zone 93*. New York: American Society of Civil Engineers, 2111–25.

Stive, M.J.F., Roelvink, D.J.A. and de Vriend, H.J. 1990. Large-scale coastal evolution concept. *Coastal Engineering: Proceedings of the Twenty-second Coastal Engineering Conference*. New York: American Society of Civil Engineers, 1962–74.

Stive, M.J.F., Nicholls, R.J. and de Vriend, H.J. 1991. Sea-level rise and shore nourishment: a discussion. *Coastal Engineering* 16: 147–63.

Stoddard, G. 1995. Coastal policy implications of right to rebuild questions. *Shore and Beach* 63 (1): 25–33.

Stone, G.W. and Finkl, C.W. Jr. editors, 1995. Impacts of Hurricane Andrew. *Journal of Coastal Research* Special Issue 21.

Stone, K.E. 1991. How to pay for coastal protection: governmental approaches. *Coastal Zone 91*. New York: American Society of Civil Engineers, 3778–86.

Stronge, W.B. 1995. The economics of government funding for beach nourishment projects: the Florida case. *Shore and Beach* 63 (3): 4–6.

Sturgess, P. 1992. Clear-felling dune plantations: studies in vegetation recovery. In Carter, R.W.G., Curtis, T.G.F. and Sheehy-Skeffington, M.J. editors, *Coastal Dunes:*

Geomorphology, Ecology and Management for Conservation. Rotterdam: A.A. Balkema, 339–49.

Stutts, A.T., Siderelis, C.D. and Rogers, S.M. 1989. Effect of ocean setback standards on the location of permanent structures. *Coastal Zone 89*. New York: American Society of Civil Engineers, 2459–67.

Sudar, R.A., Pope, J., Hillyer, T. and Crumm, J. 1995. Shore protection projects of the U.S. Army Corps of Engineers. *Shore and Beach* 63 (2): 3–16.

Swart, D.H. 1991. Beach nourishment and particle size effects. *Coastal Engineering* 16: 61–81.

Swart, D.H. and Reyneke, P.G. 1988. The role of driftsands at Waenhuiskrans, South Africa. *Journal of Coastal Research* Special Issue 3: 97–102

Tait, J.P. and Griggs, G.B. 1990. Beach response to the presence of a seawall. *Shore and Beach* 58 (2): 11–28.

Tait, L.S. editor, 1996. *The Future of Beach Nourishment*, Tallahassee, FL: Florida Shore and Beach Preservation Association.

Tait, S. 1991. Florida's comprehensive beach management law. *Shore and Beach* 59 (4): 23–6.

Taussik, J. 1995. The contribution of development plans to coastal policy with particular reference to nature conservation. In Healy, M.G. and Doody, J.P. editors, *Directions in European Coastal Management*. Cardigan, UK: Samara Publishing Ltd., 461–9.

Technische Adviescommissie voor de Waterkeringen. 1995. *Basisrapport Zandige Kust*. Delft: Drukkerij & DTP-Service Nivo.

Tekke, R. 1995. Ecoteams and greenteams: taking care of nature outside protected areas. In Healy, M.G. and Doody, J.P. editors, *Directions in European Coastal Management*. Cardigan, UK: Samara Publishing Ltd., 441–6.

Télez-Duarte, M.A. 1993. Cultural resources as a criterion in coastal zone management: the case of northwestern Baja California, Mexico. In Fermán-Almada, J.L., Gómez-Morin, L. and Fischer, D.W. editors, *Coastal Zone Management in Mexico: the Baja California Experience*. New York: American Society of Civil Engineers, 137–47.

Terchunian, A.V. and Merkert, C.L. 1995. Little Pikes Inlet, Westhampton, New York. *Journal of Coastal Research* 11: 697–703.

Terwindt, J.H.J., Kohsiek, L.H.M. and Visser, J. 1988. The Netherlands. In Walker, H.J. editor, *Artificial Structures and Shorelines*. Dordrecht: Kluwer Academic Publishers, 104–14.

Thieler, R.E. and Danforth, W.W. 1994. Historical shoreline mapping (II): application of the digital shoreline mapping and analysis system (DSMS/DSAS) to shoreline change mapping in Puerto Rico. *Journal of Coastal Research* 10: 600–20.

Thompson, J. 1995. Aesthetics and the value of nature. *Environmental Ethics* 17: 291–305.

Tiffney, W.N. and Andrews, J.C. 1989. Is there a relationship between pond opening and bluff erosion on Nantucket Island, Massachusetts? *Coastal Zone 89*. New York: American Society of Civil Engineers, 3760–72.

Tinley, K.L. 1985. *Coastal Dunes of South Africa*. Pretoria: Council for Scientific and Industrial Research.

Tippets, W.E. and Jorgensen, P.D. 1991. Restoration of coastal dune systems in southern California state parks. *Coastal Zone 91*. New York: American Society of Civil Engineers, 2994–3000.

Titus, J.G. 1990. Greenhouse effect, sea level rise, and barrier islands: case study of Long Beach Island, New Jersey. *Coastal Management* 18: 65–90.

Titus, J.G. and Narayanan, V. 1996. The risk of sea level rise. *Climatic Change* 33: 151–212.

Tomasicchio, U. 1996. Submerged breakwaters for the defence of the shoreline at Ostia: field experiences, comparison. *Coastal Engineering 1996: Proceedings of the Twenty-fifth International Conference*. New York: American Society of Civil Engineers, 2404–17.

Tooley, M.J. 1990. The chronology of coastal dune development in the United Kingdom. *Catena* Supplement 18: 81–8.

Torresani, S. 1989. Historical evolution of the coastal settlement in Italy: ancient times. In Fabbri, P. editor, *Coastlines of Italy*. New York: American Society of Civil Engineers, 150–9.

Townend, I.H. and Fleming, C.A. 1991. Beach nourishment and socio-economic aspects. *Coastal Engineering*. 16: 115–27.

Toyoshima, O. 1982. Variation of foreshore due to detached breakwaters. *Coastal Engineering: Proceedings of the Eighteenth Coastal Engineering Conference*. New York: American Society of Civil Engineers, 1873–92.

Trampenau, T., Göricke, F. and Raudkivi, A.J. 1996. Permeable pile groins. *Coastal Engineering 1996: Proceedings of the Twenty-fifth International Conference*. New York: American Society of Civil Engineers, 2142–51.

Trembanis, A.C. and Pilkey, O.H. 1998. Summary of beach nourishment along the U.S. Gulf of Mexico shoreline. *Journal of Coastal Research* 14: 407–17.

Trofimov, A.M. 1987. On the problem of geomorphological prediction. *Catena* Supplement 10: 193–7.

Trudgill, S. and Richards, K. 1997. Environmental science and policy: generalizations and context sensitivity. *Transactions of the Institute of British Geographers* 22: 5–12.

Truitt, C.L., Kraus, N.C. and Hayward, D. 1993. Beach fill performance at the Lido Beach, Florida groin. In Stauble. D.K. and Kraus, N.C. editors, *Beach Nourishment: Engineering and Management Considerations*. New York: American Society of Civil Engineers, 31–42.

Tsuchiya, Y., Kawata, Y., Yamashita, T., Shibano, T., Kawasaki, M. and Habara, S. 1992. Sandy beach stabilization: preservation of Shirarahama Beach, Wakayama. *Coastal Engineering: Proceedings of the Twenty-third Coastal Engineering Conference*. New York: American Society of Civil Engineers, 3426–39.

Turner, I.L. and Leatherman, S.P. 1997. Beach dewatering as a 'soft' engineering solution to coastal erosion – a history and critical review. *Journal of Coastal Research* 13: 1050–63.

Tye, R.S. 1983. Impact of Hurricane David and mechanical dune restoration on Folly Beach, South Carolina. *Shore and Beach* 51 (2): 3–9.

Uematsu, Y., Yamada, M., Higashiyama, H. and Orimo, T. 1992. Effects of the corner shape of high-rise buildings on the pedestrian-level wind environment with consideration for mean and fluctuating wind speeds. *Journal of Wind Engineering and Industrial Aerodynamics*. 41–44: 2289–300.

Ulrich, C.P. 1993. Selecting the optimum dune height for Panama City beaches, Florida. In Stauble. D.K. and Kraus, N.C. editors, *Beach Nourishment: Engineering and*

Management Considerations. New York: American Society of Civil Engineers, 148–61.

US Army Corps of Engineers. 1957. *Beach Erosion Control Report on Cooperative Study (Survey) of the New Jersey Coast Barnegat Inlet to the Delaware Bay Entrance to the Cape May Canal*. Philadelphia: US Army Corps of Engineers, Philadelphia District.

US Army Corps of Engineers. 1962. *Coastal Storm of 6–7 March 1962. Post Flood Report*. Philadelphia: US Army Corps of Engineers, Philadelphia District.

US Army Corps of Engineers. 1963. *Report on Operation Five-High: March 1962 Storm*. New York: US Army Corps of Engineers, North Atlantic Division.

US Army Corps of Engineers. 1985. *Post Flood Report: Coastal Storm of 28–29 March 1984*. Philadelphia: US Army Corps of Engineers, Philadelphia District.

US Army Corps of Engineers. 1989. *Benefits Re-evaluation Study: Brigantine Island, New Jersey*. Philadelphia: US Army Corps of Engineers, Philadelphia District.

US Army Corps of Engineers. 1990. *New Jersey Shore Protection Study: Report of Limited Reconnaissance Study*. Philadelphia: US Army Corps of Engineers, Philadelphia District.

US Army Corps of Engineers. 1991. *Sand Bypassing System, Engineering and Design Manual*. EM 1110–2–1616. Washington, DC: US Army Corps of Engineers.

US Army Corps of Engineers. 1993. *Post Flood Report: Coastal Storm of 11–15 December 1992*. Philadelphia: US Army Corps of Engineers, Philadelphia District.

US Congress. 1976. New Jersey Coastal Inlets and Beaches Hereford Inlet to Delaware Bay Entrance to Cape May Canal. *House Document 94–641*. Washington, DC: US Government Printing Office.

Valentini, E. and Rosman, P.C.C. 1993. Managing beach erosion in Fortaleza, Brazil. *Coastal Zone 93*. New York: American Society of Civil Engineers, 799–811.

Valiela, I., Peckol, P., D'Avanzo, C., Kremer, J., Hersh, D., Foreman, K., Lajtha, K., Seely, B., Geyer, W.R., Isaji, T. and Crawford, R. 1998. Ecological effects of major storms on coastal watersheds and coastal waters: Hurricane Bob on Cape Cod. *Journal of Coastal Research* 14: 218–38.

van Beckhoven, K. 1992. Effects of groundwater manipulation on soil processes and vegetation in wet dune slacks. In Carter, R.W.G., Curtis, T.G.F. and Sheehy-Skeffington, M.J. editors, *Coastal Dunes: Geomorphology, Ecology and Management for Conservation*. Rotterdam: A.A. Balkema, 251–63.

van Bohemen, H.D. 1996. Environmentally friendly coasts: dune breaches and tidal inlets in the foredunes. Environmental engineering and coastal management: a case study from The Netherlands. *Landscape and Urban Planning* 34: 197–213.

van Bohemen, H.D. and Meesters, H.J.N. 1992. Ecological engineering and coastal defense. In Carter, R.W.G., Curtis, T.G.F. and Sheehy-Skeffington, M.J. editors, *Coastal Dunes: Geomorphology, Ecology and Management for Conservation*. Rotterdam: A.A. Balkema, 369–78.

van Boxel, J.H., Jungerius, P.D., Kieffer, N. and Hampele, N. 1997. Ecological effects of reactivation of artificially stabilized blowouts in coastal dunes. *Journal of Coastal Conservation* 3: 57–62.

van de Graaff, J. 1994. Coastal and dune erosion under extreme conditions. *Journal of Coastal Research* Special Issue 12: 253–62.

van de Graaff, J., Niemeyer, H.D. and J. Overeem. 1991. Beach nourishment, philosophy and coastal protection policy. *Coastal Engineering* 16: 3–22.

van der Maarel, E. 1979. Environmental management of coastal dunes in The Netherlands. In Jefferies, R.L. and Davy, A.J. editors, *Ecological Processes in Coastal Environments*. Oxford: Basil Blackwell, 543–70.

van der Maarel, E. and van der Maarel-Versluys, M. 1996. Distribution and conservation status of littoral vascular plant species along the European coasts. *Journal of Coastal Conservation* 2: 73–92.

van der Meulen, F. 1990. European dunes: consequences of climate change and sea level rise. *Catena* Supplement 18, 209–23.

van der Meulen, F. and Jungerius, P.D. 1989a. Landscape development in Dutch coastal dunes: the breakdown and restoration of geomorphological and geohydrological processes. *Proceedings of the Royal Society of Edinburgh* 96B: 219–29.

van der Meulen, F. and Jungerius, P.D. 1989b. The decision environment of dynamic dune management. In van der Meulen, F., Jungerius, P.D. and Visser, J.H. editors, *Perspectives in Coastal Dune Management*, The Hague: SPB Academic Publishing, 133–40.

van der Meulen, F. and Salman, A.H.P.M. 1996. Management of Mediterranean coastal dunes. *Ocean and Coastal Management* 30: 177–95.

van der Meulen, F. and van der Maarel, E. 1989. Coastal defense alternatives and nature development perspectives. In van der Meulen, F., Jungerius, P.D. and Visser, J.H. editors, *Perspectives in Coastal Dune Management*, The Hague: SPB Academic Publishing, 183–95.

van der Putten, W.H. and Kloosterman, E.H. 1991. Large-scale establishment of *Ammophila arenaria* and quantitative assessment by remote sensing. *Journal of Coastal Research* 7: 1181–94.

van der Putten, W.H., van Dyke, C. and Peters, B.A.M. 1993. Plant specific soil borne diseases contribute to succession in foredune vegetation. *Nature* 362: 53–5.

van der Wal, D. 1996. The development of a digital terrain model for the geomorphological engineering of the "rolling" foredune of Terschelling, the Netherlands. *Journal of Coastal Conservation* 2: 55–62.

van der Wal, D. 1998. The impact of the grain-size distribution of nourishment sand on aeolian sand transport. *Journal of Coastal Research* 14: 620–31.

van Dijk, H.W.J. 1989. Ecological impact of drinking-water production in Dutch coastal dunes. In van der Meulen, F., Jungerius, P.D. and Visser, J.H. editors, *Perspectives in Coastal Dune Management*, The Hague: SPB Academic Publishing, 163–82.

van Dijk, H.W.J. 1992. Grazing domestic livestock in Dutch coastal dunes: experiments, experiences and perspectives. In Carter, R.W.G., Curtis, T.G.F. and Sheehy-Skeffington, M.J. editors, *Coastal Dunes: Geomorphology, Ecology and Management for Conservation*. Rotterdam: A.A. Balkema, 235–50.

van Dolah, R.F., Gayes, P.T., Katuna, M.P. and Devoe, M.R. 1993. Five year program to evaluate sand, mineral and hard bottom resources of the continental shelf off South Carolina. *Coastal Zone 93*. New York: American Society of Civil Engineers, 1188–96.

van Heerden, I.L. and de Rouen, K. 1997. Implementing a barrier island and barrier shoreline restoration program – the State of Louisiana's perspective. *Journal of Coastal Research* 13: 679–85.

van Huis, J. 1989. European dunes, climate and climatic change, with case studies of the
        Coto Doñana (Spain) and the Slowinski (Poland) National Parks. In van der
        Meulen, F., Jungerius, P. D. and Visser, J. H. editors, *Perspectives in Coastal Dune
        Management*, The Hague: SPB Academic Publishing, 313–26.

van Oorschot, J. H. and van Raalte, G. H. 1991. Beach nourishment: execution methods and
        developments in technology. *Coastal Engineering* 16: 23–42.

van 't Hoff, J., de Groot, M. B., Winterwerp, J. C., Verwoert, H. and Bakker, W. T. 1992.
        Artificial sand fills in water. *Coastal Engineering: Proceedings of the Twenty-third
        Coastal Engineering Conference*. New York: American Society of Civil Engineers,
        2599–612.

Verhagen, H. J. 1991. Coastal maintenance: the Netherlands. *Coastal Zone 91*. New York:
        American Society of Civil Engineers, 2734–43.

Verhagen, H. J. 1992. Method for artificial beach nourishment. *Coastal Engineering:
        Proceedings of the Twenty-third Coastal Engineering Conference*. New York: American
        Society of Civil Engineers, 2474–85.

Verhagen, H. J. 1996. Analysis of beach nourishment schemes. *Journal of Coastal Research* 12:
        179–85.

Vestergaard, P. and Alstrup, V. 1996. Loss of organic matter and nutrients from a coastal
        dune heath in northwest Denmark caused by fire. *Journal of Coastal Conservation* 2:
        33–40.

Viggóssen, G. 1990. Rubble mound breakwaters in Iceland. *Journal of Coastal Research* Special
        Issue 7: 41–61.

Viles, H. and Spencer, T. 1995. *Coastal Problems: Geomorphology, Ecology and Society at the Coast*.
        London: Edward Arnold.

Vinh, T. T., Kant, G., Huan, N. N. and Pruszak, Z. 1996. Sea dike erosion and coastal retreat
        at Nam Ha Province, Viet Nam. *Coastal Engineering 1996: Proceedings of the Twenty-
        fifth International Conference*. New York: American Society of Civil Engineers,
        2820–8.

Waks, L. J. 1996. Environmental claims and citizen rights. *Environmental Ethics* 18:
        133–48.

Waldrop, J. 1988. In the path of disaster. *American Demography* 10 (8): 6–7.

Walker, H. J. 1981a. The peopling of the coast. In Ma, L. J. C. and Noble, A. G. editors, *The
        Environment: Chinese and American Views*, New York: Methuen and Co., 91–105.

Walker, H. J. 1981b. Man and shoreline modification. In Bird, E. C. F. and Koike, K. editors,
        Coastal Dynamics and Scientific Sites. Tokyo: Komazawa University, 55–90.

Walker, H. J. 1984. Man's impact on shorelines and nearshore environments: a
        geomorphological perspective. *Geoforum*, 15: 395–417.

Walker, H. J. 1985. The shoreline: realities and perspectives. In *Coastal Planning: Realities and
        Perspectives*, Genoa: Comune di Genova, 59–90.

Walker, J. R. 1991. Downdrift effects of navigation structures on the California coast.
        Coastal Zone 91. New York: American Society of Civil Engineers, 1889–903.

Wanders, E. 1989. Perspectives in coastal dune management. In van der Meulen, F.,
        Jungerius, P. D. and Visser, J. H. editors, *Perspectives in Coastal Dune Management*, The
        Hague: SPB Academic Publishing, 141–8.

Warne, A. G. and Stanley, D. J. 1993. Late quaternary evolution of the northwest Nile delta
        and adjacent coast in the Alexandria region, Egypt. *Journal of Coastal Research* 9:
        26–64.

Warrick, R.A. and Barrow, E.M. 1991. Climate change scenarios for the UK. *Transactions of the Institute of British Geographers* 16: 387–99.

Wasyl, J., Jenkins, S.A. and Skelly, D.W. 1991. Sediment bypassing around dams – a potential beach erosion control mechanism. In Domurat, G.W. and Wakeman, T.H. editors, *The California Coastal Experience*. New York: American Society of Civil Engineers, 251–65.

Watanabe, A. and Horikawa, K. 1983. Review of coastal stabilization works in Japan. *Proceedings of the International Conference on Coastal and Port Engineering in Developing Countries*, 186–200.

Watson, I. and Finkl, C.W. Jnr. 1990. State of the art in storm-surge protection: The Netherlands Delta Project. *Journal of Coastal Research* 6: 739–64.

Watson, J.J., Kerley, G.I.H. and McLachlan, A. 1997. Nesting habitat of birds breeding in a coastal dunefield, South Africa and management implications. *Journal of Coastal Research* 13: 36–45.

Watzin, M.C. and McGilvrey, F.B. 1989. Coastal Barrier Resources Act report to Congress. In Stauble, D.K. editor, *Barrier Islands: Processes and Management*. New York: American Society of Civil Engineers, 208–22.

Webb, C.J., Stow, D.A. and Chang, H.H. 1991. Morphodynamics of southern California inlets. *Journal of Coastal Research* 7: 167–87.

Weggel, J.R. 1988. Seawalls: the need for research, dimensional considerations and a suggested classification. *Journal of Coastal Research* Special Issue 4: 29–40.

Weggel, J.R. and Farrell, S.C. 1990. The effect of a shore-parallel offshore breakwater on the beaches at Ocean City, NJ. *Coastal Engineering: Proceedings of the Twenty-second Coastal Engineering Conference*. New York: American Society of Civil Engineers, 2020–33.

Weggel, J.R. and R.M. Sorensen. 1991. Performance of the 1986 Atlantic City, New Jersey, beach nourishment project. *Shore and Beach* 59 (3): 29–36.

Weggel, J.R., Morreale, M. and Giegengack, R. 1995. The Ocean City, New Jersey, beach nourishment project: monitoring its early performance. *Shore and Beach* 63 (3): 25–32.

Wells, J.T. and McNinch, J. 1991. Beach scraping in North Carolina with special reference to its effectiveness during Hurricane Hugo. *Journal of Coastal Research* Special Issue 8: 249–61.

Western Australia Department of Planning and Urban Development. 1993. *Horrocks Beach Coastal Plan*. Perth: State of Western Australia.

Western Australia Department of Planning and Urban Development. 1994a. *Central Coast Regional Profile*. Perth: State of Western Australia.

Western Australia Department of Planning and Urban Development. 1994b. *Central Coast Regional Strategy*. Perth: State of Western Australia.

Westhoff, V. 1985. Nature management in coastal areas of Western Europe. *Vegetatio* 62: 523–32.

Westhoff, V. 1989. Dunes and dune management along the North Sea Coasts. In van der Meulen, F., Jungerius, P.D. and Visser, J.H. editors, *Perspectives in Coastal Dune Management*. The Hague: SPB Academic Publishing, 41–51.

Weston, A. 1996. Self-validating reduction: toward a theory of environmental devaluation. *Environmental Ethics* 18: 123–32.

White, A.T., Barker, V. and Tantrigama, G. 1997. Using integrated coastal management and economics to conserve coastal tourism resources in Sri-Lanka. *Ambio* 26: 335–44.

Whitlock, W. and Becker, R.H. 1991. Nature based tourism: an alternative for rural coastal economic enhancement. *Coastal Zone 91.* New York: American Society of Civil Engineers, 1046–52.

Wiedemann, A.M. 1984. *The Ecology of Pacific Northwest Coastal Sand Dunes: a Community Profile.* Washington, DC: US Department of the Interior Fish and Wildlife Service.

Wiedemann, A.M. 1990. The coastal parabola dune system at Sand Lake, Tillamook County, Oregon, USA. In Davidson-Arnott, R.G.D. editor, *Proceedings of the Symposium on Coastal Sand Dunes,* Ottawa: National Research Council Canada, 171–94.

Wiegel, R.L. 1987. Trends in coastal erosion management. *Shore and Beach* 55 (1): 3–11.

Wiegel, R.L. 1991. Protection of Galveston, Texas, from overflows by Gulf storms: grade-raising, seawall and embankment. *Shore and Beach* 59 (1): 4–10.

Wiegel, R.L. 1992a. Some complexities of coastal waves, currents, sand, and structures. *Shore and Beach* 60 (1): 21–33.

Wiegel, R.L. 1992b. Dade County, Florida, beach nourishment and hurricane surge protection project. *Shore and Beach* 60 (4): 2–27.

Wiegel, R.L. 1992c. Beach nourishment, sand by-passing, artificial beaches: bibliography of articles. *Shore and Beach* 60 (3): 3–5.

Wiegel, R.L. 1993a. Artificial beach construction with sand/gravel made by crushing rock. *Shore and Beach* 61 (4): 28–9.

Wiegel, R.L. 1993b. Dana Point Harbor, California. *Shore and Beach* 61 (3): 37–55.

Wiegel, R.L. 1994. Ocean beach nourishment on the USA Pacific coast. *Shore and Beach* 62 (1): 11–36.

Wiegel, R.L. 1995. Waikiki Beach, Oahu. *Shore and Beach* 63 (4): 34–6.

Williams, A.T. 1987. Coastal conservation policy development in England and Wales with special reference to the Heritage Coast concept. *Journal of Coastal Research* 3: 99–106.

Williams, A.T. and Davies, P. 1980. Man as a geological agent: the sea cliffs of Llantwit Major, Wales, U.K. *Zeitschrift für Geomorphologie* Supplementband 34: 129–41.

Williams, A.T. and Randerson, P.F. 1989. Nexus: ecology, recreation and management of a dune system in south Wales. In van der Meulen, F., Jungerius, P.D. and Visser, J.H. editors, *Perspectives in Coastal Dune Management,* The Hague: SPB Academic Publishing, 217–27.

Wong, P.P. 1985. Artificial coastlines: the example of Singapore. *Zeitschrift für Geomorphologie* 57: 175–92.

Wong, P.P. 1988. Singapore. In Walker, H.J. editor, *Artificial Structures and Shorelines.* Dordrecht: Kluwer Academic Publishers, 383–92.

Wong, P.P. 1990. The geomorphological basis of beach resort sites – some Malaysian examples. *Ocean and Shoreline Management* 13: 127–47.

Wong, P.P. 1993. Island tourism development in peninsular Malaysia: environmental perspective. In Wong, P.P. editor, *Tourism vs Environment: the Case for Coastal Areas.* Dordrecht: Kluwer Academic Publishers, 83–97.

Wood, J.D., Ramsey, J.S. and Weishar, L.L. 1996. Beach nourishment along Nantucket Sound: a tale of two beaches. In Tait, L.S. editor, *The Future of Beach Nourishment,* Tallahassee, FL: Florida Shore and Beach Preservation Association, 117–29.

Woodell, G. and Hollar, R. 1991. Historical changes in the beaches of Los Angeles County. *Coastal Zone 91.* New York: American Society of Civil Engineers, 1342–55.

Work, P. A. and Dean, R. G. 1995. Assessment and prediction of beach-nourishment evolution. *Journal of Waterway, Port, Coastal and Ocean Engineering*. 121: 182–9.

Yamazaki, T., Uda, T., Shinoda, T. and Koarai, M. 1991. Creation of calm sea area by large-scale offshore breakwater and planning method for use of the created area. In Nagao, Y. editor, *Coastlines of Japan*. New York: American Society of Civil Engineers, 199–211.

Yazdani, N. and Ycaza, I. D. 1995. Multi-agency integrated code for coastal construction. *Journal of Coastal Research* 11: 899–903.

Yazdani, N., Nnaji, S. and Rambo-Rodenberry, M. 1997. Conceptual breakaway swimming pool design for coastal areas. *Journal of Coastal Research* 13: 61–6.

Yesin, N. V. and Kos'yan, R. D. 1993. Ecology of Gelendzhik Bay. In Kos'yan, R. editor, *Coastlines of the Black Sea*. New York: American Society of Civil Engineers, 156–72.

Yüksek, O., Onsoy, H., Birben, A. R. and Ozolcer, I. H. 1995. Coastal erosion in eastern Black Sea region. *Coastal Engineering* 26: 225–39.

Zawadzka-Kahlau, E. 1995. Erosion/accretion bistructures of the south Baltic coast. In Healy, M. G. and Doody, J. P. editors, *Directions in European Coastal Management*. Cardigan, UK: Samara Publishing Ltd., 317–21.

Zawadzka, E. 1996. Coastal zone dynamics during artificial nourishment. *Coastal Engineering 1996: Proceedings of the Twenty-fifth International Conference*. New York: American Society of Civil Engineers, 2955–68.

Zenkovich, V. P. and Aibulatov, N. A. 1993. Specific features of the Russian Black Sea coast dynamics and morphology (Kerch Strait – Psou River mouth. In Kos'yan, R. editor, *Coastlines of the Black Sea*. New York: American Society of Civil Engineers, 278–302.

Zenkovich, V. P. and Schwartz, M. L. 1987 Predicting the Black Sea–Georgian S. S. R. gravel coast. *Journal of Coastal Research* 3: 201–9.

Zenkovich, V. P. and Schwartz, M. L. 1988. Restoration of the Georgian SSR coast. *Shore and Beach* 56 (1): 8–12.

Zunica, M. 1990. Beach behavior and defences along the Lido di Jesolo, Gulf of Venice, Italy. *Journal of Coastal Research* 6: 709–19.

Index

Printed in the United States
21558LVS00001B/343-350